# MANAGEMENT TECHNIQUES APPLIED TO THE CONSTRUCTION INDUSTRY

26.50

## Fifth edition

**R. Oxley** *MSc, MCIOB, MIMgt*

and

**J. Poskitt** *MCIOB, MIMgt*

*b*

Blackwell Science

Blackwell Science Ltd
Editorial Offices:
Osney Mead, Oxford OX2 0EL
25 John Street, London WC1N 2BL
23 Ainslie Place, Edinburgh EH3 6AJ
238 Main Street, Cambridge,
Massachusetts 02142, USA
54 University Street, Carlton
Victoria 3053, Australia

Other Editorial Offices:
Arnette Blackwell SA
224, Boulevard Saint Germain
75007 Paris, France

Blackwell Wissenschafts-Verlag GmbH
Kurfürstendamm 57
10707 Berlin, Germany

Zehetnergasse 6
A-1140 Wien
Austria



First published by Crosby Lockwood and Son Ltd 1968
Second edition published 1971
Reprinted by Granada Publishing Ltd in Crosby Lockwood Staples 1976, 1979
Third edition published by Granada Publishing 1980
Fourth edition published by Collins Professional and Technical Books 1986
Reprinted by BSP Professional Books 1989, 1990, 1992
Fifth edition published by Blackwell Science Ltd 1996

Set in 10 on 13pt Times
by Aarontype Ltd, Bristol
Printed and bound in Great Britain
by Hartnolls Ltd, Bodmin, Cornwall



DISTRIBUTORS

Marston Book Services Ltd
PO Box 269
Abingdon
Oxon OX14 4YN
(*Orders*: Tel: 01235 465500
Fax: 01235 465555)

USA
Blackwell Science, Inc.
238 Main Street
Cambridge, MA 02142
(*Orders*: Tel: 800 215-1000
617 876-7000
Fax: 617 492-5263)

Canada
Copp Clark, Ltd
2775 Matheson Blvd East
Mississauga, Ontario
Canada, L4W 4P7
(*Orders*: Tel: 800 263-4374
905 238-6074)

Australia
Blackwell Science Pty Ltd
54 University Street
Carlton, Victoria 3053
(*Orders*: Tel: 03 9347-0300
Fax: 03 9347-5001)

A catalogue record for this title is available from the British Library

ISBN 0-632-03862-4

Library of Congress
Cataloging-in-Publication Data

Oxley, R.
Management techniques applied to the construction industry/R. Oxley and J. Poskitt. – 5th ed.
p. cm.
Includes biographical references and index.
ISBN 0-632-03862-4
1. Construction industry – Management.
I. Poskitt, J. II. Title.
TH438.095 1996
690′.68–dc20

96-13801
CIP

# Contents

# Preface

This book is concerned with the application of management techniques to construction projects. It is aimed mainly at students studying construction management as part of a degree or higher diploma course or courses leading to the Chartered Institute of Building examinations.

It is assumed that readers are familiar with basic techniques, such as pre-tender and pre-contract planning and no excuse is made for omitting some of the background information in some chapters. In addition, certain topics have been omitted from this latest edition to make room for more practical examples: the chapter on statistics is no longer included as the topic is adequately covered by other texts. Operational research has been dropped since – apart from network analysis techniques, which have been retained – OR techniques are little used in the industry.

Some examples used in the text have been adapted and simplified in order to convey the important points in a clearer manner. The methods of presentation in the book have evolved from our experience with students studying the various topics.

The techniques in this book have been considered separately for convenience. It must be appreciated, however, that if a company is to operate efficiently, it needs an integrated information system that will meet the needs of management. Information that is gathered for one department, such as production control, will be relevant to other departments, such as estimating cost control. Case study 1 illustrates the integration of the estimating and production processes.

There seems little point in departments within a company collecting information or processing data identical to that collected or processed by another department, but this happens all too often in practice.

In preparing the fifth edition of this book, we have taken account of suggestions received from academics and practitioners to add more practical examples and to set some exercises, but to retain chapters that are found to be useful in encouraging students to question methods and procedures.

While some of the existing material has been retained unaltered, some has been changed considerably and most major examples have been replaced. We have also taken the opportunity to reorganise the order of presentation of the material.

Both of the authors of this book are directors of a construction company named O & P Construction Services Ltd. A selection of case studies of actual projects undertaken by this company have been included.

While this book is aimed primarily at students, it will also be of interest to staff in construction firms.

Management techniques are the tools of management, and the techniques considered in this book can and should be used by all companies, irrespective of size. They enable decisions to be taken with greater knowledge of the facts; ignorance of the techniques available to management leads to inefficient operation and may eventually lead to bankruptcy. The use of scientific methods is essential to effective management.

*R. Oxley and J. Poskitt*

# Acknowledgements

We are indebted to our friends and colleagues for their help received when writing the various editions of this book.

We wish to express our gratitude for the help and encouragement received from our wives Sheila and Jean, who are also part-time secretaries for O&P Construction Services Ltd, when preparing this edition and the previous editions. We also wish to thank Stuart Oxley and Robert Mitchell for their work on the case studies in Chapter 7.

*Trademark acknowledgements*: '1-2-3' is a registered trademark of the Lotus Development Corporation; 'Pertmaster' is a registered trademark of People in Technology Ltd.

# Chapter 1
# Construction Planning and Control

## 1.1 Introduction

Planning can be applied in varying degrees of detail, depending on the stage at which it is being carried out. For construction work it is usually divided into *pre-tender planning*, *pre-contract planning* and *short-term planning*.

Examples of the use of planning and control at all stages of a project are shown in the case studies in Chapter 7.

### 1.1.1 *Pre-tender planning*

Pre-tender planning is undertaken to allow the estimator to arrive at an estimate of cost, based on the proposed methods of working and an estimate of the time required to carry out the major operations. Programming at the pre-tender stage is usually in an outline form on large projects, to consider only the phasing of the main operations, as much information is not available at that time.

### 1.1.2 *Pre-contract planning*

Pre-contract planning is carried out when the contract has been won, and the project is considered more fully. Planning at this stage includes the master programme, labour schedule, plant schedule, and materials schedules. The master programme should not break the operations down excessively on large projects or it will become unrealistic.

The information to be used as a basis for calculating productive hours at this stage can be obtained from operational records of past contracts and from the bill of quantities, provided it is accurate. If the bill is employed, many items will have to be collected together under the heading of the programme operations (for example, the operation 'brickwork' will include such items as 'plain bands' and 'damp-proof courses').

The bill rates should be broken down to show labour and plant content so that the labour and plant content of operations can be easily ascertained.

### *1.1.3 Short-term planning*

Short-term planning is carried out in greater detail, and the programmes at this stage are broken down much further.

Short-term programming can cover a period of four to six weeks, or may cover a stage of the work. For six-week programmes a new programme is drawn up every fourth week, thus providing a two-week overlap and allowing a review to be made of the work outstanding from the previous month; any work behind schedule can then be included in the current programme.

The purpose of short-term planning is to ensure that work proceeds in accordance with the overall programme, which must be updated regularly, while a review is simultaneously made of all requirements by checking schedules.

The most suitable time for preparation of the programme is after the monthly planning meeting, which is in turn best held immediately after the Architect's monthly meeting. In this way account can be taken of any alterations or variations that are necessary.

The basis of calculations at this stage can be the bill of quantities, combined with operational records. Work study synthetics can be useful if available. There should be an integrated cost control/bonusing/planning and surveying system, which will provide the basic information for all departments. The expected outputs for bonus operations should be used in calculating the time required.

### *1.1.4 Control of progress on site*

Control must be carried out to make planning effective, as without control planning loses much of its value. It must be applied continuously to update the plans, and to enable reconsideration of the workload in the light of what has already taken place.

Control involves comparing, at regular intervals, the actual achievement with the plans and then taking any necessary corrective action to bring operations back on schedule.

If a programme is to be really effective as a control document, it must represent time and quantity of work carried out. This is particularly important on programmes where operations take several weeks or months to perform, as time alone will not clearly indicate whether work in progress is behind programme or not.

Progress can be recorded on planning charts that clearly indicate what is happening and where corrective action needs to be taken. Another method is to use pictorial diagrams, by colouring plans and elevations when certain sections of work are completed,. This method is often used on housing projects, and provides a visual impression of overall progress, but other ways are also available depending on the type of building and the amount of detail necessary.

### *1.1.5 Use of meetings*

Monthly and weekly meetings are invaluable in helping to control progress. The work that is not proceeding as planned will receive particular attention, and explanations will be required where sufficient progress is not being achieved. The action necessary for correcting underproduction will be considered, and the best solution will then be incorporated into the programme for the next period.

## 1.2 Bar charts

### *1.2.1 Introduction*

Bar charts are far less involved than network charts, and are particularly useful for small projects. Many contractors use bar charts for all their programmes, although there is an ever-increasing tendency for architects and engineers to request network presentation, either by asking for networks specifically or by asking for information that is best produced using networks: for example, requests to show critical activities and float.

In examination conditions, bar chart programmes may be the only reasonable method of presentation, because the alternative would necessitate drawing a network, analysing it, and then scheduling the project, which would involve resource allocation. This may then have to be adjusted to bring the project within the allowed duration. Consequently, unless a network is specifically requested in an examination, bar chart presentation would be more logical.

### 1.2.2 *Drawing bar charts*

As the name implies, bars are used to represent the duration over which an operation is planned to take place. They are also used to record the period over which the operation was carried out.

**Example 1.1: Roadworks project**

As an illustration, a roadworks project for a housing scheme will be used as follows:

Contract period 18 weeks

- ❑ *Clear site*
  - ■ Commence beginning on week 1.
  - ■ Complete end of week 3.
- ❑ *Site excavations*
  - ■ Commence beginning of week 1.
  - ■ Complete end of week 6.
- ❑ *Sewers*
  - ■ Commence beginning on week 3.
  - ■ Complete end of week 14.
- ❑ *Surface water drains*
  - ■ Commence beginning on week 3.
  - ■ Complete end of week 15.
- ❑ *Concrete road*
  - ■ Commence beginning on week 6.
  - ■ Complete end of week 16.

A bar chart programme for the project is shown in Fig. 1.1.

The duration of each operation must of course be realistic, and is best obtained from production data. This is clearly illustrated in the case studies in Chapter 7.

### 1.2.3 *Showing progress on bar charts*

The bars on the chart in Fig. 1.1 represent the duration of the operations, but they can also represent the quantity of work to be done: that is, the bar represents 100% of the work involved in the operation. By exploiting this, the chart can be used to present a considerable amount of information during the progress of the works.

| Programme for roadworks | | | | | | | | | | | | | | | | | | | |
|---|---|---|---|---|---|---|---|---|---|---|---|---|---|---|---|---|---|---|---|
| Week number | 1 | 2 | 3 | 4 | 5 | 6 | 7 | 8 | 9 | 10 | 11 | 12 | 13 | 14 | 15 | 16 | 17 | 18 | 19 |
| Clear site | | | | | | | | | | | | | | | | | | | |
| Site excavation | | | | | | | | | | | | | | | | | | | |
| Sewers | | | | | | | | | | | | | | | | | | | |
| Surface water drains | | | | | | | | | | | | | | | | | | | |
| Concrete roads | | | | | | | | | | | | | | | | | | | |

**Fig. 1.1** Bar chart programme for housing scheme roadworks project.

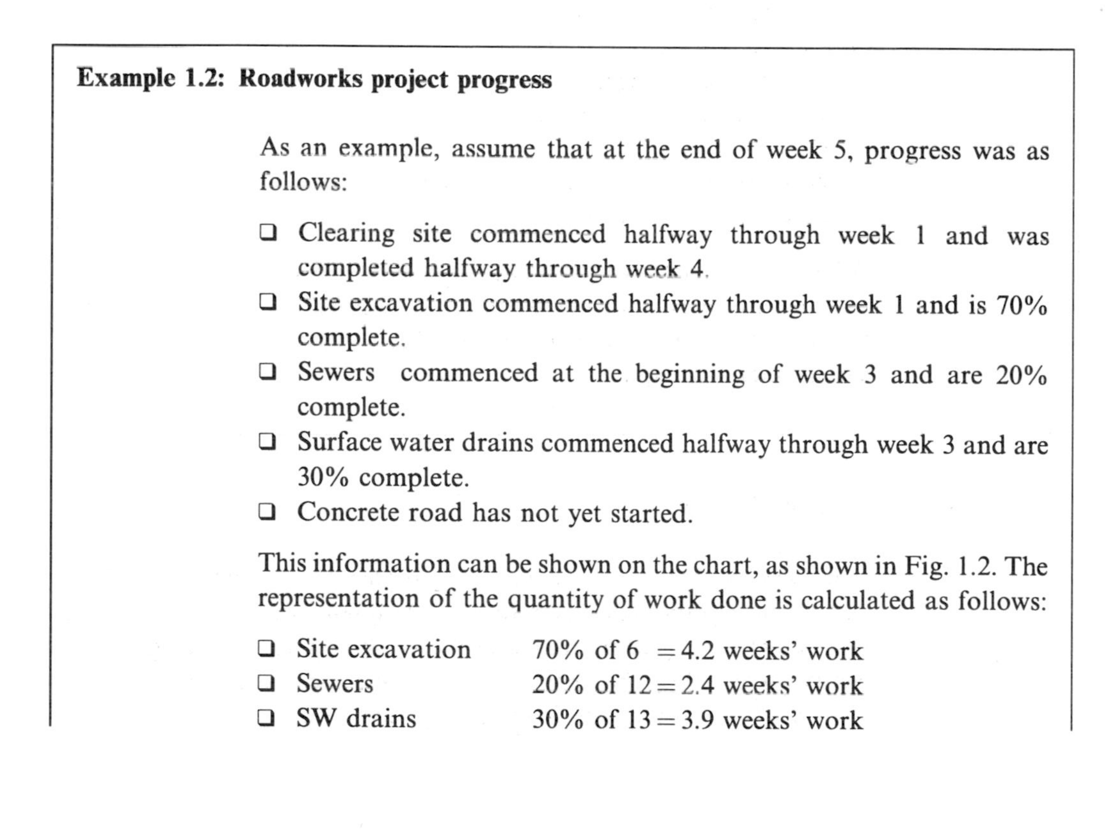

**Example 1.2: Roadworks project progress**

As an example, assume that at the end of week 5, progress was as follows:

- ❑ Clearing site commenced halfway through week 1 and was completed halfway through week 4.
- ❑ Site excavation commenced halfway through week 1 and is 70% complete.
- ❑ Sewers commenced at the beginning of week 3 and are 20% complete.
- ❑ Surface water drains commenced halfway through week 3 and are 30% complete.
- ❑ Concrete road has not yet started.

This information can be shown on the chart, as shown in Fig. 1.2. The representation of the quantity of work done is calculated as follows:

- ❑ Site excavation   70% of 6 = 4.2 weeks' work
- ❑ Sewers   20% of 12 = 2.4 weeks' work
- ❑ SW drains   30% of 13 = 3.9 weeks' work

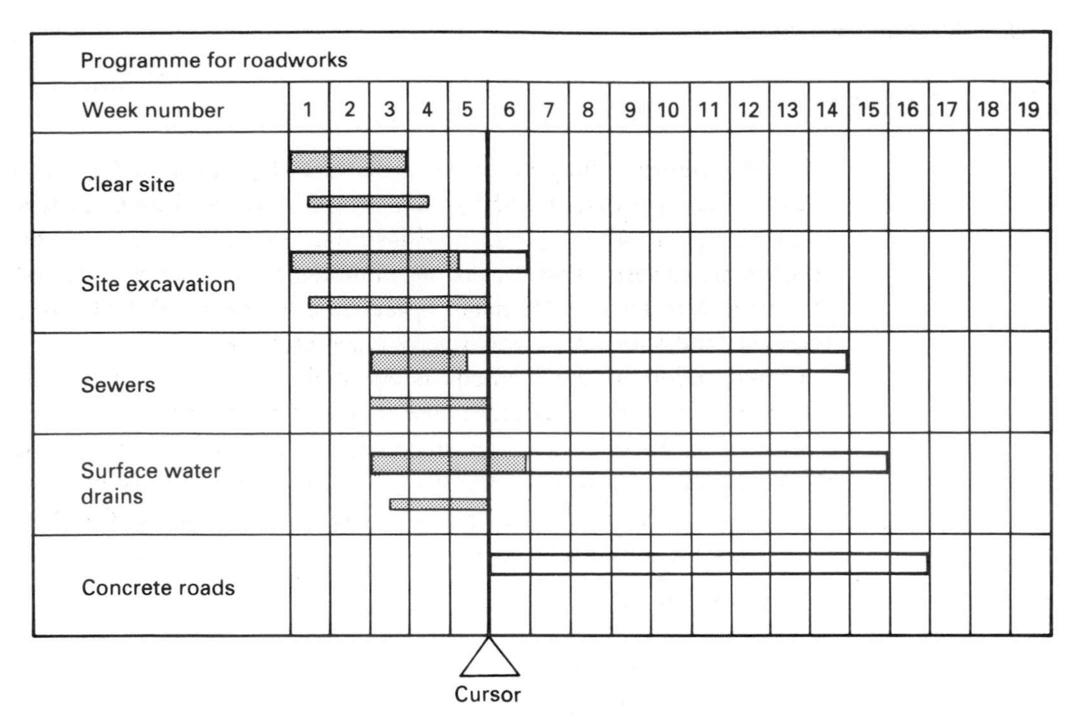


**Fig. 1.2** Roadworks project: progress at end of week 5.

The information presented can be summarised as follows:

- *Clear site*: commenced half a week late and completed half a week late. It took the amount of time programmed.
- *Site excavation*: Commenced half a week late, is still proceeding and is now 0.6 weeks behind programme. At the present rate it will finish 1 week behind programme.
- *Sewers*: commenced on time and is now 0.6 weeks behind programme. At the present rate of progress it will finish 3 weeks late. Transfer operatives from surface water drains (or increase the labour strength by 25%).
- *SW drains*: commenced 0.5 weeks late but is now 0.8 weeks in front. At the present rate of progress it will finish 4.67 weeks in front of programme. Transfer some resources to sewers.
- *Concrete road*: not yet started. Start next Monday.

## 1.3 Method statements

### 1.3.1 Introduction

Method statements may be drawn up at any stage of a project. The extent of detail is determined by the purpose of the method statement and the stage at which it is being drawn up. For example, at the pre-tender stage, decisions need to be taken on major plant to be used and the methods to be used for major operations. At the detailed planning stage, method statements would be in much more detail.

Consideration of the methods is essential before programming is carried out, as this clearly affects the operation durations.

Method statements are best presented on a standard form, and should include as much information as is appropriate to the stage at which they are being drawn up. An example of the information that might be required is as follows:

- a heading for the item being described
- description of method
- quantity if appropriate
- plant and equipment
- gang sizes if appropriate
- output if appropriate
- duration if appropriate.

Sometimes cost is added, but this cannot be ascertained for non-continuous operations involving plant until a programme is drawn up.

Whenever possible, the prospective site manager would be involved in determining methods.

Questions similar to those asked in the examination stage of work study should be asked before decisions are taken:

- What should be done?
- Where should it be done?
- When should it be done?
- Who should do it?
- How should it be done?

### 1.3.2 Pre-tender method statement

This is one of the most important pre-tender documents, because it is used by the estimator as the basis for arriving at the cost of many of the major items of work. It follows that the method statement should include all work of any consequence where alternative methods or choice of plant are possible.

The method statement takes the form of a schedule (see Fig. 1.3) stating the method to be adopted or plant to be used for such activities as excavation, formwork, and concreting. It is compiled at the same time as the pre-tender programme and site layout drawing, after the estimator has considered each problem in detail and has arrived at the best solution with regard to resources required and cost. Having arrived at the resources required, it is common practice to record these in the schedule for future reference. Together with the pre-tender programme it forms the basis for the overall programme. The schedule is also very useful for informing site staff of the method that the estimator has used when arriving at a price.

Pre-tender method statement
Office block, Manchester Road, Sheffield    Date:

| Item | Quantity | Remarks | Labour and plant | Time required |
|---|---|---|---|---|
| Excavation | 110 m$^3$ | Reduced level dig - Front bucket of JCB Sitemaster - easy going | JCB Sitemaster operator<br>3 labourers | 1 day |
| | 135 m$^3$ | Ground beam trenches - Rear backacter bucket of JCB Sitemaster )<br>Internal drain trenches - Rear backacter bucket of JCB Sitemaster ) | JCB Sitemaster operator<br>3 labourers | 1 week |
| | | Disposal of surplus spoil - Remove to tip 2 miles | 3 lorries<br>2 drivers | 1 week |
| General hoisting and transportation | | 750 kg platform hoist placed adjacent to front entrance | 750 kg platform hoist | 15 weeks |
| Concreting | 98 m$^3$ | Ready mixed concrete | | |
| | 98 m$^3$ | Transportation - Placed directly into trenches for foundations. Placed directly into GF slab using hit and miss method of construction | Ready mix lorry | |
| | | Mobile crane for first floor | Lorry mounted mobile crane | 1 day |
| | 98 m$^3$ | Placing - Poker vibrators for foundations and ground beams. | Vibrating pokers | 1 week |
| | | Vibrating screed for all slabs | Vibrating screed | GF - 1 week<br>FF - 1 week |

**Fig. 1.3** Part of a pre-tender method statement.

Time spent in the preparation of a well-thought-out method statement usually results in a more competitive tender, and when considered with a programme it can result in the solution of some potential problems before they arise.

Part of a pre-tender method statement is shown in Fig. 1.3.

### 1.3.3 Detailed method statement

Method statements can be compiled along with stage or short-term programmes, recording in detail the methods selected for completing a particular stage of the work on a project. The information in this type of method statement would be primarily for the benefit of the site staff, but it could be fed back to the estimator for his future reference.

Part of a detailed method statement is shown in Fig. 1.4.

Detailed method statement
Amenity centre, Bradford

| Item | Quantity | | Method | Output | Plant | Labour | Time required |
|---|---|---|---|---|---|---|---|
| Underpinning | 40 piles | 1 | Piles driven 2 m from face of existing building | - | Sub-contract | | 5 days |
| | 30 m$^3$ | 2 | Excavate preliminary trench down to base of existing foundation by hand-excavated material deposited at side of trench | 3 h/m$^3$ | | | |
| | 150 m$^3$ | 3 | Excavate trenches in short lengths to 600 mm below level of column bases in basement. Wall divided into six equal sections and sequence of excavation to be (1 and 10), (3 and 7), (5), (2 and 8), (4 and 6). Sections being excavated not to exceed quarter of length of wall at any time | 5 h/m$^3$ | | 3 labs | 32 days |
| | 54 m$^2$ | 4 | Support to side of excavation under building to be provided by precast concrete slabs as work proceeds, grouted up at back and left in-strutted off sheet piles | 1 h/m$^2$ | | | |
| | 6 m$^3$ | 5 | Concrete to foundations to be mixed on site and to be rebated at both ends | Mixer 1 h/m$^3$ | 150/100 mixer | | |
| | 45 m$^2$ | 6 | Brickwork to underside of existing foundation to be toothed both sides | 40 bks/h | 150/100 mixer | 1 blr 1 lab | |
| | | 7 | Weak concrete fill to back of wall as brickwork proceeds | To keep pace with Op. 6 | 150/100 mixer | 1 blr 1 lab | 31 days |
| | 3 m$^2$ | 8 | Pin up brickwork | 7.5 h/m$^2$ | | 1 blr 1 lab | |
| | 1.5 m$^3$ | 9 | Hack off existing foundations | 15 h/m$^3$ | Compressor 2 No drills | 3 labs | 1 day |
| Basement excavation | 84 piles | 10 | Piles driven around basement, 600 mm outside outer face of wall on NW, SW and SE sides | - | Sub-contract | | 12 days |
| | 545 m$^3$ | 11 | Commence excavation at SW side of site and work back towards car park. Dispose of all excavated material to tip. Compact bottom of excavation using plate vibrator | 0.1 h/m$^3$ | JCB Sitemaster 3 No 4 m$^3$ lorries | 3 labs | 7 days |

**Fig. 1.4** Part of a detailed method statement.

### *1.3.4 Further examples of method statements*

Further examples of method statements are shown in the case studies in Chapter 7.

## 1.4 Sequence studies

### *1.4.1 Introduction*

It is necessary to compile sequence studies when closely related operations have to be coordinated: for example, the construction of the columns, beams and slabs in a reinforced concrete framed building, or the finishing trades in a building such as a multi-storey office block, which would have much repetitive work on the various floors.

### *1.4.2 Achieving continuity*

To achieve continuity of work throughout these repetitive cycles, it is necessary to balance the gangs of workmen.

**Example 1.3: Procedure for preparing a sequence study**

Assume the frame of a 12-storey reinforced concrete building is to be erected and the work content per floor is as follows:

- *Columns and walls*
  - reinforcement 128 man-hours
  - concreting 20 man-hours
  - formwork 512 man-hours
- *Floors (including beams and slab)*
  - reinforcement 160 man-hours
  - concreting 80 man-hours
  - formwork 640 man-hours

An 8 hour day is used in the calculations. The hardening period for columns and walls is 2 days and for floors and beams 7 days, the props being left in. The preparation of the sequence study is then carried out as follows:

(1) Calculate the work content of steelfixers and carpenters

| | *Reinforcement* | *Formwork* |
|---|---|---|
| Columns and walls | 128 | 512 |
| Floors | 160 | 640 |
| Total | 288 | 1152 |

(2) Balance the gangs

Ratio of carpenters to steelfixers $\frac{1152}{288} = 4 : 1$

Therefore use 8 carpenters and 2 steelfixers (could have used other combinations giving a 4 : 1 ratio).

*Note*: Concrete gang will be used intermittently as required and will be working on other operations when not required on the frame.

The sequence will be arranged to reuse formwork as much as possible. In this example sufficient formwork for one complete floor will be required.

(3) Calculate operations times:

Time required for columns and walls:

reinforcement $\frac{128}{2 \times 8} = 8$ days (half floor = 4 days)

formwork $\frac{512}{8 \times 8} = 8$ days – assume 6 erect and 2 strip
(half floor assume 3 erect and 1 strip)

Time required for floors and beams:

reinforcement $\frac{160}{2 \times 8} = 10$ days

formwork $\frac{640}{8 \times 8} = 10$ days – assume $7\frac{1}{2}$ erect $2\frac{1}{2}$ strip
(half floor asume $3\frac{3}{4}$ erect $1\frac{1}{4}$ strip)

(4) Prepare the sequence study providing continuity on the repetitive work cycles on each floor (see Fig. 1.5). Of the three trades involved, the carpenters have the least flexibility because all their work takes place on the building itself. Steelfixers have a reasonable amount of flexibility because they can fabricate columns and beams when they are not actually fixing steel in position. When sufficient column cages have been erected it is possible to start the formwork, and care must be taken to allow the required hardening periods before the various formwork elements are stripped. The sequence of formwork erection and stripping is shown in Fig. 1.5.

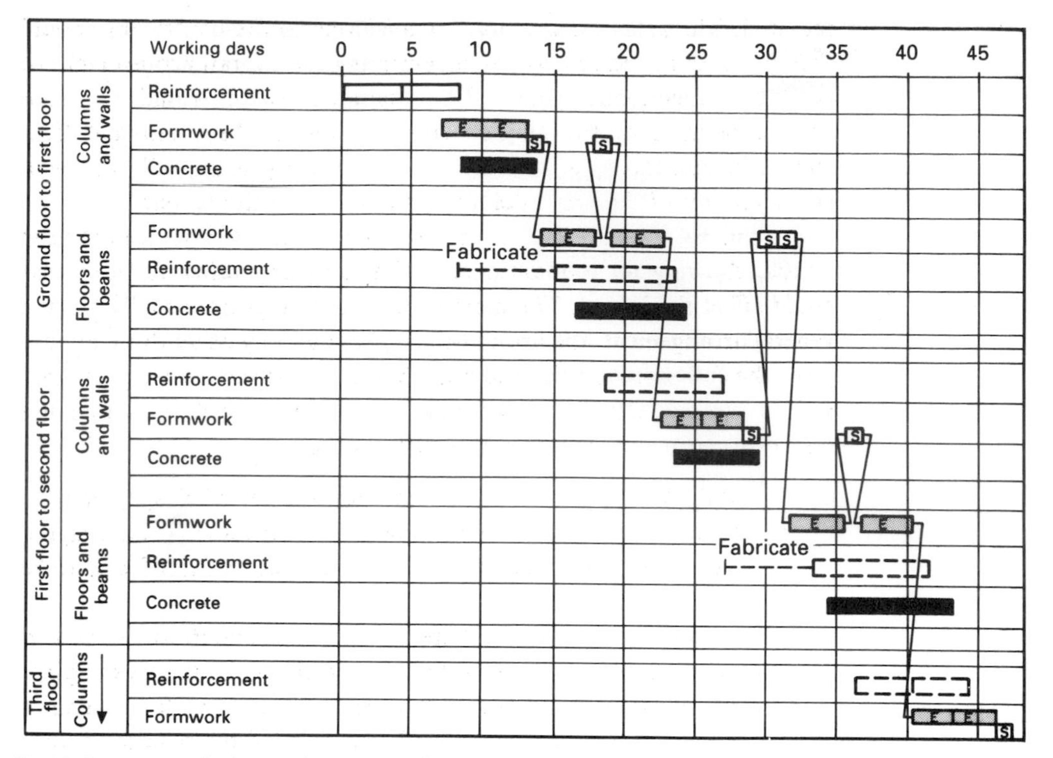


**Fig. 1.5** Sequence study: formwork erection and stripping.

As the formwork and steelfixing gangs have been balanced it is not necessary to show the fixing of steel in great detail, but merely to indicate the period over which it would be carried out. As stated previously, the concrete would be placed intermittently as required.

If more detailed planning is to be carried out, this would take place at the short-term planning stage.

## 1.5 Site layout

The physical factors of the site often affect the method and sequence adopted in the construction programme, so the programme and the site layout are usually prepared together.

### *1.5.1 Factors affecting site layout*

Before attempting to construct a site layout drawing, a list should be compiled of all accommodation, plant and material storage areas required. The areas on site that are available to the contractor should then be investigated to assist in determining the overall arrangement of the items mentioned above. For example, where would the staff accommodation or the major items of plant be best situated? This should be considered in overall terms at an early stage, followed by a consideration of the more detailed aspects of the site layout.

For this work a detailed site drawing is necessary to a scale of at least 1:100, showing all foundations, drains, and any other features that could affect the layout. The most suitable drawings are usually the site general arrangement, the ground-floor plan and any other drawing that gives the full extent of the site and shows all obstructions. As changes will be made frequently in arriving at the site layout, two common methods may be used to facilitate alterations:

- The scale drawing can be overlaid with a sheet of clear plastic on which the planner will draw his tentative layout with a wax crayon, which can easily be erased if parts of the layout are later found to be unsuitable.
- Silhouettes or models of all items of accommodation, plant and storage areas are produced to the same scale as the layout drawing, and placed on the scale drawing in the areas set aside for particular activities. They can then be moved around until the required layout is achieved. This method is a work study technique, which is very useful in obtaining the best layout.

Temporary roads may or may not be necessary, depending upon the type of subsoil experienced on site. When they *are* used they should be planned to serve all major items of plant and material storage compounds; hardcore or railway sleepers may be employed as construction material. In some cases the site service roads may be completed before building work commences, in which case they could be used to provide access about the site.

A number of factors can influence the location of accommodation, plant, storage areas and temporary roads, and these are considered in the example of a site layout shown in the case studies in Chapter 7.

## 1.6 The line of balance method (elemental trend analysis)

### *1.6.1 The advantages of the method*

One of the main advantages of this method is that it gives a better indication of the dependence of one activity on another. It is very useful when progressing because it is immediately obvious when corrective action needs to be taken.

**Example 1.4: Use of line of balance method**

Normally the line of balance method is used only when large numbers of units are to be built, but in order to illustrate the method a smaller project will be used.

Assume that ten identical units are to be constructed. The work content and sequence for some of the internal activities is shown in Table 1.1. No overlap is to be used between units. A 5 day week is to be worked.

**Table 1.1** Work content and sequences.

| Activity | Time required |
|---|---|
| Plumbing carcass | 4 gang-days |
| Electrical carcass | 3 gang-days |
| Carpenter first fix | 2 gang-days |
| Plastering | 4 gang-days |
| Plumbing sanitary fittings | 3 gang-days |
| Electrical fittings | 1 gang-day |
| Carpenter second fix | 5 gang-days |

In all types of repetitive construction, it is necessary to balance the gangs of operatives if non-productive time and project duration are to be kept to a minimum.

To illustrate this point, assume a line of balance schedule is prepared using one gang on each activity. Each gang is to be continuously occupied once they start on site, and no buffer is to be used between the activities. The line of balance schedule would be as shown in Fig. 1.6. It is clear that the project duration is excessive, which results in high overhead costs.

If a schedule is produced using one gang on each activity, as in Fig. 1.6, but starting each activity as soon as the previous one is completed on each unit, the line of balance schedule would be as shown in Fig. 1.7. This would reduce the project duration, but would result in excessive non-productive time, which in turn would result in high direct costs.

In order to prepare a schedule to give continuity of work and to complete the project in reasonable time, more gangs would be required. To clarify the method it is assumed that the completion rate for each unit is to be one per day. The number of gangs required to achieve this is shown in Table 1.2.

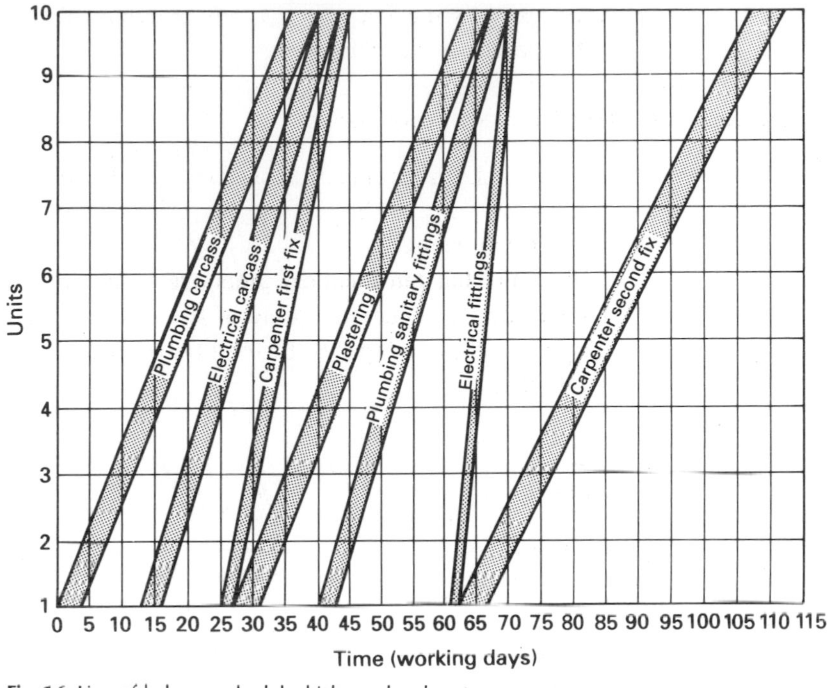


**Fig. 1.6** Line of balance schedule: high overhead costs.

**Table 1.2** Number of gangs required.

| Activity | Time required | Number of gangs |
|---|---|---|
| Plumbing carcass | 4 gang-days | 4 |
| Electrical carcass | 3 gang-days | 3 |
| Carpenter first fix | 2 gang-days | 2 |
| Plastering | 4 gang-days | 4 |
| Plumbing sanitary fittings | 3 gang-days | 3 |
| Electrical fittings | 1 gang-day | 1 |
| Carpenter second fix | 5 gang-days | 5 |

The work could be carried out using one gang in each unit (i.e. if two gangs are required, each gang would work on alternate units), or by using more than one gang in each unit, provided this does not result in overcrowding.

A line of balance schedule using one gang in each unit is shown in Fig. 1.8, which also shows the movement of some of the gangs. A 2 day buffer has been included to provide some flexibility.

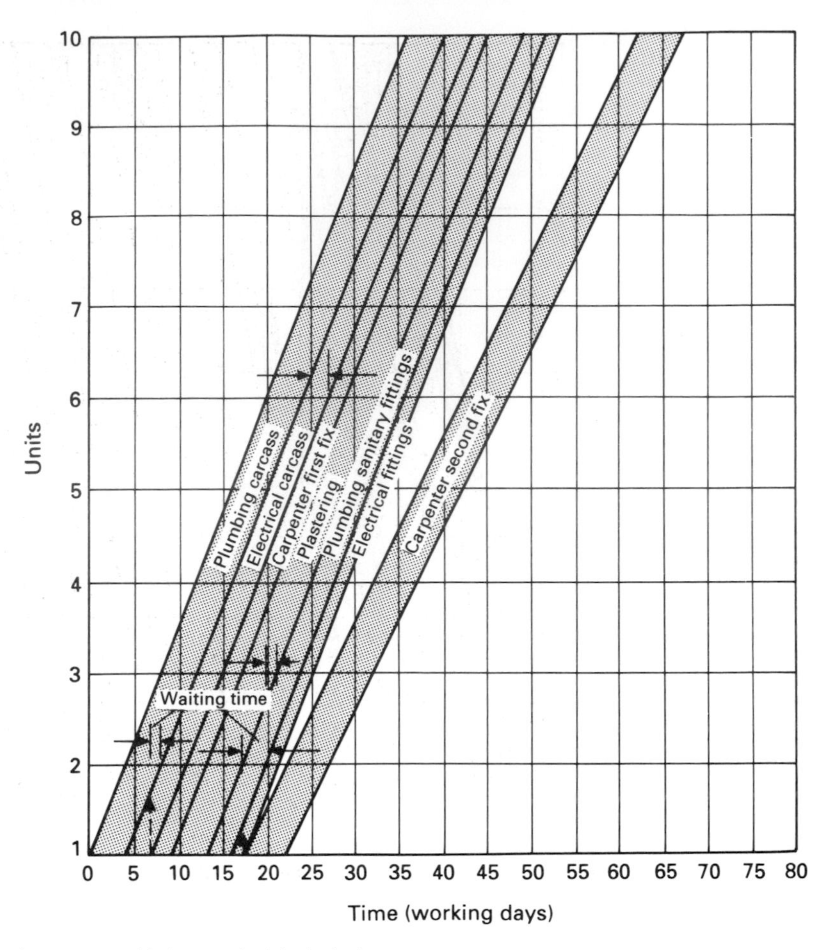


**Fig. 1.7** Line of balance schedule: high direct costs.

### 1.6.2 *Introducing costs to the line of balance method*

The type of construction for repetitive projects can vary from traditional labour-intensive projects to system-built capital-intensive projects, and this will have an influence on the type of line of balance schedule used.

Basically line of balance schedules can be produced by one of two methods:

- *Method 1: Parallel scheduling*. In this method the lines of balance are drawn parallel to each other, progressing at the required rate of production, often with no buffer between them. To ensure that all activities should be able to progress at this rate, the number of gangs

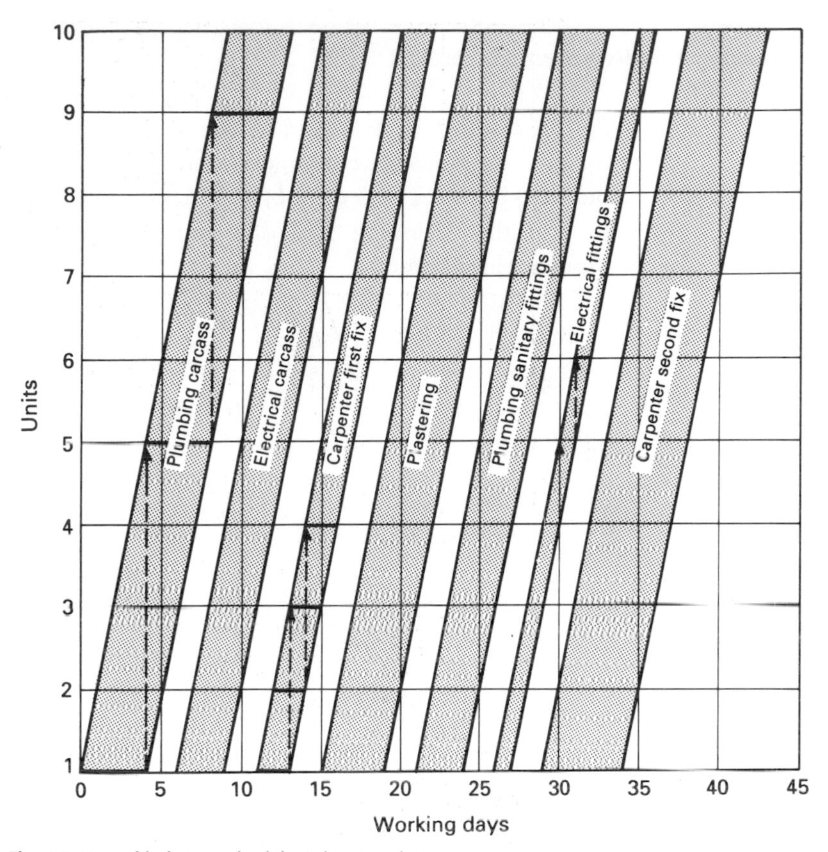


**Fig. 1.8** Line of balance schedule: a better solution.

required is always rounded upwards. This method of scheduling would be used on capital-intensive projects, where site plant and site overheads are expensive. The cost of loss of production caused by overmanning is more than saved by the reduction in the overall planned duration of the project.

- *Method 2: Resource scheduling*. In this method the schedule is determined by the progress of each activity, and interference between activities is avoided at the planning stage. This results in a longer project duration and consequently greater project overheads. This method is used on labour-intensive projects. To reduce the project duration, some of the activities can be carried out in stages, using a different number of gangs in each stage.

The method of scheduling is determined mainly on economic grounds.

**Example 1.5: Refurbishment project**

Part of a refurbishment project consists of the erection of the shells for 100 extensions. The completion rate is to be 10 units per week and the working week is 5 days of 8 hours per day. All activities are to start at the beginning of a day, and a buffer of 2 days must be provided at all times.

The cost of labour is £72 per man per day, and site overheads are to be charged at £360 per day.

The activity data are shown in Table 1.3. As the construction consists of small extensions, no overlapping of activities will be allowed.

**Table 1.3** Activity data.

| Activity no. | Activity | Total man-hours per unit | Gang size |
|---|---|---|---|
| 1 | Demolition of existing lean-to and load rubble | 24 | 2 |
| 2 | Construct foundation and make drain connection | 16 | 2 |
| 3 | Brickwork to DPC level | 15 | 3 (2 + 1) |
| 4 | Oversite concrete and hardcore | 11 | 2 |
| 5 | Brickwork to roof level | 69 | 3 (2 + 1) |

***Method 1: parallel scheduling***

*Procedure for programming (Table 1.4)*

(1) Using the minimum optimum number of operatives, calculate the time required by this gang ($G$) (Col 4).

(2) The handover rate required is ten units per week. Based on a 40 hour week one unit must be handed over every 4 hours after completion of the first unit. Calculate the number of gangs required to give this handover rate, i.e. gang-hours/4. Round *up* the number of gangs to the nearest whole number. This is because of the high capital costs to ensure all operations proceed at a minimum of 10 units per week. This will balance the gangs as nearly as possible (Col 5).

(3) Calculate the number of units per week, i.e. $R = 40g/G$ (Col 6).

(4) Determine the number of gangs per unit, i.e. $P$ (Col 7).

(5) Calculate the duration required per unit, i.e. $T = G/8P$ (Col 8).

**Table 1.4** Parallel scheduling.

| Activity no. | 1<br>Activity | 2<br>Total man hours per unit | 3<br>Min. gang size | 4<br>Time reqd by min. gang, $G$ | 5<br>Number of gangs, $g = G/t$ | 6<br>No. of units per week, $R = 40g/G$ | 7<br>No. of gangs per unit, $P$ | 8<br>Time per unit (days), $T = G/8P$ | 9<br>Time from start first to start last (days), $5(N-1)/R$ |
|---|---|---|---|---|---|---|---|---|---|
| 1 | Demolition | 24 | 2 | 12 | 3 | 10 | 1 | 1.5 | 49.5 |
| 2 | Foundations and drains | 16 | 2 | 8 | 2 | 10 | 1 | 1 | 49.5 |
| 3 | Brickwork to DPC | 15 | 3 (2 + 1) | 5 | 2 | 16 (10) | 1 | 0.625 | 30.94 (49.5) |
| 4 | Oversite concrete | 11 | 2 | $5\frac{1}{2}$ | 2 | 14.55 (10) | 1 | 0.7 | 34.02 (49.5) |
| 5 | Brickwork to roof | 69 | 3 (2 − 1) | 23 | 6 | 10.43 | 1 | 2.075 | 47.46 (49.5) |

**Table 1.5** Cost calculations for parallel scheduling.

| | | |
|---|---|---|
| Demolition | 6 × £72 = £ 432 per day | Cost = 500/10 × £ 432 = £ 21 600 |
| Foundations | 4 × £72 = £ 288 per day | Cost = 500/10 × £ 288 = £ 14 400 |
| Brickwork to DPC | 6 × £72 = £ 432 per day | Cost = 500/10 × £ 432 = £ 21 600 |
| Oversite concrete | 4 × £72 = £ 288 per day | Cost = 500/10 × £ 288 = £ 14 400 |
| Brickwork to roof | 18 × £72 = £1296 per day | Cost = 500/10 × £1296 = £ 64 800 |
| | | £136 800 |
| | | Overheads 66 × 360 £ 23 760 |
| | | £160 560 |
| | Delay start of brickwork to roof | Cost = 500/10.43 × £1296 = £ 62 128.50 |
| | | Saving £ 2 671.50 |

**Table 1.6** Additional calculations for resource scheduling with complete continuity for all activities.

| Activity no. | Activity | Total man-hours per unit | Min. gang size | Time reqd by min. gang | Number of gangs | No. of units per week | No. of gangs per unit | Time per unit (days) | Time from start first to start last |
|---|---|---|---|---|---|---|---|---|---|
| 3 | Brickwork to DPC | 15 | 3 (2 + 1) | 5 | 1 | 8 | 1 | 0.625 | 61.88 |
| 4 | Oversite concrete | 11 | 2 | $5\frac{1}{2}$ | 1 | 7.27 | 1 | 0.7 | 68.09 |

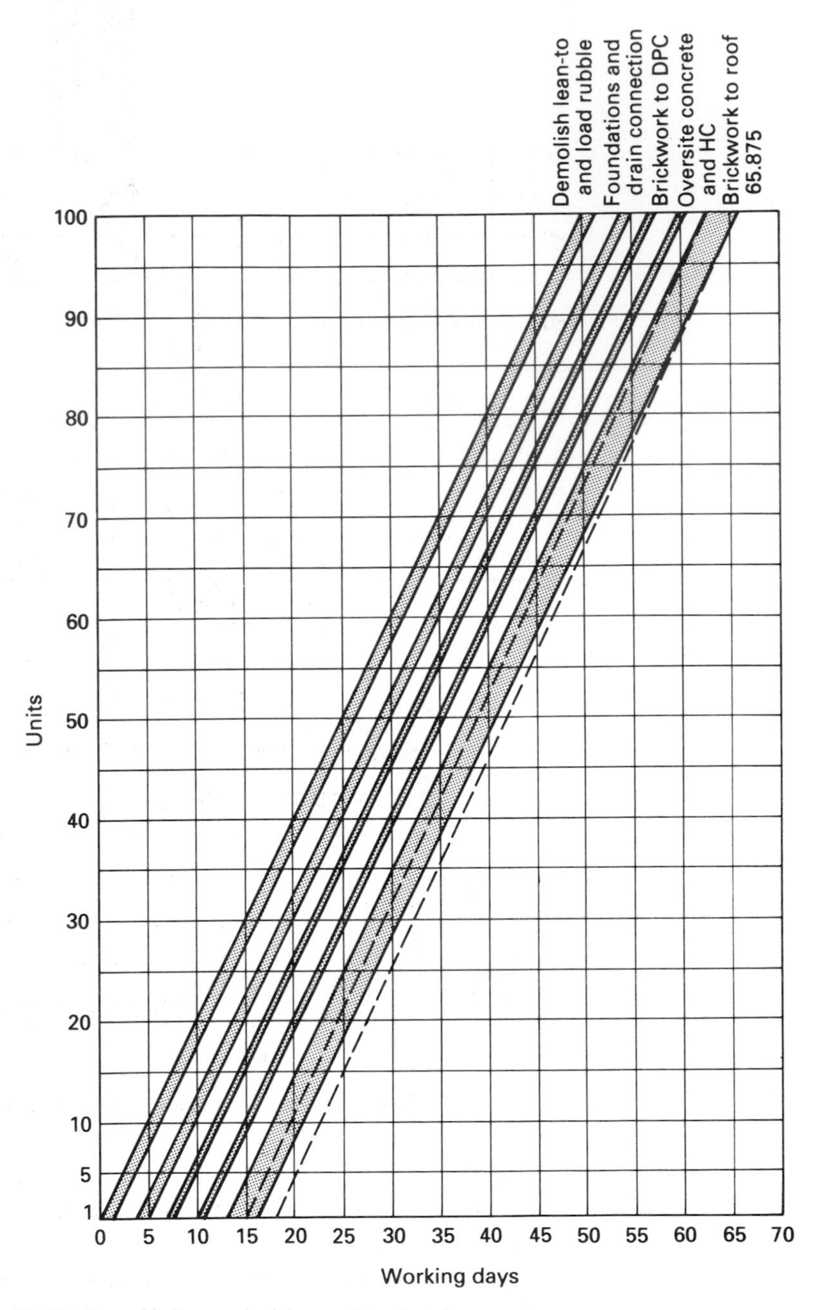


**Fig. 1.9** Line of balance schedule: parallel scheduling.

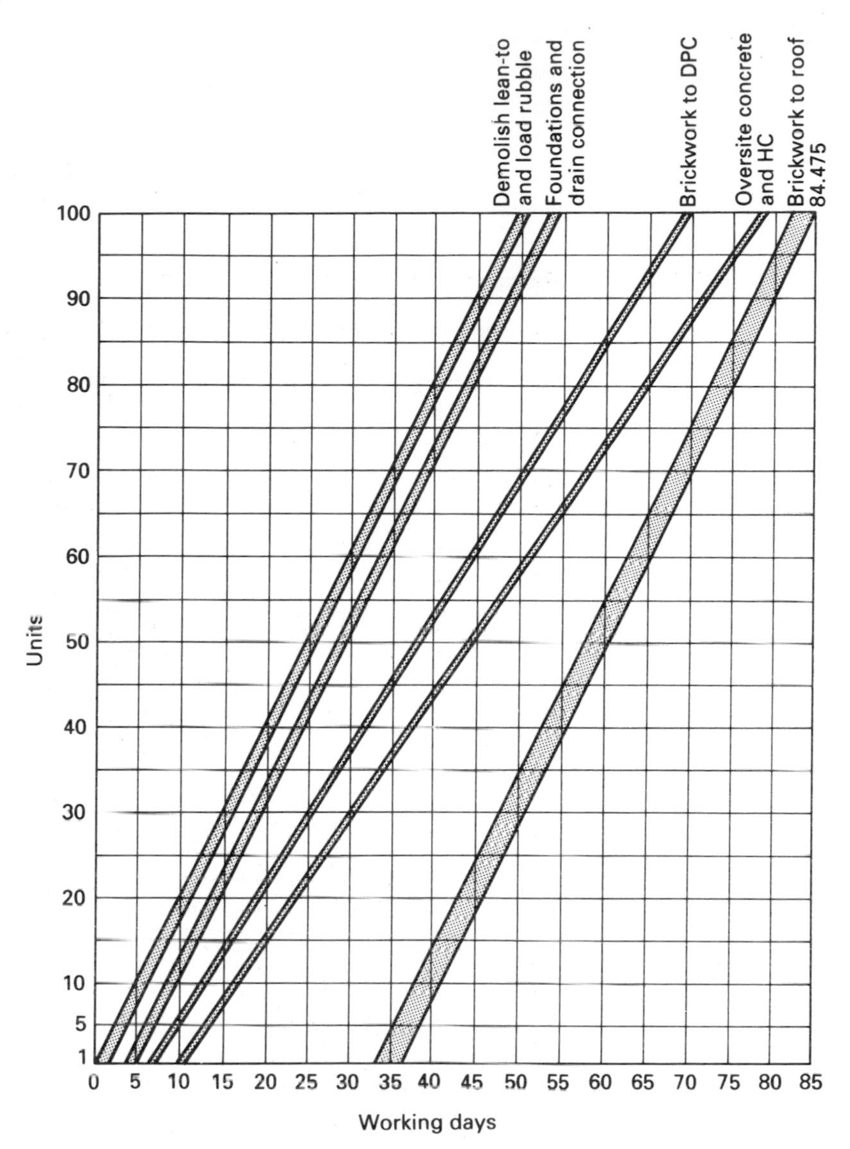

**Fig. 1.10** Line of balance schedule: resource scheduling with complete continuity for all activities.

(6) Calculate the overall duration from the start of the first unit to the start of the last unit, i.e. $5(N-1)/R$, (Col 9) when $N$ is the number of units.

(7) Plot the line of balance schedule based on the above information (Fig. 1.9).

The line of balance schedule is shown in Fig. 1.9, and the cost calculations are shown in Table 1.5. The start time for brickwork to roof could be delayed, and the activity could still be completed by day 66. This would result in a saving of £2671.50 as shown. This would not, however, be possible if the roof construction was to follow immediately after. The cost could of course be reduced by omitting the buffers, but in a project of this type this would result in considerable interference between gangs, which could outweigh any theoretical savings.

### *Method 2: resource scheduling*

*Resource scheduling with complete continuity for all activities*

The additional calculations are shown in Table 1.6, but the number of gangs is rounded to the nearest number. The line of balance schedule is shown in Fig. 1.10, and the cost calculations in Table 1.7. It can be seen that the project duration is extensive, owing to the different rates of progress of the activities, even though the gangs have been balanced as nearly as possible. If the gangs had been rounded upwards, the times from the start of the first unit to the start of the last one would have been as shown in Table 1.4 for each activity. The project duration would have been reduced by two days to 83 days, resulting in a saving of 2 × £360 = £720.

**Table 1.7** Cost calculations for resource scheduling.

| | | |
|---|---|---|
| Demolition | 6 × £72 = £432 per day | |
| Foundations | 4 × £72 = £288 per day | |
| Brickwork to DPC | 3 × £72 = £216 per day | |
| Oversite concrete | 2 × £72 = £144 per day | |
| Brickwork to roof | 18 × £72 = £1296 per day | |
| Cost = 500/10 | × £ 432 = | £ 21 600.00 |
| Cost = 500/10 | × £ 288 = | £ 14 400.00 |
| Cost = 500/8 | × £ 216 = | £ 13 500.00 |
| Cost = 500/7.27 | × £ 144 = | £ 9 904.50 |
| Cost = 500/10.43 | × £1296 = | £ 62 128.50 |
| | | £121 533.00 |
| Overheads 85 × £360 | | £ 30 600.00 |
| | | £152 133.00 |

*Resource scheduling with some activities carried out in stages*

A number of solutions are possible when adapting the schedule, and the optimum solution will depend on a number of factors, such as site overheads, cost of labour, the extent to which the activities are out of parallel, or the effects of lack of continuity. All the relevant factors must be taken into account when developing a solution, but because of all these factors determination of the optimum solution is extremely complex. In practical terms it is more appropriate to

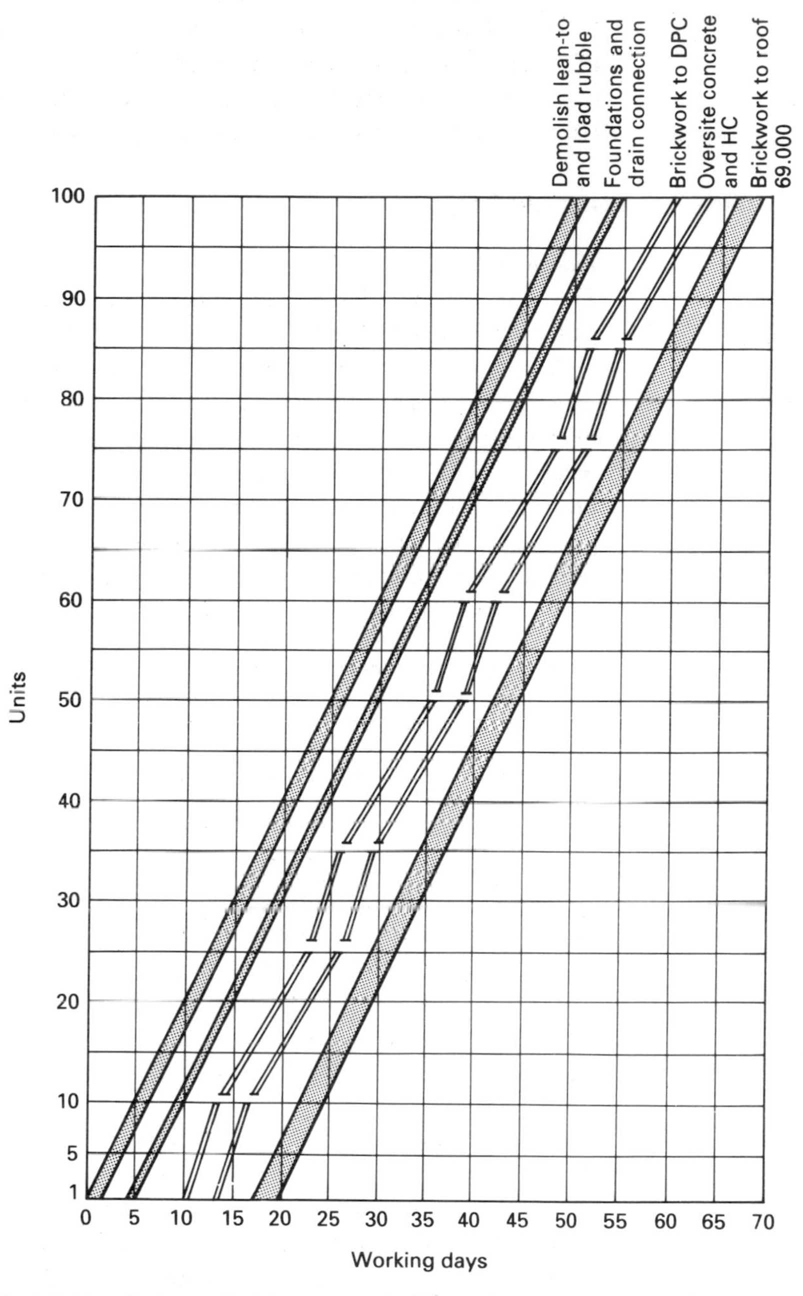


**Fig. 1.11** Line of balance schedule: resource scheduling with some activities carried out in stages.

analyse each schedule logically and determine an economic solution based on this approach.

To simulate the effects of non-continuity of work, a charge of £45 per man will be made each time resources are brought back to site. In this example one solution is to alternate the number of gangs on 'Brickwork to DPC' and on 'Oversite concrete'.

Considering the activity 'Brickwork to DPC', to achieve a 'start first' to 'start last' duration of 49.5 required for 10 units per week, it is necessary to alternate between two gangs and one gang. The number of units to be completed in each case can be calculated as follows:

activity duration in each unit using one gang = 0.625 days
average activity duration in each unit using two gangs = 0.3125 days

$$x \times 0.625 + (100 - x) \times 0.3125 = 50$$
$$0.625x - 0.3125x = 50 - 31.25 = 18.75$$
$$x = 18.75 \div 0.3125$$
$$= 60 \text{ units}$$

Therefore 60 units are to be built using one gang and 40 units using two gangs.

As units are to be built in stages, build 15 units with one gang and 10 units with two gangs in each stage.

Additional calculations are necessary to determine the time from 'start first' to 'start last' in each stage as follows:

Fifteen units using one gang $5(15 - 1) \div 8 = 8.75$
Ten units using two gangs $5(10 - 1) \div 16 = 2.81$

Similar calculations for oversite concrete would not produce an economic programme, and it has therefore been decided to programme this activity in parallel with brickwork to DPC.

The line of balance schedule is shown in Fig. 1.11.

The adjustment to the price relative to the resource scheduling example shown in Fig. 1.10 is set out below:

| | | |
|---|---|---|
| *Saving in cost* | | |
| Overheads (85 – 69) × £360 | = £5760 | £5760 |
| *Additional costs* | | |
| Brickwork to DPC | | |
| 3 (men) × 3 (returns to site) × £45 | = £465 | |
| Oversite concrete | | |
| 2 × 3 × £45 | = £270 | |
| | £675 | £ 675 |
| | Savings = | £5085 |

Chapter 2

# Project Network Techniques

## 2.1 Introduction

Project network techniques cover a number of techniques, one example being the Critical Path Method. Network techniques are particularly applicable to one-off projects, and hence are of considerable use for many construction projects.

For small projects, networks can be successfully analysed by hand, but on larger projects computers can be useful, and save time in analysis, re-analysis and updating. This applies particularly when cost optimisation and/or resource allocation is being undertaken.

Networks can be presented as arrow diagrams (activity on arrow) or precedence diagrams (activity on node).

## 2.2 Advantages of network techniques over the bar chart

When using network techniques, the interrelationship of all operations is clearly shown. The normal bar chart does not do this, and consequently requires the dependence of one operation on another to be remembered by the planner: this is extremely difficult with large projects, and in addition the site manager (who must carry out the work) has to be informed how dependent one operation is on another.

When a delay occurs, and networks are being used, critical operations will stand out as requiring particular attention. When bar charts are used on a large project many operations tend to be 'crashed' unnecessarily, as it is almost impossible to remember which operations are interdependent.

It is far easier for anyone taking over a partially completed project to become familiar with the progress when networks are employed.

When using networks it is essential to study the sequence of operations very carefully, leading to a closer understanding of the project.

Planning, analysing and scheduling are separated when using networks, which allows a greater concentration on the planning aspect.

## 2.3 The preparation of network diagrams

The first stage in the preparation of network diagrams is to make a list of the activities to be used. The amount of detail required in the breakdown depends on many factors, such as the size of the project or the stage of planning, i.e. pre-tender, pre-contact, or short term. One activity in a network drawn at the pre-tender stage may be broken down into a number of activities at a later stage. To avoid misunderstanding, diagrams are arranged so that time flows left to right. When drawing network diagrams it is important to remember that off-site activities such as delivery of plasterboard can be critical, and must therefore be included on the diagram.

As each activity is considered, the following questions should be asked:

- ❑ Which activity must be completed before this activity can start?
- ❑ Which other activities cannot start until this activity is completed?

A common mistake made here is for the planner to show an activity where he thinks it will be done, rather than the earliest time at which it can be done.

## 2.4 The preparation of arrow diagrams

### *2.4.1 Activities and events*

An arrow diagram consists of activities and events. An activity is an operation or process. All activities start and finish at an event, which is a point in time and may be the junction of two or more activities (Fig. 2.1).

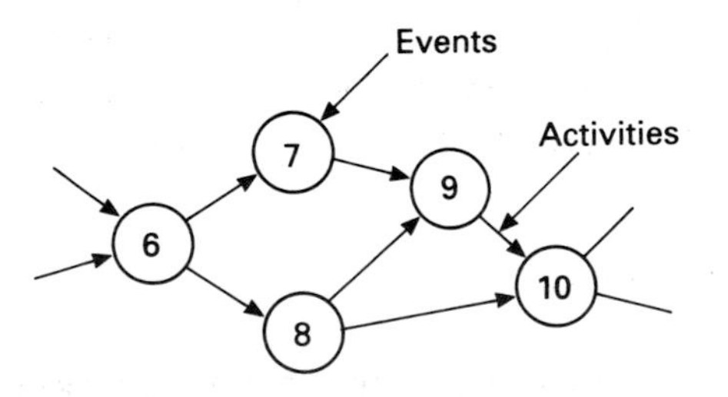


**Fig. 2.1** Activity 9–10 cannot start until activities 7–9 and 8–9 are completed.

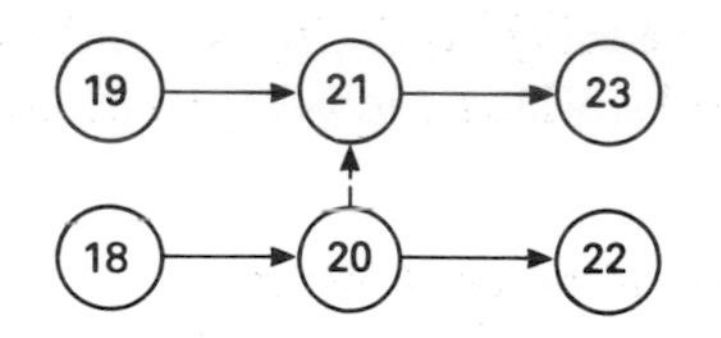


**Fig. 2.2** Activity 21–23 cannot start until activities 19–21 and 18–20 are completed, but activity 20–22 can start when activity 18–20 is completed, and is not dependent on activity 19–21.

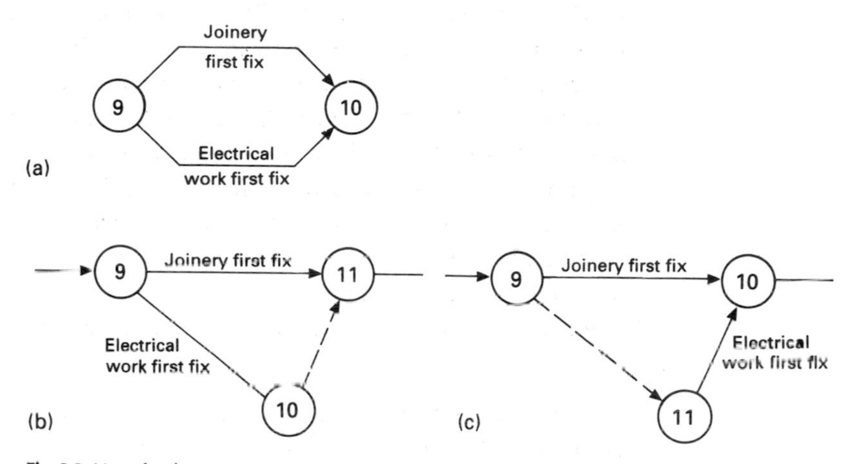


**Fig. 2.3** Use of a dummy activity to give a unique numbering system: (a), (b) and (c) represent the same situation.

Activities take time whereas events do not, and an activity cannot be started until all the activities leading into its preceding event are completed.

Activities are represented by arrows, and the sequence of the arrows represents the sequence of activities. Events are normally represented by circles.

Dummy activities are sometimes necessary in an arrow diagram. These do not take time to perform, and are used either to make the sequence clear (Fig. 2.2) or to give a unique numbering system (Figs 2.3(a), (b) and (c)). If a dummy is not used here, two activities would have the same reference: for example, joinery first fix and electrical work first fix (Fig. 2.3(a)) would have the same preceding and succeeding event. This is not acceptable when using a computer for analysis, and it is therefore essential that a unique numbering system be used (Figs 2.3(b) and (c)).

### 2.4.2 *Overlap of activities*

Many activities in a building project can overlap, one activity starting before the previous activity is completed. For example, in a large

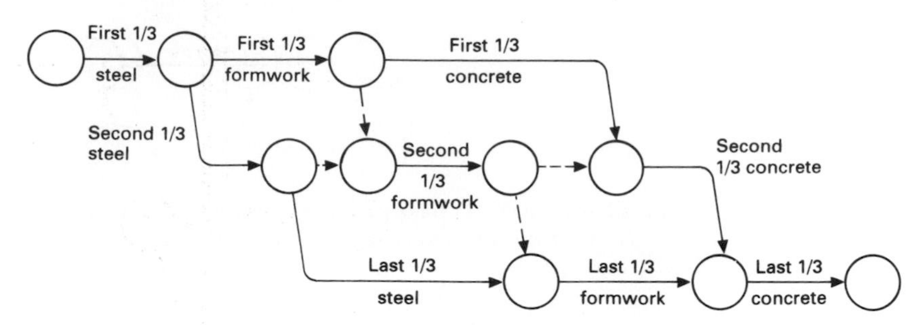


**Fig. 2.4** Overlap of activities: division of each activity into smaller portions.

concreting operation involving numerous columns, steelfixing would begin first and then formwork could be started when perhaps two columns had been completed; concreting would then follow when the formwork was sufficiently advanced. To show this in a network it is necessary to split each activity.

(1) Divide each activity into smaller equal portions (Fig. 2.4). This method of dividing each activity into the same number of equal parts is the only method that truly portrays the situation, but if

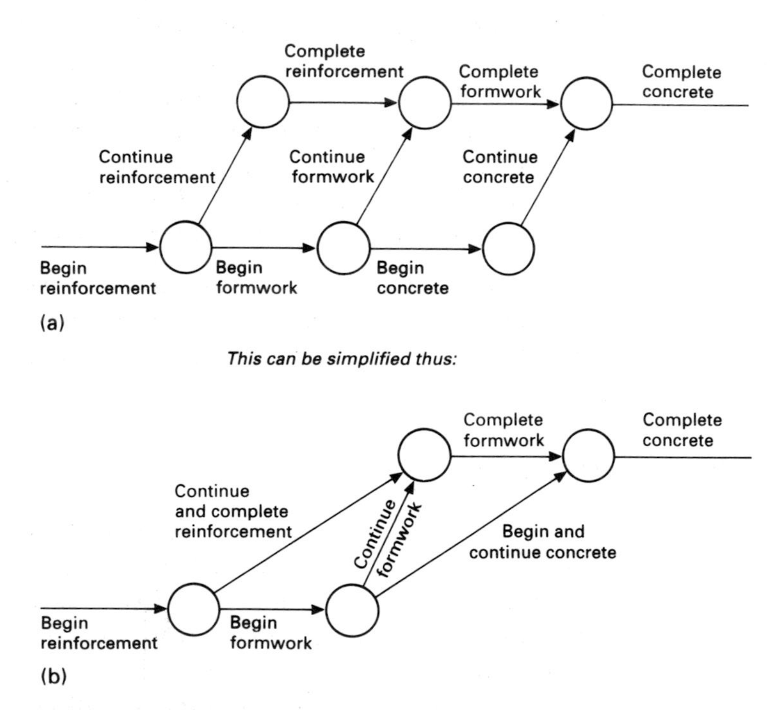


**Fig. 2.5** Overlap of activities: division of each activity into three sections (begin, continue, complete).

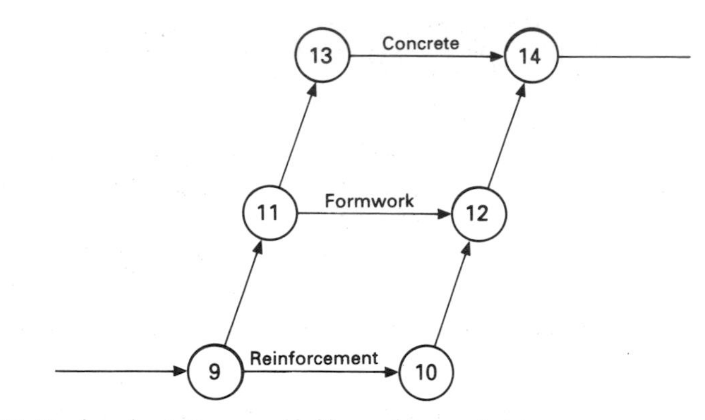


**Fig. 2.6** Overlap of activities: use of ladders and time restraints.

each activity is divided into many parts the diagram becomes unduly complicated, and in practice one of the following methods is often used as an alternative.

(2) Divide each activity into three sections – begin, continue and complete – and again show the sequence in the normal way (Figs 2.5(a) and (b)).

(3) Use *ladders* and *time restraints* (Fig. 2.6). Activities 9–11 and 11–13 are known as *lead time* and 10–12 and 12–14 as *lag time.* Activity 9–11 represents the amount of reinforcement that must be fixed before formwork can start. Activity 11–13 represents the amount of formwork to be erected before concrete can be poured. Activity 10–12 represents the last of the formwork to be erected after steelfixing is completed. Activity 12–14 represents the last of the concrete to be poured after the formwork is completed.

Methods 2 and 3 will give the correct event times, but the relationship between activities may not be truly represented when the analysis is tabulated. Care is therefore necessary when using these techniques.

**Example 2.1: Village hall project**

Figure 2.7 shows a village hall, which is attached to an existing building by means of a link corridor. The new building is to be constructed using timber portal frames with side elevations clad with brickwork and end elevations clad with timber frame units incorporating windows and doors. In this example the activities are limited to those listed below, and for simplification no overlap has been allowed.

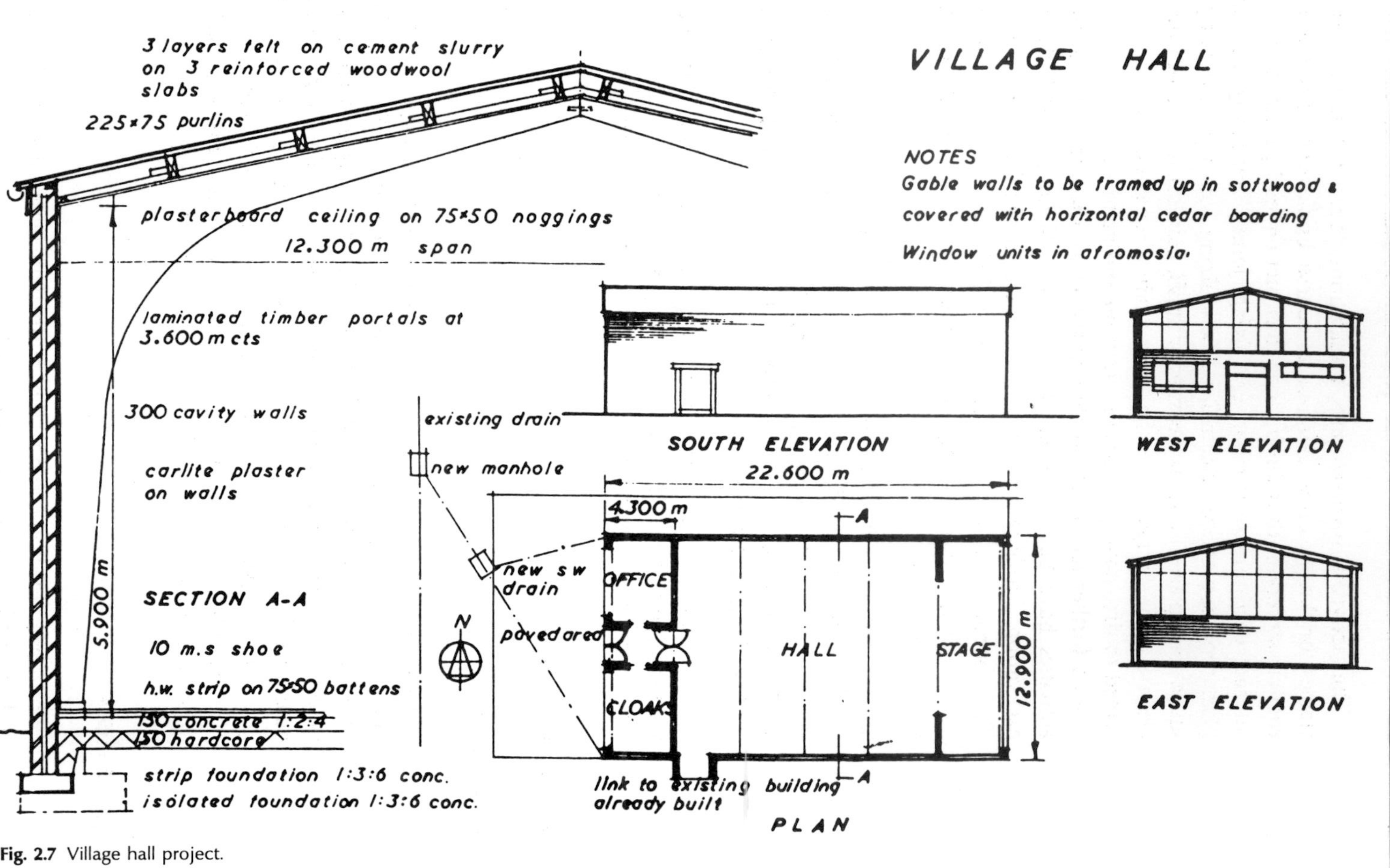


**Fig. 2.7** Village hall project.

The activities involved are: foundations, deliver frames, erect frames, brickwork up to damp-proof course, brickwork up to eaves, roof construction, roof finish, floor construction, end panels, glazing, internal partitions, electrician first fixing, joiner first fixing, deliver plasterboard, plasterer, electrician second fixing, joiner second fixing, make good plaster, floor screeds, floor finish, painting and decorating, electrician finishings, joiner finishings, rainwater goods, allow floor screed to dry, paving, drainage, clean up and hand over.

As a further means of simplifying the arrow diagram some of the activities have been grouped. For example, 'foundations' includes excavation, levelling and ramming, and concrete bed. The arrow diagram based on these activities and the drawing (Fig. 2.7) is shown in Fig. 2.8; the following assumptions have been made in its construction:

- ❑ The end panels are built off the concrete slab.
- ❑ The internal partitions cannot be built until the external brickwork and end panels are complete.
- ❑ The building must be weatherproof before joiner and electrician first fixings are carried out.

The sequence of activities shown in Fig. 2.8 is not unique, and other planners might draw this network slightly differently.

## 2.5 Analysis of arrow diagrams

### *2.5.1 Activity duration*

Activity durations (shown on the arrows) can be in days, weeks or months, depending on how detailed the diagram is to be. The durations for contractors' activities would be arrived at by using past records of outputs, synthetic times based on work study, and data from bills of quantities. The gang sizes and plant to be used would be the ones that give optimum performance at this stage irrespective of the requirements of other activities. Subcontractors and suppliers should be approached in order to obtain realistic durations for their activities.

### *2.5.2 Project duration*

The project duration is the minimum time in which it can be completed with the activity times assigned to it.

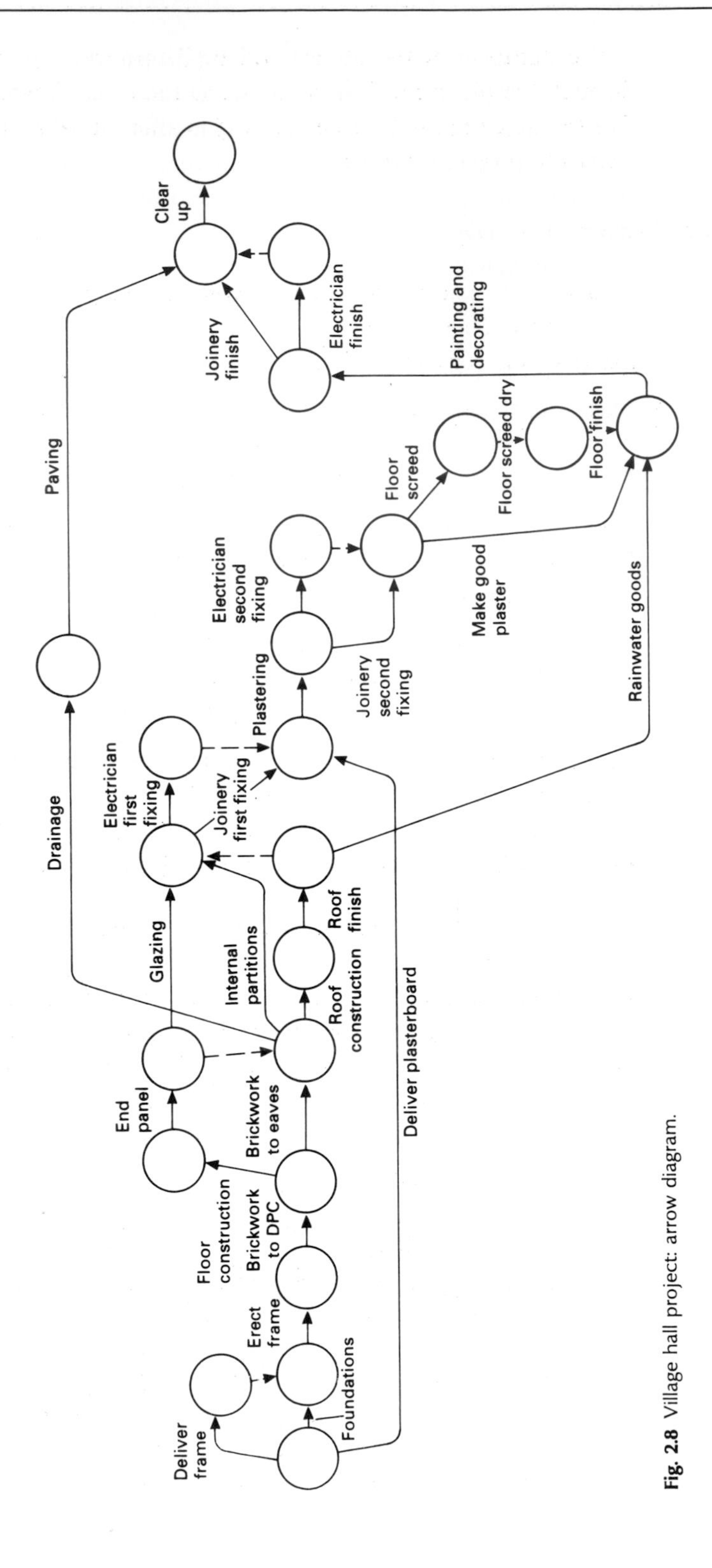


**Fig. 2.8** Village hall project: arrow diagram.

The duration of the project will be determined by the longest path through the diagram. This is known as the *critical path*.

If the time required for any activity is affected, this will automatically affect the project duration.

### 2.5.3 Method of analysis (Fig. 2.9)

*Calculate the earliest event times by working through the diagram and selecting the longest path*

Earliest time for event 1 is day 0:

Therefore the earliest time for event 2 is $0 + 3 = 3$ weeks
for event 3 is $0 + 2 = 2$ weeks
and for event 4 is $0 + 5 = 5$ weeks

Event 5 has three activities leading into it, and the earliest time for event 5 is determined by the longest path. The alternative paths give the following results:

event 2 to event 5 would give $3 + 0 = 3$ weeks
event 3 to event 5 would give $2 + 4 = 6$ weeks
event 4 to event 5 would give $5 + 3 = 8$ weeks

The longest path to event 5 is therefore from event 4 to event 5, which means that the earliest time for event 5 is 8 weeks.

The remainder of the earliest event times are calculated in this way, and the earliest time for the final event gives the project duration.

*Calculate the latest event times by working backwards through the diagram again, selecting the longest path*

The latest event time is the latest time by which an event can be completed without affecting the project duration, so that the latest time for the final event is the same as its earliest time (the project duration).

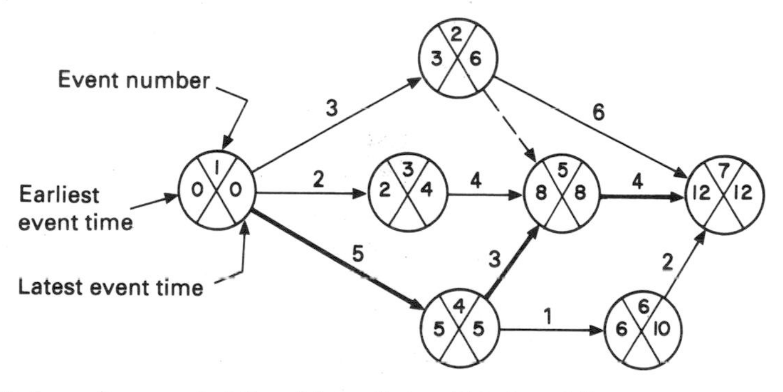


**Fig. 2.9** Arrow diagram: calculation of the earliest and latest event times.

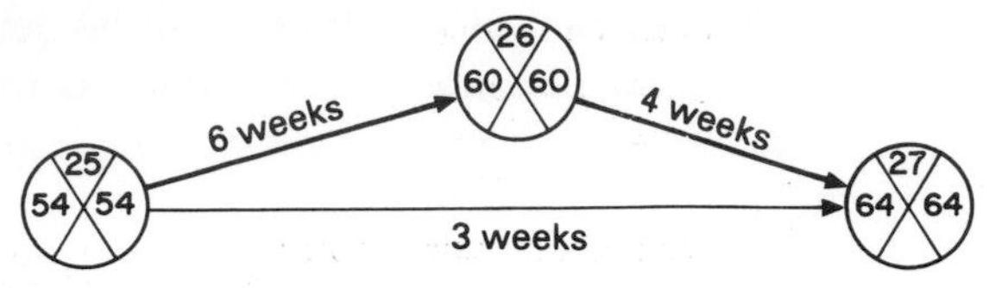


**Fig. 2.10** Arrow diagram: locating critical path.

Therefore the latest time for event 7 is week 12
for event 6 is 12 – 2 = 10 weeks
and for event 5 is 12 – 4 = 8 weeks

Event 4 has two activities leaving it, so the latest time for event 4 is determined by the longest path. The alternative paths give the following results:

event 5 to event 4 would give 8 – 3 = 5 weeks
event 6 to event 4 would give 10 – 1 = 9 weeks

The longest path to event 4 is therefore from event 5 to event 4, which means that the latest time for event 4 is 5 weeks.

*Critical events* have earliest times equal to latest times. The critical path passes through these events, along the activities whose duration is equal to the difference between the preceding and succeeding event times (Fig. 2.10). The critical path passes along activities 25–26 and 26–27, but 25–27, is not critical, as activity duration of three weeks is not equal to the difference between the times for events 25 and 27, i.e. 10 weeks.

### 2.5.4 Float

Float is the name given to the spare time available on non-critical activities. There are various types of float (Fig. 2.11), as follows.

- *Total float* is the total amount of spare time available in an activity. It is calculated by (latest succeeding event time) minus (earliest preceding event time) minus (activity duration). Note that total float can be the amount of float in a chain of activities.
- *Free float* is a part of total float, and represents the amount of spare time that can be used without affecting subsequent activities, provided the activity starts at its earliest time. It is calculated by: (earliest succeeding event time) minus (earliest preceding event time) minus (activity duration).

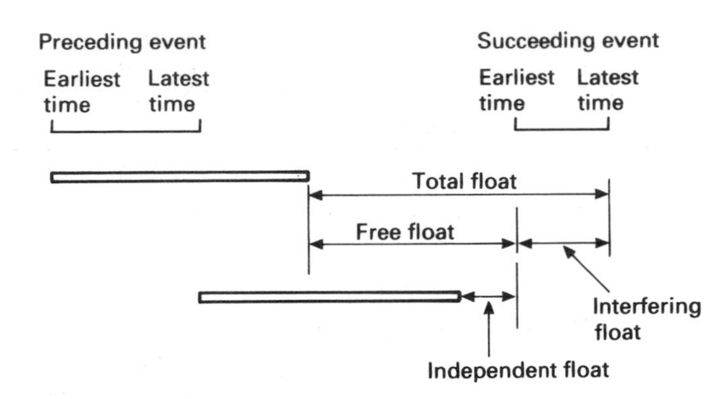


**Fig. 2.11** Arrow diagram: the various types of float.

- *Interfering float* is the amount of spare time available that, if used, will affect subsequent activities, and may be calculated by: (total float) minus (free float).
- *Independent float* is the amount of spare time available that can be used without affecting any succeeding activity, and which cannot be affected by any preceding activity. The calculation is: (earliest succeeding event time) minus (latest preceding event time) minus (activity duration).

### 2.5.5 *Significance of the various types of float*

If additional resources were required on a critical activity they could be obtained from:

- activities with independent float without affecting the rest of the network at all;
- activities with free float without affecting the float of subsequent activities;
- activities with interfering float only, which will affect the float of previous and subsequent activities.

If only the total float is calculated, this may include independent, free and interfering float.

### 2.5.6 *Tabular presentation of arrow diagrams*

The analysis of the diagram can be presented in tabular form, and when computers are used in analysis this is the way in which the results will normally be presented. Table 2.1 presents the information in terms of

Table 2.1 Tabular presentation of arrow diagram.

| Activity number | Duration | Earliest | | Latest | | Total float |
|---|---|---|---|---|---|---|
| | | Start | Finish | Start | Finish | |
| 1–2 | 3 | 0 | 3 | 3 | 6 | 3 |
| 1–3 | 2 | 0 | 2 | 2 | 4 | 2 |
| 1–4 | 5 | 0 | 5 | 0 | 5 | 0 |
| 2–5 | 0 | 3 | 3 | 8 | 8 | 5 |
| 2–7 | 6 | 3 | 9 | 6 | 12 | 3 |
| 3–5 | 4 | 2 | 6 | 4 | 8 | 2 |
| 4–5 | 3 | 5 | 8 | 5 | 8 | 0 |
| 4–6 | 1 | 5 | 6 | 9 | 10 | 4 |
| 5–7 | 4 | 8 | 12 | 8 | 12 | 0 |
| 6–7 | 2 | 6 | 8 | 10 | 12 | 4 |

activities in numerical order, but the sequence can be varied in many ways (for example, by arranging in the order of least total float). The table can also be used as a means of controlling the project.

### 2.5.7 Method of analysis

(1) List the activities in numerical order.
(2) Fill in the earliest start time, which is the earliest time for the preceding event (obtainable from the network; Fig. 2.9).
(3) Calculate the earliest finish time, which is the earliest start time plus activity duration.
(4) Fill in the latest finish time, which is the latest time for the succeeding event.
(5) Calculate the latest start time, which is the latest finish time minus the activity duration.
(6) Calculate the total float, which is either the latest start time minus the earliest start time or the latest finish time minus the earliest finish time. Both methods give the same result.

Note that critical activities have no float.

**Example 2.2: Analysis of village hall project**

Table 2.2 shows the activity durations and resources for the village hall in Fig. 2.7. The analysed arrow diagram for this project is shown in Fig. 2.12 and the completed table of results with total float in Table 2.3.

**Table 2.2** Activity durations and resources for village hall project.

| Activity | Man-hours | Plant-hours | Resources | Time req'd (days) |
|---|---|---|---|---|
| Foundations | | | | |
| Excavate oversite strip } Excavate for foundations } | 40 | 20 | JCB 4 + operator + 2 labourers } | 5 |
| Concrete foundations | 160 | 20 | 8 labourers 400/300 litre mixer } | |
| Frame | | | | |
| Delivery | | | | 10 |
| Erect | 16 | 8 | Crane + operator + 2 labourers | 1 |
| Brickwork to DPC | 96 | 16 | 4 bricklayers 2 labourers 150/100 litre mixer | 2 |
| Floor construction | | | | |
| Hardcore } Concrete } | 208 | 24 | 8 labourers 400/300 litre mixer | 3 |
| Brickwork to eaves | 528 | 88 | 4 bricklayers 2 labourers 150/100 litre mixer | 11 |
| End panels | 208 | | 2 joiners | 13 |
| Glazing | 112 | | Plumber and mate | 7 |
| Roof construction | | | | |
| Carcass } Woodwool } | 240 | | 2 joiners | 15 |
| Internal partitions | 96 | 16 | 4 bricklayers 2 labourers 150/100 litre mixer | 2 |
| Drainage | | | | |
| Excavation and backfill } Concrete bed } | 136 | 6 | 8 labourers 400/300 litre mixer } | 4 |
| Lay drains } Manholes } | 32 | 16 | Bricklayer and mate 150/100 litre mixer } | |
| Roof finish | 144 | | 2 roofers | 9 |
| Rainwater goods | 32 | | Plumber and mate | 2 |
| Electrician first fixing | 112 | | Electrician and mate | 7 |
| Joiner first fixing | 96 | | 2 joiners | 6 |
| Deliver plasterboard | | | | 70 |
| Plastering | 410 | 80 | 4 plasterers 2 labourers 150/100 litre mixer | 10 |
| Electrician second fixing | 48 | | Electrician and mate | 3 |

**Table 2.2** *Continued.*

| Activity | Man-hours | Plant-hours | Resources | Time req'd (days) |
|---|---|---|---|---|
| Joinery second fixing | 80 | | 2 joiners | 5 |
| Make good plaster | 8 | | Plasterers | 1 |
| Floor screed | 160 | 32 | 4 plasterers<br>1 labourer<br>150/100 litre mixer | 4 |
| Floor screed dry | | | | 10 |
| Floor finish | 192 | | 2 joiners | 12 |
| Painting and decorating | 256 | | 4 painters | 8 |
| Paving | 144 | | Bricklayer and mate | 9 |
| Electrical finishing | 16 | | Electrician and mate | 1 |
| Joinery finishing | 48 | | 2 joiners | 3 |
| Clean and hand over | 40 | | 1 labourer | 5 |

**Table 2.3** Completed results table for village hall project.

| Activity | Activity number | Duration | Earliest | | Latest | | Total float |
|---|---|---|---|---|---|---|---|
| | | | Start | Finish | Start | Finish | |
| Deliver frame | 1–2 | 10 | 0 | 10 | 10 | 20 | 10 |
| Foundations | 1–3 | 5 | 0 | 5 | 15 | 20 | 15 |
| Deliver plasterboard | 1–13 | 70 | 0 | 70 | 0 | 70 | 0 |
| Dummy | 2–3 | 0 | 10 | 10 | 20 | 20 | 10 |
| Erect frame | 3–4 | 1 | 10 | 11 | 20 | 21 | 10 |
| Brickwork to DPC | 4–5 | 2 | 11 | 13 | 21 | 23 | 10 |
| Floor construction | 5–6 | 3 | 13 | 16 | 23 | 26 | 10 |
| Brickwork to eaves | 5–8 | 11 | 13 | 24 | 28 | 39 | 15 |
| End panels | 6–7 | 13 | 16 | 29 | 26 | 39 | 10 |
| Dummy | 7–8 | 0 | 29 | 29 | 39 | 39 | 10 |
| Glazing | 7–11 | 7 | 29 | 36 | 56 | 63 | 27 |
| Roof construction | 8–9 | 15 | 29 | 44 | 39 | 54 | 10 |
| Internal partitions | 8–11 | 2 | 29 | 31 | 61 | 63 | 32 |
| Drainage | 8–21 | 4 | 29 | 33 | 109 | 113 | 80 |
| Roof finish | 9–10 | 9 | 44 | 53 | 54 | 63 | 10 |
| Dummy | 10–11 | 0 | 53 | 53 | 63 | 63 | 10 |
| Rainwater goods | 10–20 | 2 | 53 | 55 | 109 | 111 | 56 |
| Electrician first fixing | 11–12 | 7 | 53 | 60 | 63 | 70 | 10 |
| Joiner first fixing | 11–13 | 6 | 53 | 59 | 64 | 70 | 11 |
| Dummy | 12–13 | 0 | 60 | 60 | 70 | 70 | 10 |
| Plastering | 13–14 | 10 | 70 | 80 | 70 | 80 | 0 |
| Electrician second fixing | 14–15 | 3 | 80 | 83 | 82 | 85 | 2 |
| Joiner second fixing | 14–16 | 5 | 80 | 85 | 80 | 85 | 0 |

**Table 2.3** *Continued.*

| Activity | Activity number | Duration | Earliest | | Latest | | Total float |
|---|---|---|---|---|---|---|---|
| | | | Start | Finish | Start | Finish | |
| Dummy | 15–16 | 0 | 83 | 83 | 85 | 85 | 2 |
| Make good plaster | 16–20 | 1 | 85 | 86 | 110 | 111 | 25 |
| Floor screed | 16–18 | 4 | 85 | 89 | 85 | 89 | 0 |
| Floor screed dry | 18–19 | 10 | 89 | 99 | 89 | 99 | 0 |
| Floor finish | 19–20 | 12 | 99 | 111 | 99 | 111 | 0 |
| Painting and decorating | 20–22 | 8 | 111 | 119 | 111 | 119 | 0 |
| Paving | 21–24 | 9 | 33 | 42 | 113 | 122 | 80 |
| Electrician finish | 22–23 | 1 | 119 | 120 | 121 | 122 | 2 |
| Joiner finish | 22–24 | 3 | 119 | 122 | 119 | 122 | 0 |
| Dummy | 23–24 | 0 | 120 | 120 | 122 | 122 | 2 |
| Clean and hand over | 24–25 | 5 | 122 | 127 | 122 | 127 | 0 |

## 2.6 Scheduling

Scheduling is the process of determining the actual time periods during which the activities are planned to take place: that is, start and finish dates for each activity.

### 2.6.1 *Resource allocation*

Resource allocation is concerned with scheduling activities and their resources within predetermined constraints. Initially, when a network is drawn, no account is taken of the limit of availability of resources for any particular activity, and it is assumed that resources are always available when required. In some cases, however, the resources may be required on different activities, which are on parallel paths in the network.

### 2.6.2 *Resource levelling*

If availability of resources is a critical factor, then the project duration may well be influenced by insufficient resources. Resource levelling is used to ensure that availability is not exceeded.

### 2.6.3 *Resource smoothing*

If the project duration is of prime importance, the first analysis will often show an excessive duration. The critical path method is extremely useful

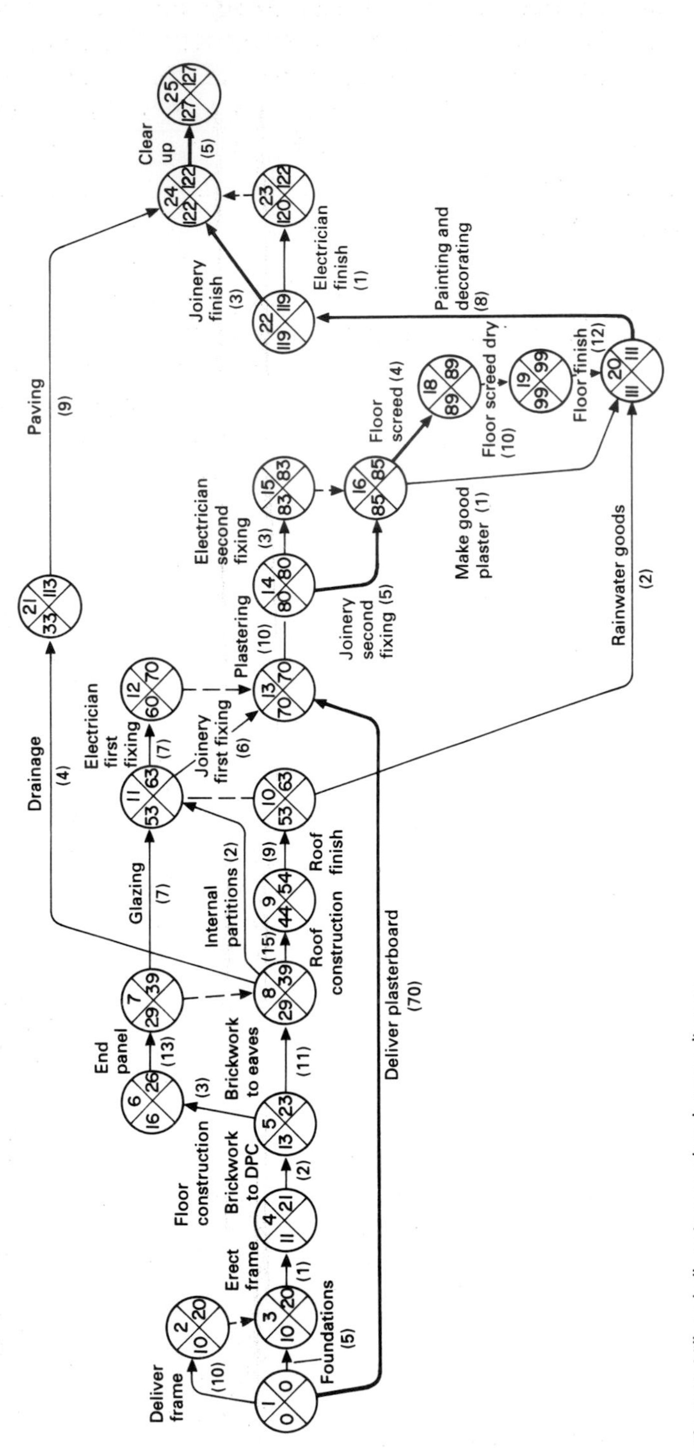


**Fig. 2.12** Village hall project: analysed arrow diagram.

in these circumstances, as it highlights those activities that must be examined in order to reduce the project duration. The aim will be to reduce the time required for those activities that will result in least cost overall.

Resource smoothing is used when some smoothing of resources is carried out within the activity floats to limit fluctuations in demand.

## Example 2.3: Resource levelling

Figure 2.13 represents a small project. The duration of each activity and labour required are given in the first three columns in Table 2.4. From this information the critical path and the project duration can be found. For simplicity it is assumed that the operatives are totally interchangeable, although the number of operatives on the activities must be as stated. The maximum number of operatives available is 10.

**Table 2.4** Analysis of Fig. 2.13.

| Activity | Duration | Gang size | Earliest | | Latest | | Total float |
|---|---|---|---|---|---|---|---|
| | | | Start | Finish | Start | Finish | |
| 1–2 | 2 | 10 | 0 | 2 | 0 | 2 | 0 |
| 2–3 | 2 | 2 | 2 | 4 | 7 | 9 | 5 |
| 2–4 | 3 | 5 | 2 | 5 | 2 | 5 | 0 |
| 3–5 | 4 | 6 | 4 | 8 | 9 | 13 | 5 |
| 4–5 | 8 | 3 | 5 | 13 | 5 | 13 | 0 |
| 4–6 | 4 | 4 | 5 | 9 | 13 | 17 | 8 |
| 5–6 | 1 | 2 | 13 | 14 | 16 | 17 | 3 |
| 5–7 | 8 | 8 | 13 | 21 | 13 | 21 | 0 |
| 6–7 | 4 | 4 | 14 | 18 | 17 | 21 | 3 |

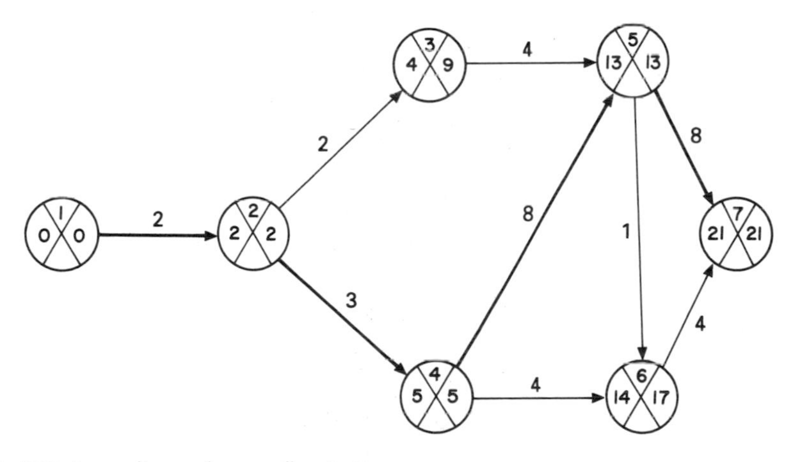


**Fig. 2.13** Arrow diagram for a small project.

| ACT. | 0 | 1 | 2 | 3 | 4 | 5 | 6 | 7 | 8 | 9 | 10 | 11 | 12 | 13 | 14 | 15 | 16 | 17 | 18 | 19 | 20 | 21 |
|---|---|---|---|---|---|---|---|---|---|---|---|---|---|---|---|---|---|---|---|---|---|---|
| 1–2 | 10 | 10 | | | | | | | | | | | | | | | | | | | | |
| 2–3 | | | 2 | 2 | | | | | | | | | | | | | | | | | | |
| 2–4 | | | 5 | 5 | 5 | | | | | | | | | | | | | | | | | |
| 3–5 | | | | | 6 | 6 | 6 | 6 | | | | | | | | | | | | | | |
| 4–5 | | | | | | 3 | 3 | 3 | 3 | 3 | 3 | 3 | 3 | | | | | | | | | |
| 4–6 | | | | | | 4 | 4 | 4 | 4 | | | | | | | | | | | | | |
| 5–6 | | | | | | | | | | | | | | 2 | | | | | | | | |
| 5–7 | | | | | | | | | | | | | | 8 | 8 | 8 | 8 | 8 | 8 | 8 | 8 | |
| 6–7 | | | | | | | | | | | | | | | 4 | 4 | 4 | 4 | | | | |
| | 10 | 10 | 7 | 7 | 11 | 13 | 13 | 13 | 7 | 3 | 3 | 3 | 3 | 10 | 12 | 12 | 12 | 12 | 8 | 8 | 8 | |

**Fig. 2.14** Bar chart showing all activities at their earliest times.

| ACT. | 0 | 1 | 2 | 3 | 4 | 5 | 6 | 7 | 8 | 9 | 10 | 11 | 12 | 13 | 14 | 15 | 16 | 17 | 18 | 19 | 20 | 21 |
|---|---|---|---|---|---|---|---|---|---|---|---|---|---|---|---|---|---|---|---|---|---|---|
| 1–2 | 10 | 10 | | | | | | | | | | | | | | | | | | | | |
| 2–3 | | | 2 | 2 | | | | | | | | | | | | | | | | | | |
| 2–4 | | | 5 | 5 | 5 | | | | | | | | | | | | | | | | | |
| 3–5 | | | | | | | | | | 6 | 6 | 6 | 6 | | | | | | | | | |
| 4–5 | | | | | | 3 | 3 | 3 | 3 | 3 | 3 | 3 | 3 | | | | | | | | | |
| 4–6 | | | | | | 4 | 4 | 4 | 4 | | | | | | | | | | | | | |
| 5–6 | | | | | | | | | | | | | | 2 | | | | | | | | |
| 5–7 | | | | | | | | | | | | | | 8 | 8 | 8 | 8 | 8 | 8 | 8 | 8 | |
| 6–7 | | | | | | | | | | | | | | | 4 | 4 | 4 | 4 | | | | |
| | 10 | 10 | 7 | 7 | 5 | 7 | 7 | 7 | 7 | 9 | 9 | 9 | 9 | 10 | 12 | 12 | 12 | 12 | 8 | 8 | 8 | |

**Fig. 2.15** Resource levelling: activities moved within their float.

***Solution***

It can be seen that the diagram as analysed gives a project duration of 21 weeks.

A bar chart showing all activities at their earliest times drawn first is as shown in Fig. 2.14. The activities are then moved within their float to level out requirements and to avoid exceeding the number available, as shown in Fig. 2.15.

| ACT. | 0 | 1 | 2 | 3 | 4 | 5 | 6 | 7 | 8 | 9 | 10 | 11 | 12 | 13 | 14 | 15 | 16 | 17 | 18 | 19 | 20 | 21 | 22 | 23 | 24 | 25 |
|---|---|---|---|---|---|---|---|---|---|---|---|---|---|---|---|---|---|---|---|---|---|---|---|---|---|---|
| 1–2 | 10 | 10 | | | | | | | | | | | | | | | | | | | | | | | | |
| 2–3 | | | 2 | 2 | | | | | | | | | | | | | | | | | | | | | | |
| 2–4 | | | 5 | 5 | 5 | | | | | | | | | | | | | | | | | | | | | |
| 3–5 | | | | | | | | | | 6 | 6 | 6 | 6 | | | | | | | | | | | | | |
| 4–5 | | | | | | 3 | 3 | 3 | 3 | 3 | 3 | 3 | 3 | | | | | | | | | | | | | |
| 4–6 | | | | | | 4 | 4 | 4 | 4 | | | | | | | | | | | | | | | | | |
| 5–6 | | | | | | | | | | | | | | 2 | | | | | | | | | | | | |
| 5–7 | | | | | | | | | | | | | | 8 | 8 | 8 | 8 | 8 | 8 | 8 | 8 | | | | | |
| 6–7 | | | | | | | | | | | | | | | | | | | | | | 4 | 4 | 4 | 4 | |
| | 10 | 10 | 7 | 7 | 5 | 7 | 7 | 7 | 7 | 9 | 9 | 9 | 9 | 10 | 8 | 8 | 8 | 8 | 8 | 8 | 8 | 4 | 4 | 4 | 4 | |

**Fig. 2.16** Resource levelling: activity 6–7 moved on 4 weeks to keep within available manning level.

As critical activities have no float these are drawn first. As the maximum number of men available is 10, some adjustment to Fig. 2.15 must be made. In this case the project duration has been increased to 25 weeks by moving on activity 6–7, as shown in Fig. 2.16.

**Example 2.4: Resource smoothing**

Figure 2.17 represents a small project. The duration of each activity and the number of operatives involved are shown in Table 2.5.

**Table 2.5** Analysis of Fig. 2.17.

| Activity | Duration (weeks) | Gang size |
|---|---|---|
| 1–2 | 2 | 12 |
| 2–3 | 5 | 4 |
| 2–4 | 8 | 2 |
| 2–5 | 4 | 3 |
| 3–8 | 2 | 4 |
| 3–4 | 0 | 0 |
| 4–7 | 1 | 4 |
| 4–5 | 3 | 3 |
| 5–6 | 2 | 6 |
| 6–7 | 1 | 5 |
| 6–11 | 3 | 2 |
| 7–10 | 2 | 8 |
| 8–9 | 4 | 3 |
| 9–10 | 0 | 0 |
| 9–11 | 7 | 6 |
| 10–11 | 4 | 3 |

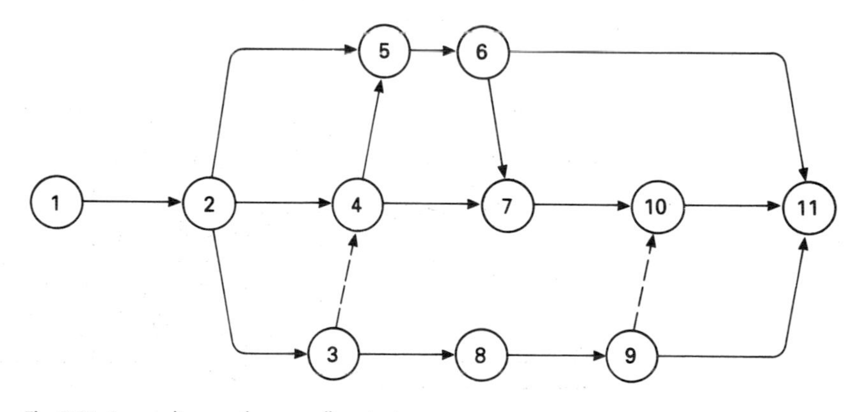


**Fig. 2.17** Arrow diagram for a small project.

***General information***

The contract has been planned to employ local casual labour working a 5 day week. The labour force can be recruited as required and employed on a week-to-week basis at a weekly rate of £360, with an 'engagement' establishment charge of £96 per person for each new appointment. Although no working is permitted on Saturday or Sunday, overtime can be worked in order to meet the requirement of the employer, which is to complete the project within 20 weeks. The employer also wishes to know the cost of carrying out the work as set out in the arrow diagram and without overtime.

The rules for working overtime are:

- The overtime can be worked in periods of 1 week, and will account for $33\frac{1}{3}\%$ increase in output on the relevant activities.
- The increased cost of overtime working is calculated at the rate of £180 per week per operative.
- Site overheads are calculated at £1050 per week.

*Note*: Each activity must be continuous: that is, once started it must be completed without a break. The gang sizes cannot be altered.

*Solution*

It can be seen that the diagram as analysed gives a project duration of 22 weeks (see Fig. 2.18). Figure 2.19 shows the network converted to a bar chart with all the activities at their earliest times, and Fig. 2.20 shows the bar chart after smoothing the resources. The cost of the project as shown would be:

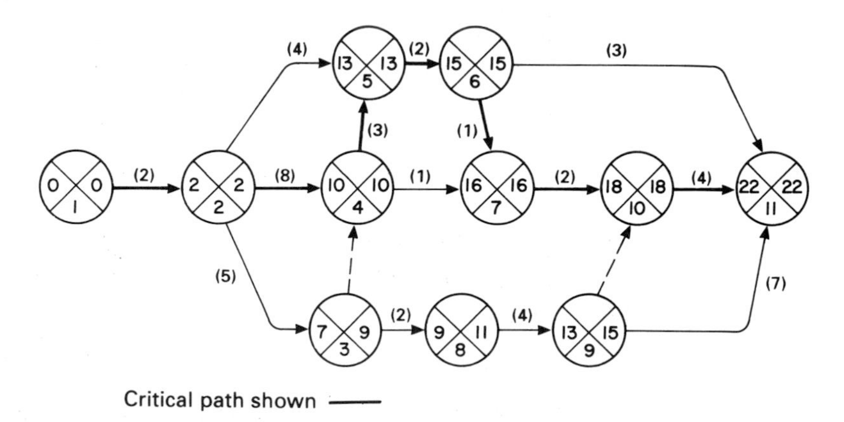


**Fig. 2.18** Analysed arrow diagram gives a project duration of 22 weeks.

*Activities shown at earliest times*

| Activity \ Week | 1 | 2 | 3 | 4 | 5 | 6 | 7 | 8 | 9 | 10 | 11 | 12 | 13 | 14 | 15 | 16 | 17 | 18 | 19 | 20 | 21 | 22 |
|---|---|---|---|---|---|---|---|---|---|---|---|---|---|---|---|---|---|---|---|---|---|---|
| 1–2 | 12 | 12 | | | | | | | | | | | | | | | | | | | | |
| 2–3 | | | 4 | 4 | 4 | 4 | 4 | | | | | | | | | | | | | | | |
| 2–4 | | | 2 | 2 | 2 | 2 | 2 | 2 | 2 | 2 | | | | | | | | | | | | |
| 2–5 | | | 3 | 3 | 3 | 3 | | | | | | | | | | | | | | | | |
| 3–8 | | | | | | | | 4 | 4 | | | | | | | | | | | | | |
| 3–4 | | | | | | | | | Dummy | | | | | | | | | | | | | |
| 4–7 | | | | | | | | | | | 4 | | | | | | | | | | | |
| 4–5 | | | | | | | | | | | 3 | 3 | 3 | | | | | | | | | |
| 5–6 | | | | | | | | | | | | | | 6 | 6 | | | | | | | |
| 6–7 | | | | | | | | | | | | | | | | 5 | | | | | | |
| 6–11 | | | | | | | | | | | | | | | | 2 | 2 | 2 | | | | |
| 7–10 | | | | | | | | | | | | | | | | | 8 | 8 | | | | |
| 8–9 | | | | | | | | | | 3 | 3 | 3 | 3 | | | | | | | | | |
| 9–10 | | | | | | | | | | | | | | | Dummy | | | | | | | |
| 9–11 | | | | | | | | | | | | | | 6 | 6 | 6 | 6 | 6 | 6 | 6 | | |
| 10–11 | | | | | | | | | | | | | | | | | | | 3 | 3 | 3 | 3 |
| | 12 | 12 | 9 | 9 | 9 | 9 | 6 | 6 | 6 | 5 | 10 | 6 | 6 | 12 | 12 | 13 | 16 | 16 | 9 | 9 | 3 | 3 |

**Fig. 2.19** Network converted to a bar chart; all activities shown at their earliest times.

*Activities adjusted using float*

| Activity \ Week | 1 | 2 | 3 | 4 | 5 | 6 | 7 | 8 | 9 | 10 | 11 | 12 | 13 | 14 | 15 | 16 | 17 | 18 | 19 | 20 | 21 | 22 |
|---|---|---|---|---|---|---|---|---|---|---|---|---|---|---|---|---|---|---|---|---|---|---|
| 1–2 | 12 | 12 | | | | | | | | | | | | | | | | | | | | |
| 2–3 | | | 4 | 4 | 4 | 4 | 4 | | | | | | | | | | | | | | | |
| 2–4 | | | 2 | 2 | 2 | 2 | 2 | 2 | 2 | 2 | | | | | | | | | | | | |
| 2–5 | | | 3 | 3 | 3 | 3 | | | | | | | | | | | | | | | | |
| 3–8 | | | | | | | | 4 | 4 | | | | | | | | | | | | | |
| 3–4 | | | | | | | | | Dummy | | | | | | | | | | | | | |
| 4–7 | | | | | | | | | | | | | | | 4 | | | | | | | |
| 4–5 | | | | | | | | | | | 3 | 3 | 3 | | | | | | | | | |
| 5–6 | | | | | | | | | | | | | | 6 | 6 | | | | | | | |
| 6–7 | | | | | | | | | | | | | | | | 5 | | | | | | |
| 6–11 | | | | | | | | | | | | | | | | | | | 2 | 2 | 2 | |
| 7–10 | | | | | | | | | | | | | | | | | 8 | 8 | | | | |
| 8–9 | | | | | | | | | | 3 | 3 | 3 | 3 | | | | | | | | | |
| 9–10 | | | | | | | | | | | | | | | Dummy | | | | | | | |
| 9–11 | | | | | | | | | | | | | | | | 6 | 6 | 6 | 6 | 6 | 6 | 6 |
| 10–11 | | | | | | | | | | | | | | | | | | | 3 | 3 | 3 | 3 |
| | 12 | 12 | 9 | 9 | 9 | 9 | 6 | 6 | 6 | 5 | 6 | 6 | 6 | 6 | 10 | 11 | 14 | 14 | 11 | 11 | 11 | 9 |

**Fig. 2.20** Bar chart showing activities adjusted using float to smooth the resources.

| | |
|---|---|
| Engagement 21 × £96 | £ 2016 |
| Labour costs 198 × £360 | £71 280 |
| Overheads 22 × £1050 | £23 100 |
| Cost | £96 396 |

As the employer requires the project to be completed in 20 weeks, the duration must be reduced by 2 weeks. To do this, overtime will have

*Activities shown at earliest times*

| Week / Activity | 0 | 1 | 2 | 3 | 4 | 5 | 6 | 7 | 8 | 9 | 10 | 11 | 12 | 13 | 14 | 15 | 16 | 17 | 18 | 19 | 20 | 21 22 |
|---|---|---|---|---|---|---|---|---|---|---|---|---|---|---|---|---|---|---|---|---|---|---|
| 1–2 | 12 | 12 | | | | | | | | | | | | | | | | | | | | |
| 2–3 | | | 4 | 4 | 4 | 4 | 4 | | | | | | | | | | | | | | | |
| 2–4 | | | 2 | 2 | 2 | 2 | 2 | 2 | | | | | | | | | | | | | | |
| 2–5 | | | 3 | 3 | 3 | 3 | | | | | | | | | | | | | | | | |
| 3–8 | | | | | | | | 4 | 4 | | | | | | | | | | | | | |
| 3–4 | | | | | | | | Dummy | | | | | | | | | | | | | | |
| 4–7 | | | | | | | | | 4 | | | | | | | | | | | | | |
| 4–5 | | | | | | | | | 3 | 3 | 3 | | | | | | | | | | | |
| 5–6 | | | | | | | | | | | | 6 | 6 | | | | | | | | | |
| 6–7 | | | | | | | | | | | | | | 5 | | | | | | | | |
| 6–11 | | | | | | | | | | | | | | 2 | 2 | 2 | | | | | | |
| 7–10 | | | | | | | | | | | | | | | 8 | 8 | | | | | | |
| 8–9 | | | | | | | | | | 3 | 3 | 3 | 3 | | | | | | | | | |
| 9–10 | | | | | | | | | | | | | | Dummy | | | | | | | | |
| 9–11 | | | | | | | | | | | | | | 6 | 6 | 6 | 6 | 6 | 6 | 6 | | |
| 10–11 | | | | | | | | | | | | | | | | | 3 | 3 | 3 | 3 | | |
| | 12 | 12 | 9 | 9 | 9 | 9 | 6 | 6 | 11 | 6 | 6 | 9 | 9 | 13 | 16 | 16 | 9 | 9 | 9 | 9 | | |

**Fig. 2.21** Project duration reduced by 2 weeks; all activities shown at their earliest times.

to be worked, and in order to keep the cost to a minimum it will be worked on the critical activity that will cost the least: that is, the activity with the least labour (in this case 2–4). If overtime is worked for 6 weeks, 8 weeks' work will be done (owing to $33\frac{1}{3}\%$ increase in output).

*Activities adjusted using float*

| Week / Activity | 0 | 1 | 2 | 3 | 4 | 5 | 6 | 7 | 8 | 9 | 10 | 11 | 12 | 13 | 14 | 15 | 16 | 17 | 18 | 19 | 20 | 21 22 |
|---|---|---|---|---|---|---|---|---|---|---|---|---|---|---|---|---|---|---|---|---|---|---|
| 1–2 | 12 | 12 | | | | | | | | | | | | | | | | | | | | |
| 2–3 | | | 4 | 4 | 4 | 4 | 4 | | | | | | | | | | | | | | | |
| 2–4 | | | 2 | 2 | 2 | 2 | 2 | 2 | | | | | | | | | | | | | | |
| 2–5 | | | 3 | 3 | 3 | 3 | | | | | | | | | | | | | | | | |
| 3–8 | | | | | | | | 4 | 4 | | | | | | | | | | | | | |
| 3–4 | | | | | | | | Dummy | | | | | | | | | | | | | | |
| 4–7 | | | | | | | | | | | 4 | | | | | | | | | | | |
| 4–5 | | | | | | | | | 3 | 3 | 3 | | | | | | | | | | | |
| 5–6 | | | | | | | | | | | | 6 | 6 | | | | | | | | | |
| 6–7 | | | | | | | | | | | | | | 5 | | | | | | | | |
| 6–11 | | | | | | | | | | | | | | | | | 2 | 2 | 2 | | | |
| 7–10 | | | | | | | | | | | | | | | 8 | 8 | | | | | | |
| 8–9 | | | | | | | | | | 3 | 3 | 3 | 3 | | | | | | | | | |
| 9–10 | | | | | | | | | | | | | | Dummy | | | | | | | | |
| 9–11 | | | | | | | | | | | | | | 6 | 6 | 6 | 6 | 6 | 6 | 6 | | |
| 10–11 | | | | | | | | | | | | | | | | | 3 | 3 | 3 | 3 | | |
| | 12 | 12 | 9 | 9 | 9 | 9 | 6 | 6 | 7 | 6 | 10 | 9 | 9 | 11 | 14 | 14 | 11 | 11 | 11 | 9 | | |

**Fig. 2.22** Amended bar chart after resource smoothing using float to adjust activities.

Figure 2.21 shows the amended bar chart with all activities at their earliest times, and Fig. 2.22 shows the amended bar chart after smoothing the resources. The cost of the project as shown would be:

| | |
|---|---|
| Engagements 22 × £96 | £ 2112 |
| Labour costs 194 × £360 | £69 840 |
| Overtime 2 × 6 × £180 | £ 2160 |
| Overheads 20 × £1050 | £21 000 |
| Cost | £95 112 |

It can be seen that both project duration and cost are reduced in this example.

## 2.7 Control with arrow diagrams

### *2.7.1 Introduction*

Because a network draws attention to those activities that affect the project duration, it is an excellent basis for control. More attention can be given to those activities that are critical and near critical, although this does not mean that activities with a reasonable amount of float can be ignored altogether, or they themselves may then become critical.

For site use the arrow diagram is usually converted into a bar chart, which is used for short-term control and shows operations at scheduled times; progress is then marked onto this, and the information from the bar chart can be transferred back onto the arrow diagram for re-analysis or updating of the network for the whole project.

An alternative method of presentation is to draw the arrow diagram on a time scale showing activities at scheduled times, and to mark the progress directly onto this (see section 2.11). It is possible to use the diagram in its original form and to enter the actual durations on the diagram for re-analysis: if the results are tabulated and the actual durations are entered on the schedule this will form a method of control.

Re-analysis of the diagram should be carried out at regular intervals (say fortnightly) so that any change in critical activities is brought to the attention of the management, and corrections can be made in reasonable time.

If the arrow diagram is used to show progress, and it is not drawn on a time scale, elasticated string can be used to give an impression of progress. By inserting pins into activities in progress (the position being proportionate to the amount of work done on each activity) and then stretching the string from pin to pin, it would be obvious that all activities to the right of the string are still outstanding.

### 2.7.2 *Re-analysis of a project partially completed*

One method of re-analysing a project partially completed is as follows:

(1) Insert 'lead time' to represent the amount of time passed.
(2) Insert duration zero on all completed activities.
(3) Insert estimated time to complete partially completed activities.
(4) Carry out analysis in the normal way.

For example, assume that in the diagram in Fig. 2.9 a period of 5 weeks has passed and the position is:

- Activities 1–2 and 1–4 are completed.
- Activity 1–3 requires an estimated 1 week to complete.
- Activity 2–7 requires an estimated 5 weeks to complete.
- All other activities are not yet started.

The re-analysed diagram would now appear as shown in Fig. 2.23. It can be seen that the critical path has changed, and the project duration is now 14 weeks. If it is necessary to complete in 12 weeks, the activities requiring attention (i.e. critical activities) are clearly shown. All activities running behind time can be demonstrated by inserting the original project duration as the latest finish time for the end event before calculating the latest finish times for the other events. From this the negative float can be calculated, which will indicate the activities that are in arrears.

A second method of showing progress on the diagram is to remove all completed activities, leaving only partially completed activities and those not yet started. Real time dummies are then inserted and given a value equal to the time that has passed. This replaces 1 and 2 in the

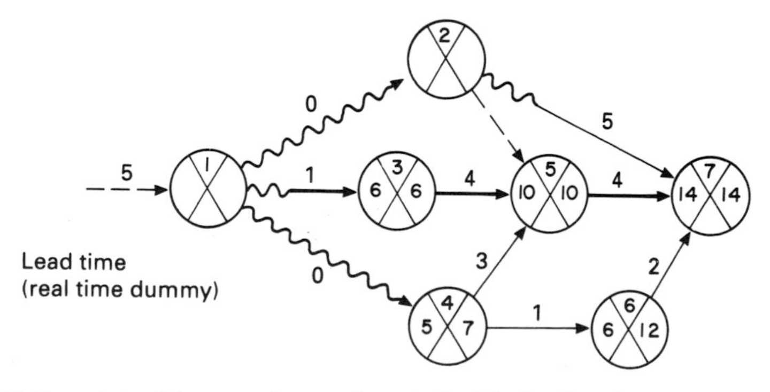


**Fig. 2.23** Re-analysis of the arrow diagram shown in Fig. 2.9, after 5 weeks.

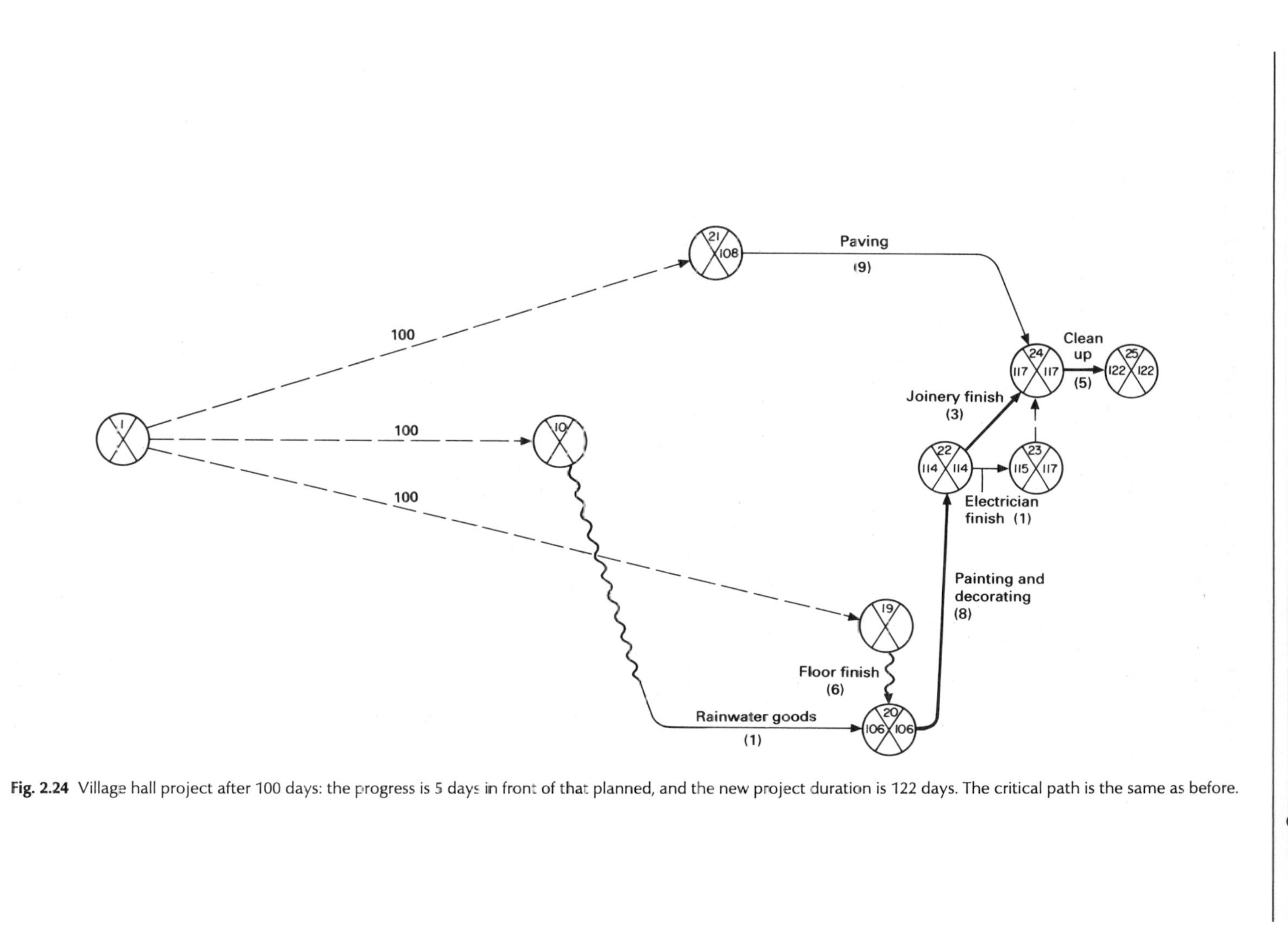


**Fig. 2.24** Village hall project after 100 days: the progress is 5 days in front of that planned, and the new project duration is 122 days. The critical path is the same as before.

preceding method, and constitutes a particularly useful alternative when the project is well advanced. For example, assume in the village hall project that after 100 days the following activities are partially completed and the estimated time necessary to complete is as shown:

- Floor finish, 6 days.
- Rainwater goods, 1 day.
- All subsequent activities have not yet started.

Using the alternative method, the arrow diagram will be as in Fig. 2.24.

## 2.8 The preparation of precedence diagrams

### 2.8.1 *Activities*

Activities are represented by boxes, and are linked together by lines of dependency. The general direction of time flow is from left to right. Activities are assumed to start at the left-hand end of the box and finish at the right-hand end of the box. When the start of an activity has more than one line of dependency it is dependent on *all* the activities to which it is connected, and therefore all the preceding activities must be completed before it can start.

### 2.8.2 *Dummy activities*

These are activities used for convenience: for example, at the beginning and end of the project.

### 2.8.3 *Delays*

When there is to be a delay between activities – hardening of concrete for example – this is shown on the link between them (see Fig. 2.25).

### 2.8.4 *Overlap*

There are two methods for dealing with overlap. The first is to divide each activity into the same number of equal parts (Fig. 2.26). This is the

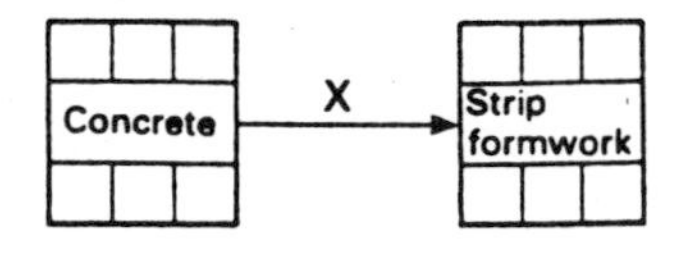


**Fig. 2.25** Precedence diagram: delay between activities.

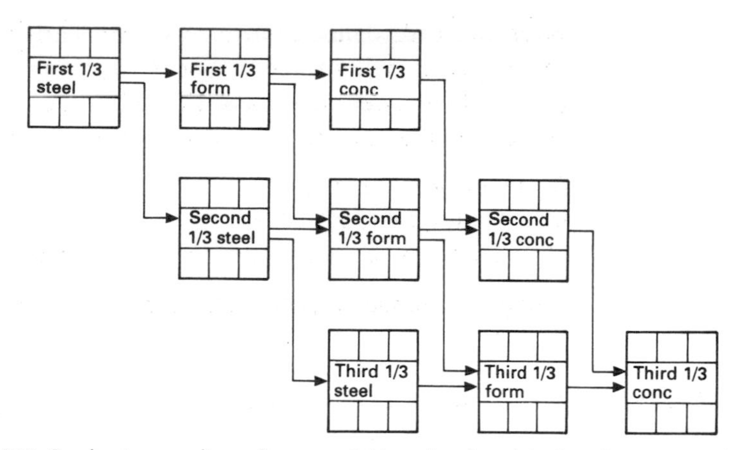


**Fig. 2.26** Overlap in precedence diagrams: division of each activity into the same number of equal parts.

only method that truly portrays the situation, but it can increase the size of the network unnecessarily in practice.

The second method is to use lead and lag restraints (Fig. 2.27). 'Begin steel' represents the time period required between start of steel and start of formwork. 'Begin formwork' represents the time period required between start of formwork and start of concrete.

'Complete formwork' represents the time period required to complete formwork after the completion of steel, and 'Complete concrete' represents the time period required to complete concrete after the completion of formwork. If no time period is necessary between the start of one activity and the start of another, or between the completion of one activity and the completion of another, the diagram is similar to Fig. 2.27 but no time period is inserted on the link.

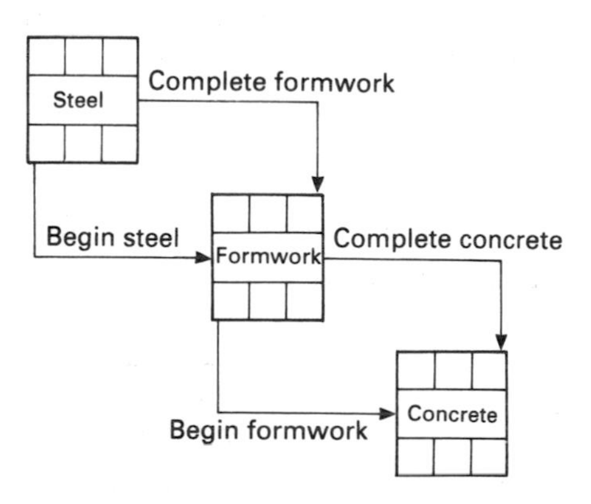


**Fig. 2.27** Overlap of precedence diagrams: use of lead and lag restraints.

## 2.9 Analysis of precedence diagrams

### 2.9.1 *Method of analysis where no overlap or delay is present (Fig. 2.28)*

*(1) Calculate the earliest activity times by working through the diagram and selecting the longest path*

In order that the diagram can start and end with a single activity, a dummy has been introduced at the beginning and at the end.

The earliest start time of an activity is the highest earliest finish of preceding activities. The earliest finish time is (earliest start) + (duration).

*(2) Calculate the latest activity times by working backwards through the diagram, again selecting the longest path*

The latest finish time is the latest time by which the activity can be completed without affecting the project duration. The latest finish time for the end activity is therefore the same as the earliest finish time.

The latest finish time for other activities is the lowest latest start time of their succeeding activities. The latest start time = (latest finish) − (duration).

*(3) Critical activities*

For critical activities, the earliest and latest times are the same.

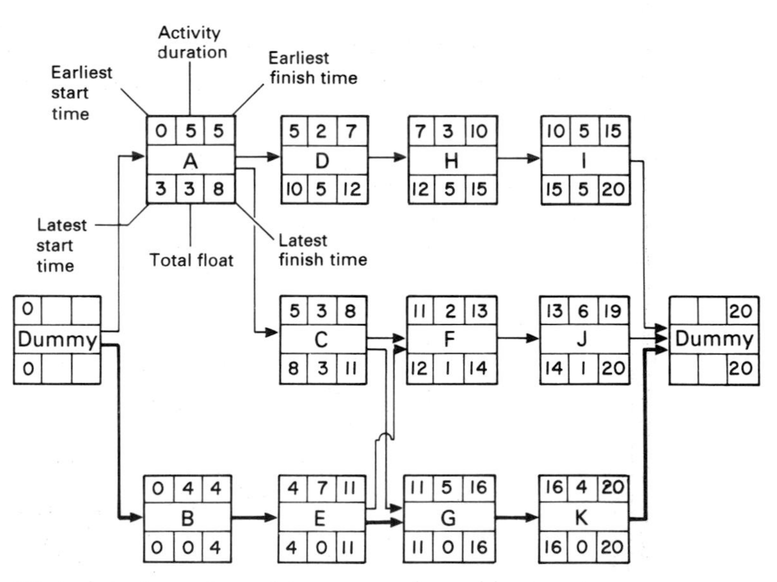


**Fig. 2.28** Analysis of precedence diagrams: no overlap or delay present.

*(4) Float*
Total float = (latest start time) − (earliest start time).

or

= (Latest finish time) − (earliest finish time).

Free float activity $x$ = (lowest earliest start of succeeding activities) − (earliest finish of activity $x$)

Interfering float = (Total float) − (free float)

Independent float on activity $x$ = (lowest earliest start of succeeding activities) − (highest latest finish of preceding activities) − (duration of activity $x$)

*(5) Analysis of delayed activities*
Earliest start of a delayed activity = (earliest finish of preceding activity) + (duration of delay)

or

= highest earliest finish of preceding activities, whichever is the greater

*(6) Analysis of overlapping activities*
Figure 2.29 shows the analysis of part of a project consisting of the erection of concrete columns. The procedure for analysis when using this method of overlap is the same as that shown in step (1) above.

Figure 2.30 shows the analysis using lead and lag restraints of 1 day, with the exception of the overlap on concrete, which is 3 days to allow for hardening. In the example the earliest and latest times and the float

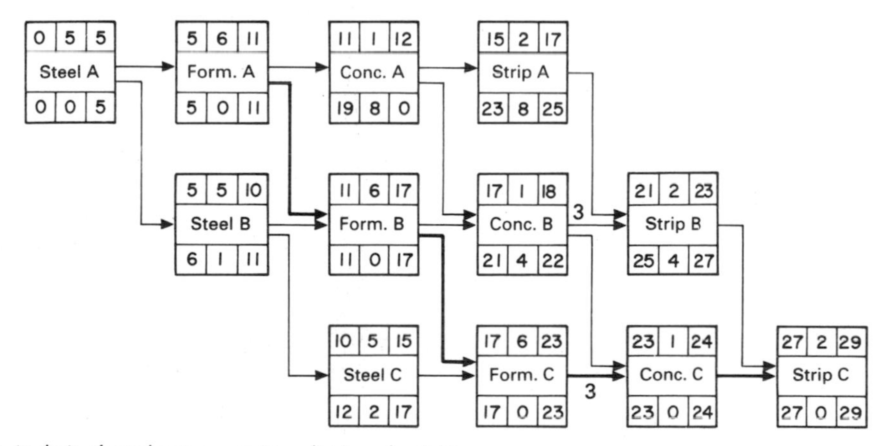


**Fig. 2.29** Analysis of overlapping activities: division of activities.

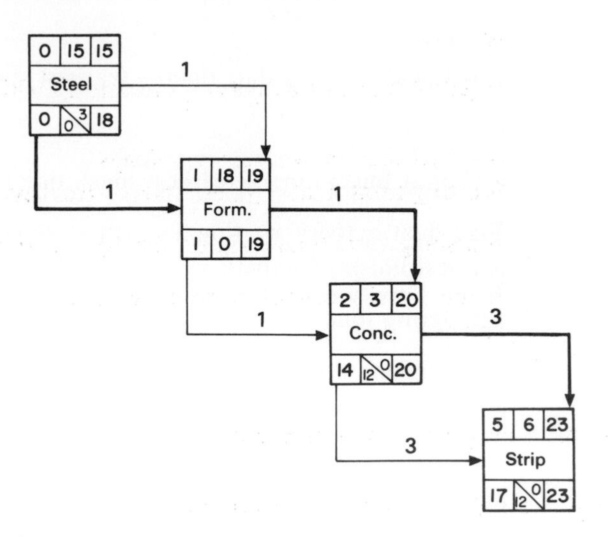


**Fig. 2.30** Analysis of overlapping activities: use of lead and lag restraints.

are influenced by the overlap, and the calculations have to be carried out as follows:

Earliest start of an activity = (earliest start of preceding activity) + (duration of lead)

or

= highest earliest finish of other preceding activities, whichever is the greater

Earliest finish of an activity = (earliest finish of preceding activity) + (duration of lag)

or

= (earliest start of activity) + (duration of activity), whichever is the greater

Latest finish of an activity = (latest finish of succeeding activity) − (duration of lag)

or

= lowest latest start time for other succeeding activities, whichever is the lesser

Latest start of an activity = (latest start of succeeding activity) − (duration of lead)

or

= (latest finish of activity) – (duration), whichever is the lesser

*(7) Total float on overlapping activities*
Starting total float = (latest start) – (earliest start)

Finishing total float = (latest finish) – (earliest finish)

If the activity cannot be split, the float has to be the lesser of the two values.

## 2.10 Control with precedence diagrams

### *2.10.1 Re-analysis of a partially completed project*

Assume that after a period of 12 weeks the progress on the project shown in Fig. 2.31 is as follows:

- ❑ Activities A, B, C, D and E are complete.
- ❑ Activity H requires 1 week to complete.
- ❑ Activity G requires 3 weeks to complete.
- ❑ All other activities have not yet started.

The period of 12 weeks is put into the dummy activity. All completed activities are given a duration of zero. Partially completed activities are given the estimated duration required to complete them, and activities

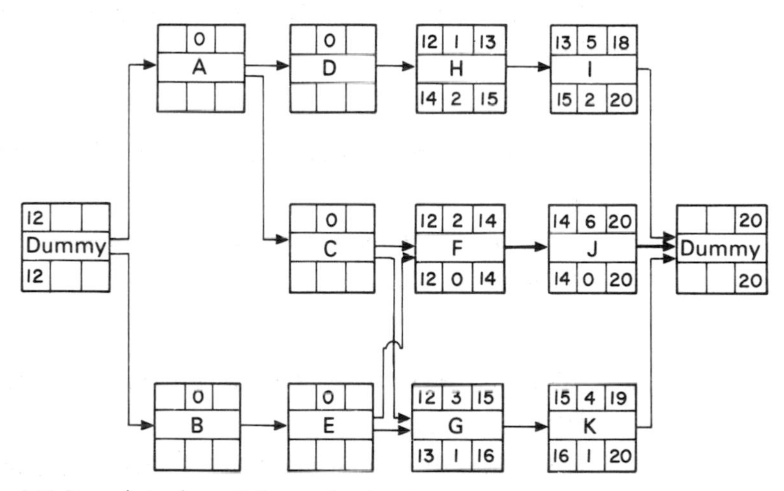


**Fig. 2.31** Re-analysis of a partially completed project.

not started are given their original duration. The analysis is then carried out as set out in section 2.9.1. It can be seen that the critical path has now changed.

An alternative method of showing progress is to remove all completed activities, leaving only partially completed activities and those not started. The analysis is then carried out as above.

**Example 2.5: Application of precedence diagrams to village hall project**

Figure 2.32 shows the analysed precedence diagram for the village hall project.

Figure 2.33 shows the diagram updated after 100 days' progress. All completed activities have been removed, and the following activities are partially completed. The estimated duration required to complete them is also shown below:

- Floor finish, 6 days.
- Rainwater goods, 1 day.
- All subsequent activities have not yet started.

## 2.11 Time scale presentation of arrow diagrams

### *2.11.1 Introduction*

Networks can be drawn in the form of a bar chart or time bar diagram, in which case the activities are normally drawn as parallel lines to a suitable time scale. This form of presentation defines the time limitations of each activity and the relationship between activities.

The first stage when constructing an activity progress chart is to compile an activity sheet, which records the sequence in which the work is to be carried out and the duration of the individual activities. The information on this sheet is then transferred to the time bar diagram for diagrammatic presentation.

**Example 2.6: Time scale presentation of village hall project**

*Activity sheet*

Table 2.6 shows an activity sheet taken from the village hall project (Table 2.3). It is a list of all the activities, their durations, and any restraints involved in the project.

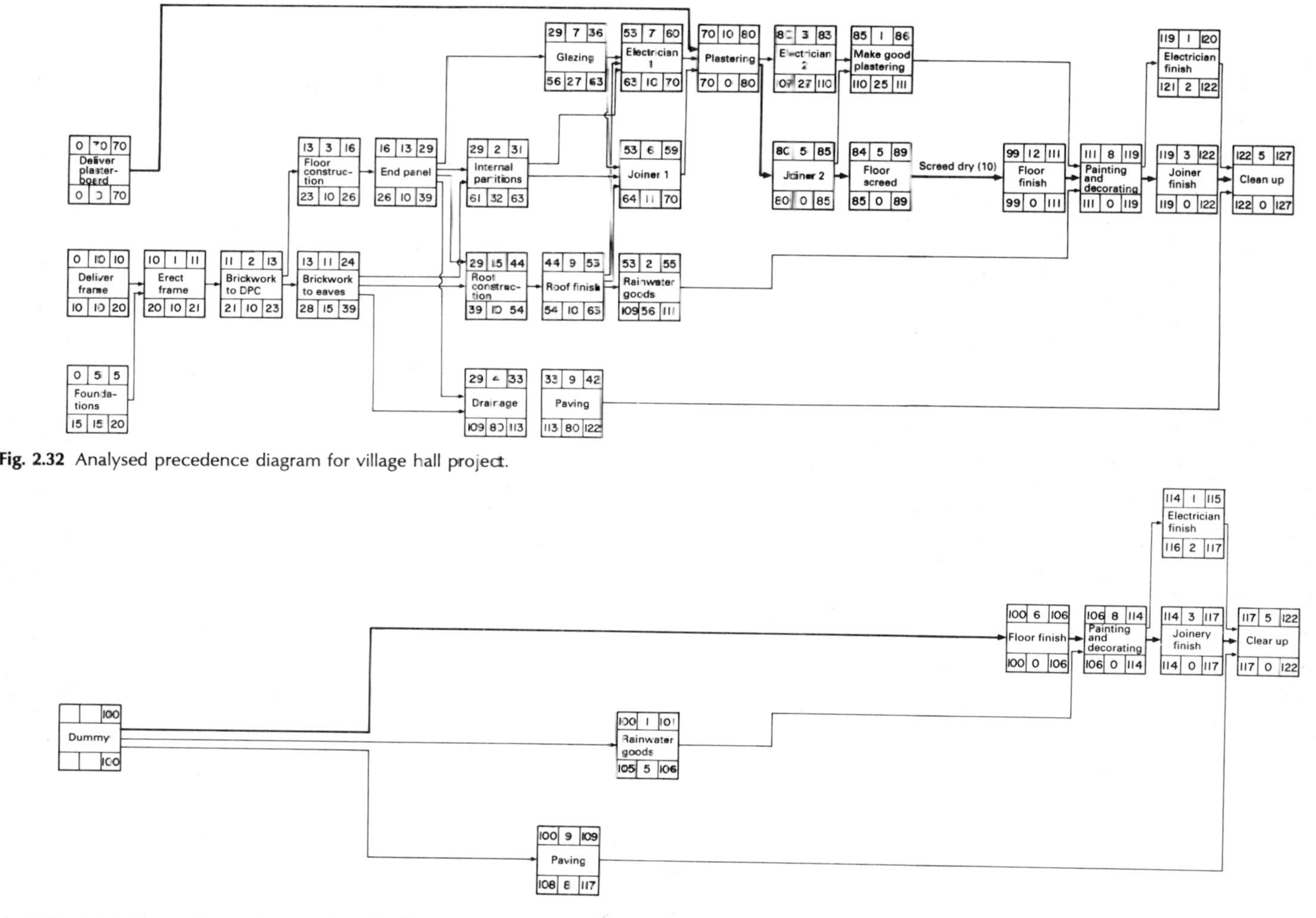


Fig. 2.32 Analysed precedence diagram for village hall project.



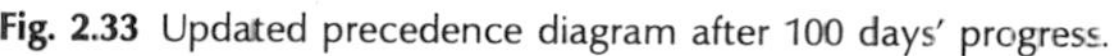

Fig. 2.33 Updated precedence diagram after 100 days' progress.

**Table 2.6** Activity sheet: village hall.

| Activity number | Activity | Activity sequence | Activity duration (days) |
|---|---|---|---|
| 1 | Deliver frame | 0;1 | 10 |
| 2 | Foundations | 0;2 | 5 |
| 3 | Deliver plasterboard | 0;3 | 70 |
| 4 | Erect frame | 1, 2;4 | 1 |
| 5 | Brickwork to DPC | 4;5 | 2 |
| 6 | Floor construction | 5;6 | 3 |
| 7 | Brickwork to eaves | 5;7 | 11 |
| 8 | End panels | 6;8 | 13 |
| 9 | Glazing | 8;9 | 7 |
| 10 | Roof construction | 7, 8;10 | 15 |
| 11 | Internal partitions | 7, 8;11 | 2 |
| 12 | Drainage | 7, 8;12 | 4 |
| 13 | Roof finish | 10;13 | 9 |
| 14 | Rainwater goods | 13;14 | 2 |
| 15 | Electrician first fixing | 9, 11, 13;15 | 7 |
| 16 | Joiner first fixing | 9, 11, 13;16 | 6 |
| 17 | Plastering | 3, 16, 15;17 | 10 |
| 18 | Electrician second fixing | 17;18 | 3 |
| 19 | Joiner second fixing | 17;19 | 5 |
| 20 | Make good plaster | 18, 19;20 | 1 |
| 21 | Floor screed | 18, 19;21 | 4 |
| 22 | Floor screed dry | 21;22 | 10 |
| 23 | Floor finish | 22;23 | 12 |
| 24 | Painting and decorating | 14, 20, 23;24 | 8 |
| 25 | Paving | 12;25 | 9 |
| 26 | Electrician finish | 24;26 | 1 |
| 27 | Joiner finish | 24;27 | 3 |
| 28 | Clean and hand over | 25, 27, 26;28 | 5 |

The first step is to list all the activities involved in the project (preferably in the order in which they are to be carried out, but this is not essential). These activities are then numbered. The next step is to provide the information regarding the relationship between activities by means of a sequence of numbers alongside the activity number. These numbers indicate which activities must be completed before the activity under consideration can be started. For example, in Table 2.6 in the sequence column for activity 1 are the figures 0;1. This indicates that there are no other activities to be completed before activity 1 can start. Another example is activity 4. In the sequence column are the figures 1, 2;4, which indicates that activity 1 (deliver frame) and activity 2 (foundations) must be completed before activity 4 (erect frame) can start. The other column records the duration of each activity, which would be arrived at via the work content and resources applied to it. This information can now be recorded diagrammatically in time bar diagram form, and the critical path ascertained.

### *Time bar diagram*

See Fig. 2.34. In this chart, each activity is represented by a straight line drawn to a suitable scale. The length of each line is proportional to the duration of the activity being represented. The chart can be drawn either vertically or horizontally. If vertically, then the chart is started at the top of the sheet; if horizontally, then the chart is drawn from left to right, as is common with most forms of bar chart.

Assuming that the chart is to be drawn horizontally, then the first step is to set up a suitable time scale on the horizontal axis and then to erect a perpendicular at time zero. This line is called the *start line*. The job lines are now drawn on the chart in the sequence indicated in the activity sheet (see Table 2.6). Each activity is given its respective number for identification purposes. A short vertical line is introduced at the end of each activity to indicate the end of that particular activity and to separate it from the next in sequence. The relationship with activities in other parts of the chart is indicated by a longer vertical line. If there is a place in the project where a number of activities must be completed so that another activity can be started, then the vertical line is drawn at the end of the activity line that is latest in time, and produced to connect all activities involved. These vertical lines can also be extended when a number of activities are to be started at the same time. The vertical lines are usually called *time lines*. The float on activities is indicated by a dotted line at the end of the activity line. This dotted line is extended to the next time line indicated on the activity sheet. The critical path is indicated by a continuous set of activities, which can be arranged to fall in one straight line if required, but this is not essential.

When the last activity has been drawn on the chart, another long vertical line is drawn at this point. This is called the *finish line*. It represents the completion of the project. The project duration is therefore the dimension between the start and finish lines.

The advantage of this chart is that it shows clearly the relationship that exists between all activities in the project, the sequence of all activities and those which are concurrent.

Following these basic principles, the construction of the time bar diagram (Fig. 2.34) is therefore as follows. Having calibrated the

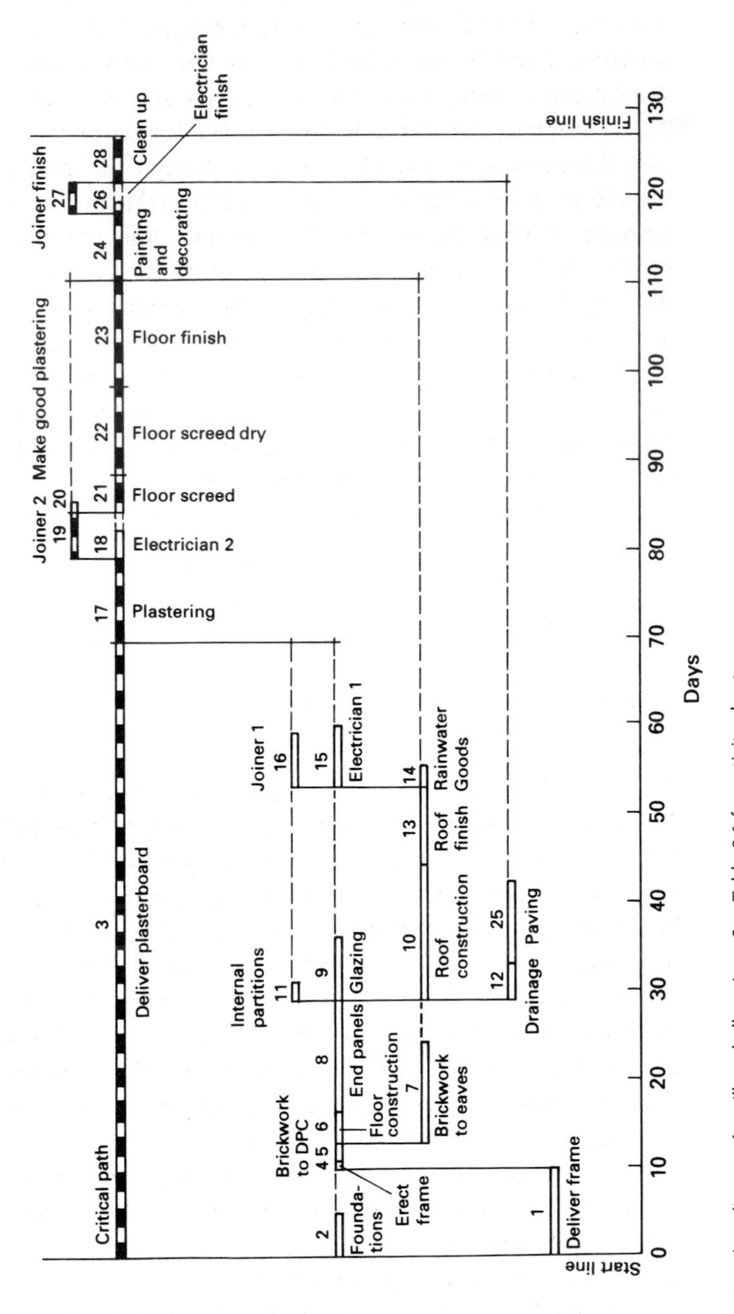


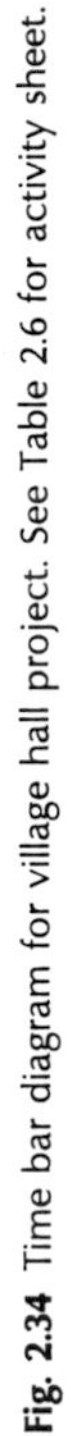
**Fig. 2.34** Time bar diagram for village hall project. See Table 2.6 for activity sheet.

horizontal axis (time scale) and drawn the start line, the next step is to draw on activity 1, commencing the line at the start line (time zero) and extending it 10 days to scale to the right. The same is done for activities 2 and 3, which both start at time zero. From the activity sheet it can be seen that activity 4 can start only when activities 1 and 2 are complete. Because activity 1 has the longer duration, a vertical time line must be erected at the end of activity 1 and extended to meet the end of activity 2 produced by dotted line (the dotted line being the amount of float on activity 2). The time line prevents activity 4 starting before activity 1 is completed. Activity 4 can now be drawn. Activity 5 can be drawn straight after activity 4 as activity 5 is dependent only upon activity 4. Activity 6 is dependent upon activity 5: therefore the straight line continues. Activity 7 is dependent upon the completion of activity 5: therefore a time line must be introduced to show the relationship between activities 5 and 7. From here on each activity is considered in a similar fashion.

## 2.12 Effects of variations and delays

### *2.12.1 Settling variations*

The recognised procedure for settling variations in accordance with the JCT Standard Form of Contract does not take all factors into account. Certain variations have a more far-reaching effect than those taken into account by adjusting the rate for measured work in the contract bills. When, for example, additional work is carried out, this will sometimes have the effect of lengthening the contract period. In these circumstances the contractor is entitled to reimbursement for the extra costs involved. Even when variations of omission are issued, it may not be a simple matter of deducting the prices of the respective items of measured work in the bill, because the omission may disrupt the continuity of work for certain labour and plant. For example, a gang of men or an item of plant may have a carefully arranged programme of work, and the omission may result in men and plant being idle awaiting the next activity. In this case the labour and plant content of the omission should still be paid for by the client.

Provision is made for the architect to allow payment to the contractor in these circumstances. It will of course be necessary for the contractor to prove his case, and to do this a programme is extremely useful.

### 2.12.2 *Other factors causing delays*

Besides variations, many other factors can cause delays to the contractor, for which he does not in many cases receive full payment, owing to his inability to substantiate his claim, even though provision is made for payment under the JCT Form of Contract. Examples of these factors are:

- late arrival of drawings, details and levels even when prior notice of these requirements has been given;
- opening up work for inspection or testing;
- discrepancies between drawings and contract bills;
- delays caused by persons employed directly by the client.

The contractor can claim payments for losses caused by these factors where his progress has been disturbed.

### 2.12.3 *Delays where no payment is warranted*

Some delays do not of course warrant extra payment to the contractor, but it is nevertheless useful to have a means available by which the amount of extension of time can be justified. Typical delays in this category are exceptionally inclement weather, fire, strikes, shortage of materials or labour which could not have been foreseen. Extension of time can be granted for these and other reasons under the JCT Form of Contract.

### 2.12.4 *Assessing the effects of delays*

It is often difficult to assess realistically the effect of delays on costs to the contractor, particularly when extension of time is involved. Without a programme it is virtually impossible. When a programme has been prepared and updated, the problem of price adjustment (where appropriate), is made much easier. This particularly applies when network techniques have been used. If the network is a contract document, it will almost certainly be used as a basis for calculating these adjustments. Even when the programme is not a contract document, provided the contractor updates it, he can present it to the quantity surveyor or architect as appropriate, whenever a variation order has the effect of interrupting the programme, and it will prove to be formidable evidence in support of a claim.

On a network programme the effect can be clearly seen, and variation orders or other causes that affect activities on the critical path will often warrant an extension of time and, possibly, subsequent additions to the contract sum.

**Example 2.7: Settling a variation**

Figure 2.12, which is the analysed arrow diagram for the village hall project, will be used to illustrate the effects of a variation order on costs and contract period.

Assume a variation order is issued for additional screed (activity 16–18) and floor finish (activity 19–20) under the stage area. (This extra flooring is considered necessary because the client will require the stage to be moved on some occasions and therefore wishes to have the whole floor covered.)

The additional screed and floor finish will increase the total amount in the bill by one fifth, and direct costs to the contractor will also increase by one fifth. These costs will be covered by adjusting the price included in the contract bills, which will also increase by one fifth. However, both these activities are on the critical path, and the time required for each will also increase by one fifth. This will result in an extension of time of $(\frac{1}{5} \times 4) + (\frac{1}{5} \times 12) = 3\frac{1}{5}$ days. The contractor may therefore require this extension of time plus the extra cost of overheads included in the preliminaries and related to the contract period.

It may of course be possible to avoid extending the contract period by overtime working or bringing in extra men. This will increase costs, however, and if the contractor agrees to take this action he is entitled to payment for the extra costs. This would probably be a matter for negotiation between the contractor and the quantity surveyor.

The example given is of course a very simple one, and in practice the problem may be more difficult. However, the above procedure can be followed, and differences of opinion should be much easier to resolve.

### *2.12.5 Assessing costs using bar charts*

Bar charts can be used in a similar manner to that shown above but it may be more difficult to 'prove' that a particular activity is on the critical path and will therefore affect the project duration. However, a good updated bar chart is far superior to no programme at all.

## 2.13 Cost optimisation

It may be possible to vary the duration of some activities in a network, and this will normally affect the cost of the activities and consequently the cost of the project.

### *2.13.1 The relationship between project duration and cost*

Generally speaking, certain assumptions can be made about the relationship between project duration and cost as follows:

- Direct costs rise as project duration decreases.
- Indirect costs rise as project duration increases.
- There is a minimum duration for any project beyond which further reduction is not feasible.

### *2.13.2 Determining optimum project cost*

At each stage in network optimisation calculations the following procedure can be adopted.

(1) Prepare a table of activities on the critical path showing normal activity duration and cost, and minimum (crash) activity duration and cost.
(2) Calculate the cost slope of each activity in the table.
(3) List the activities in order of minimum cost slope.
(4) Omit activities that cannot be compressed (and those fully crashed from previous compressions).
(5) Compress the activity (or activities if two are being compressed) with the least cost slope, the maximum amount possible, or until some other activity becomes critical.
(6) Calculate the new project duration and direct cost.
(7) When minimum duration is reached, calculate indirect costs for each project duration.
(8) Add indirect costs to direct costs and calculate total costs for each project duration.
(9) Determine optimum project cost.

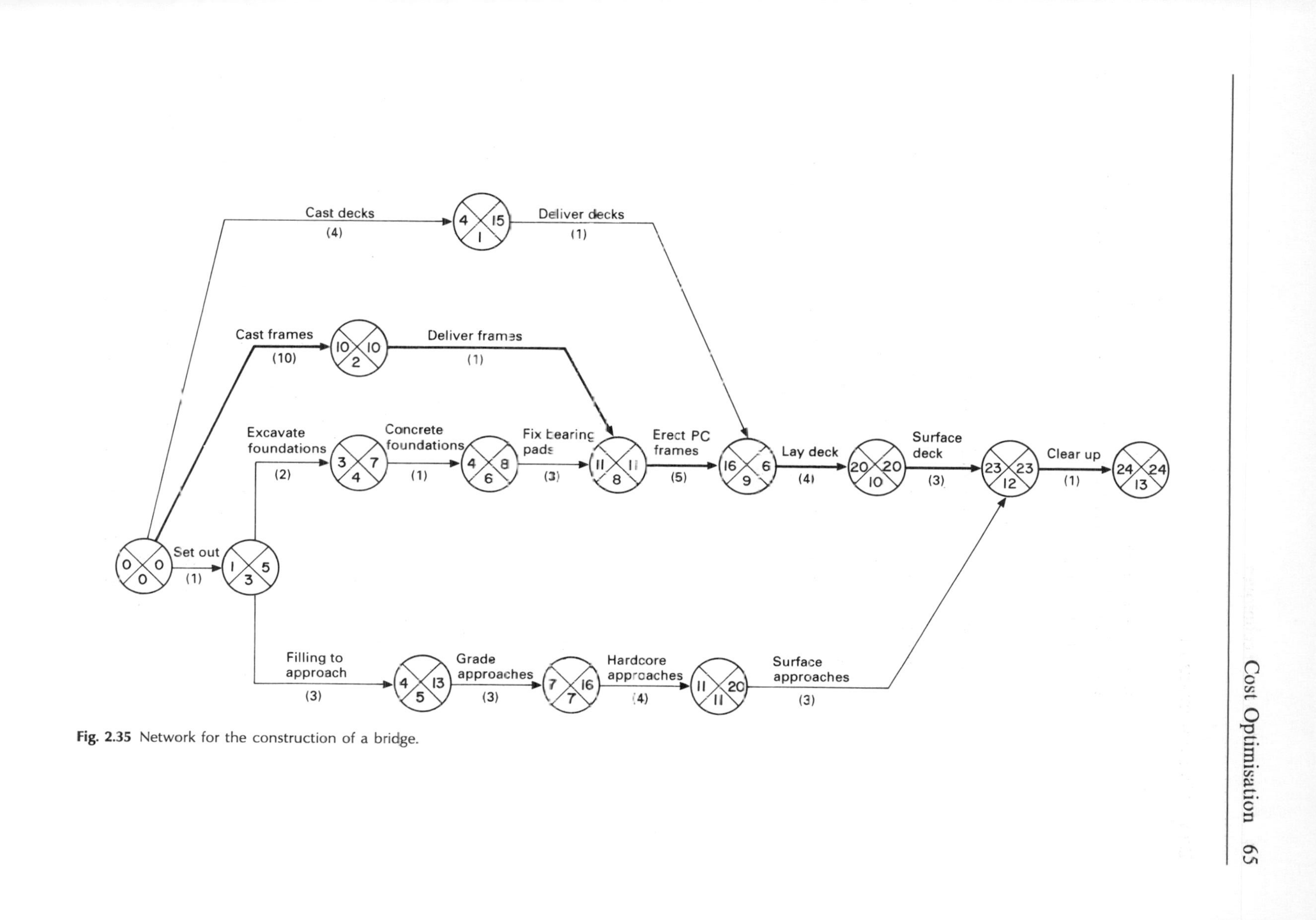


**Fig. 2.35** Network for the construction of a bridge.

**Example 2.8: Simple cost optimisation**

Figure 2.35 shows a network for the construction of a bridge. The normal cost and duration and the crash cost and duration for each activity are shown in Table 2.7. The indirect costs are £400.00 per week.

**Table 2.7** Costs and durations for Fig. 2.35.

| Activity reference | Normal | | Crash | |
|---|---|---|---|---|
| | Duration | Cost (£) | Duration | Cost (£) |
| 0–1 | 4 | 3000 | 3 | 3300 |
| 0–2 | 10 | 9000 | 5 | 9750 |
| 0–3 | 1 | 100 | 1 | 100 |
| 1–9 | 1 | 120 | 1 | 120 |
| 2–8 | 1 | 100 | 1 | 100 |
| 3–4 | 2 | 1500 | 2 | 1500 |
| 3–5 | 3 | 500 | 2 | 600 |
| 4–6 | 1 | 500 | 1 | 500 |
| 5–7 | 3 | 350 | 3 | 350 |
| 6–8 | 3 | 450 | 2 | 525 |
| 7–11 | 4 | 550 | 3 | 600 |
| 8–9 | 5 | 1500 | 3 | 2500 |
| 9–10 | 4 | 1200 | 3 | 1400 |
| 10–12 | 3 | 750 | 2 | 875 |
| 11–12 | 3 | 800 | 2 | 1100 |
| 12–13 | 1 | 500 | 1 | 500 |

The all-normal solution is given in Fig. 2.35. The critical activities are 0–2, 2–8, 8–9, 9–10, 10–12 and 12–13. The project duration is 24 weeks and the cost is £20 920.

The list of critical activities in order of minimum cost slope, after omitting activities that cannot be compressed, is shown in Table 2.8.

**Table 2.8** Critical activities.

| Activity | Maximum compression | Cost per week |
|---|---|---|
| 10–12 | 1 | £125 |
| 0–2 | 5 | £150 |
| 9–10 | 1 | £200 |
| 8–9 | 2 | £500 |

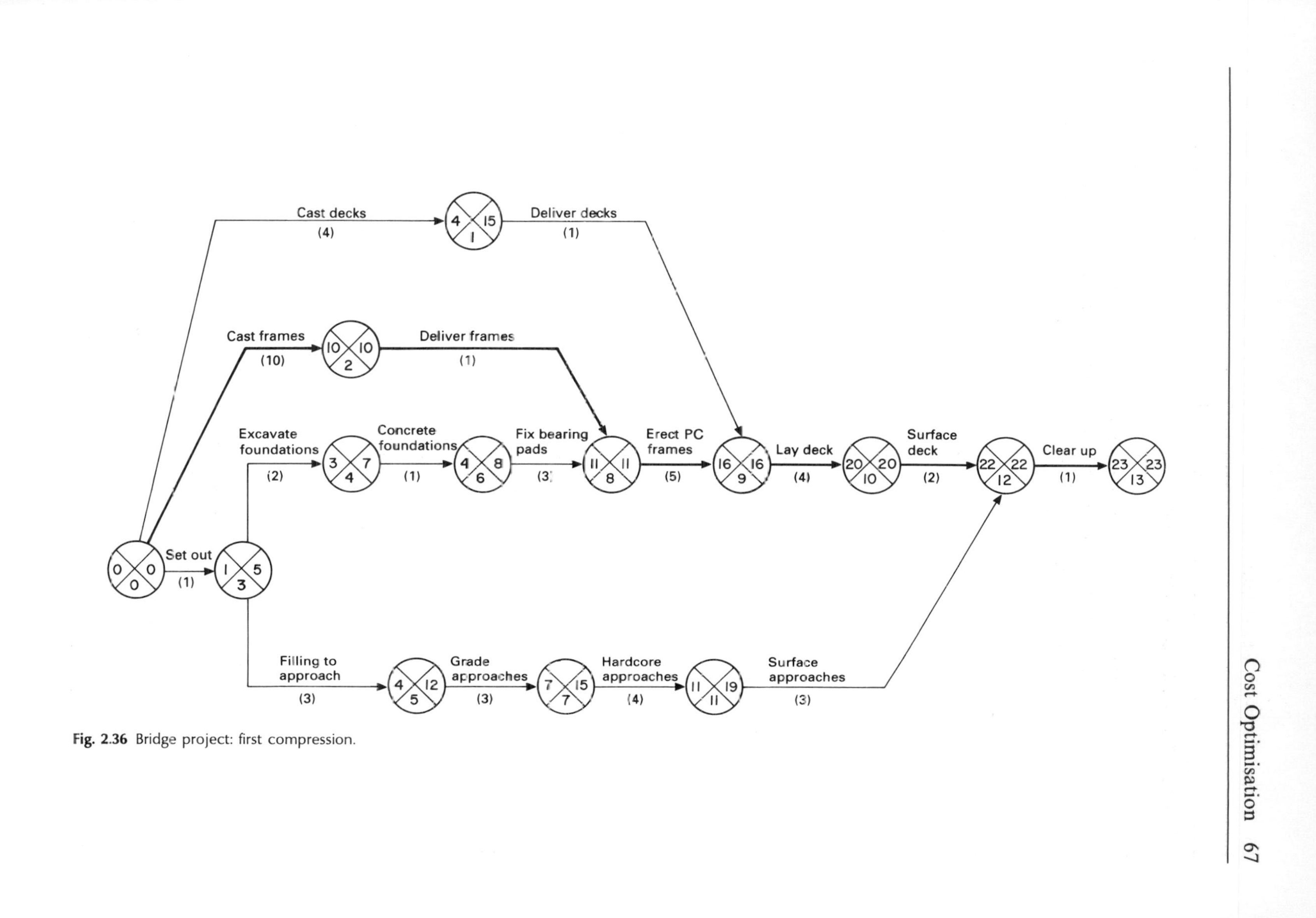


**Fig. 2.36** Bridge project: first compression.

### *First compression*

Compress activity 10–12 by 1 week at a cost of £125. The project duration is now 23 weeks and the direct cost is £20 920 + £125 = £21 045. Figure 2.36 shows the network after the first compression. The shared float on activities 3–5, 5–7, 7–11 and 11–12 is reduced to 8 weeks.

### *Second compression*

Compress activity 0–2 by 4 weeks at a cost of 4 × £150 = £600. The project duration is now 19 weeks and the direct cost is £21 045 + £600 = £21 645. Figure 2.37 shows the network after the second compression. Activities 0–3, 3–4, 4–6 and 6–8 are now also critical.

The shared float on activities 3–5, 5–7, 7–11 and 11–12 is further reduced to 4 weeks and the shared float in activities 0–1 and 1–9 is reduced to 7 weeks.

As two critical paths have now emerged, a revised list of critical activities is required in order of minimum cost slope, omitting activities that have been fully compressed and omitting activities that cannot be compressed, as shown in Table 2.9.

**Table 2.9** Revised list of critical activities.

| Activity | Maximum compression | Cost per week |
|---|---|---|
| 6–8 | 1 | £75 |
| 0–2 | 1 | £150 |
| 9–10 | 1 | £200 |
| 8–9 | 2 | £500 |

### *Third compression*

Although 6–8 has the least cost slope of the activities under consideration, compressing this activity will have no effect on the project duration unless activity 0–2 is also compressed. The combined effect of compressing both of these activities is £75 + £150 = £225 per week. It is therefore cheaper to compress activity 9–10.

Compress activity 9–10 by one week at a cost of £200. The project duration is now 18 weeks and the direct cost is £21 645 + £200 = £21 845.

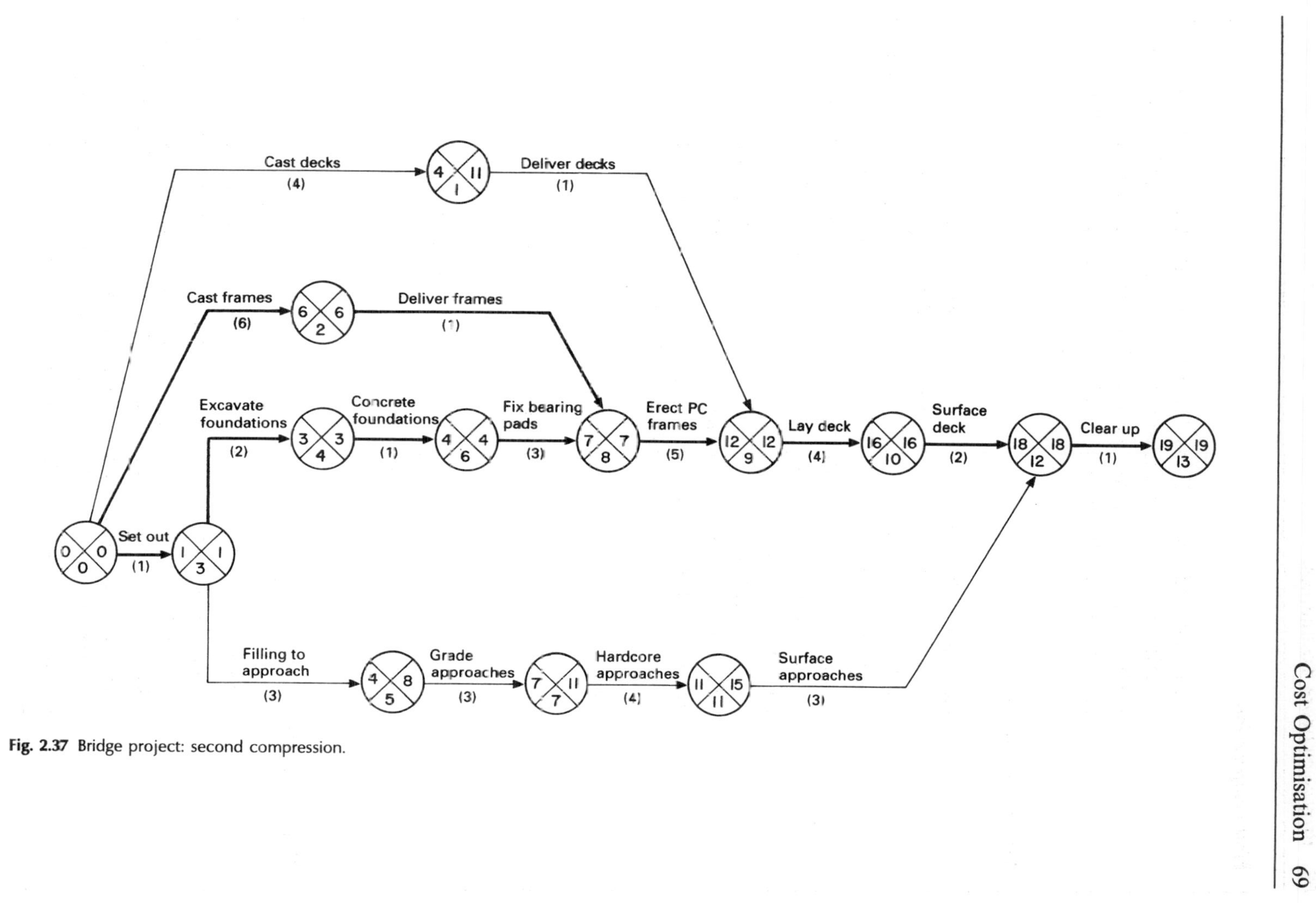


**Fig. 2.37** Bridge project: second compression.

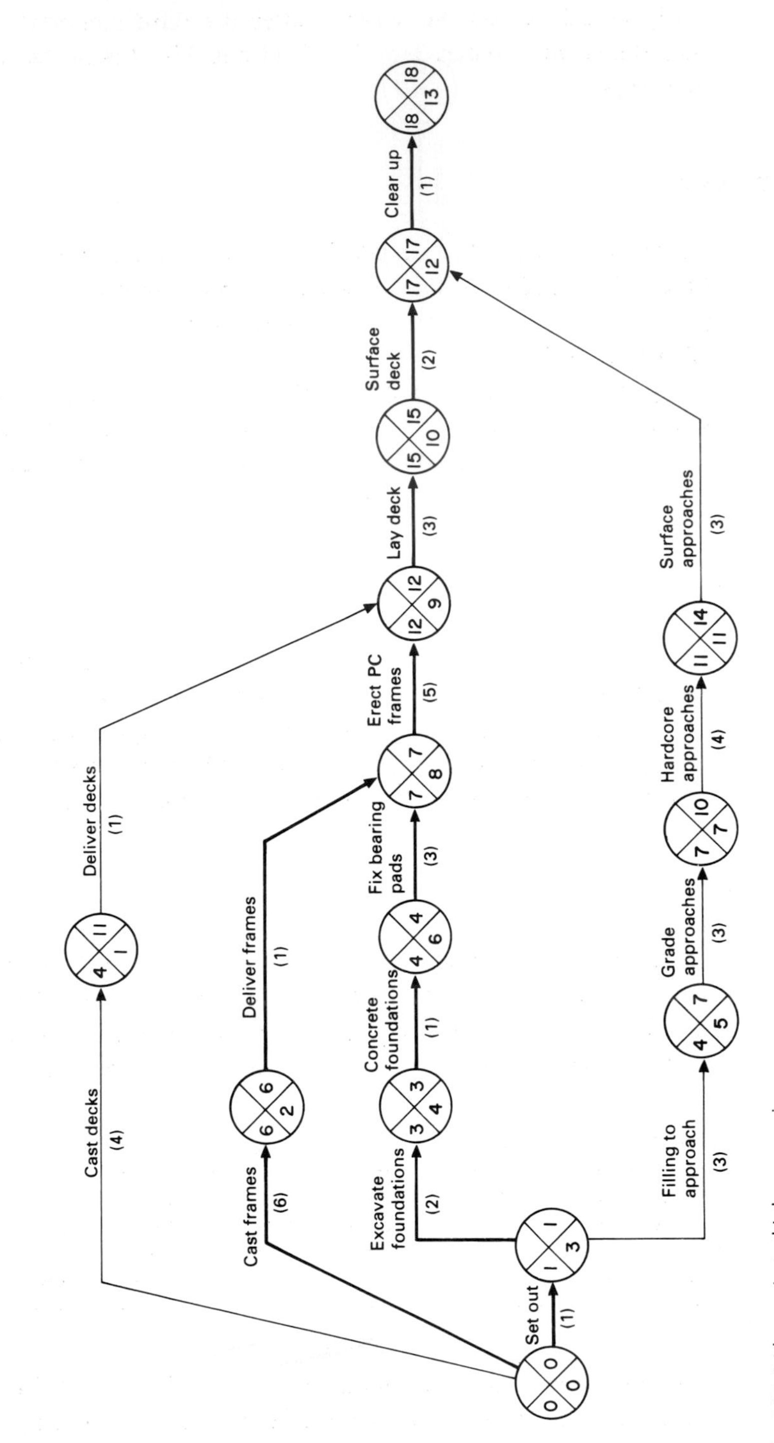


**Fig. 2.38** Bridge project: third compression.

Figure 2.38 shows the network after the third compression. The shared float in activities 3–5, 5–7, 7–11 and 11–12 is further reduced to 3 weeks.

### *Fourth compression*

Compress activities 0–2 and 6–8 by 1 week at a combined cost of £150 + £75 = £225 (as this is cheaper than compressing 8–9). The project duration is now 17 weeks and the direct cost is £21 845 + £225 = £22 070.

Figure 2.39 shows the network after the fourth compression. The shared float on activities 3–5, 5–7, 7–11 and 11–12 is further reduced to 2 weeks, and the shared float on 0–1 and 1–9 is further reduced to 6 weeks.

### *Fifth compression*

Compress activity 8–9 by 2 weeks at a cost of £500 × 2 = £1000. The project duration is now 15 weeks and the direct cost is £22 070 + 1000 = £23 070. Figure 2.40 shows the network after the fifth compression. Activities 3–5, 5–7, 7–11 and 11–12 are now critical, and the shared float on activities 0–1 and 1–9 is reduced to 4 weeks.

### *Optimising cost*

To determine the optimum cost of the project, indirect cost must be added as shown in Table 2.10. The optimum cost is £28 870 at a duration of 17 weeks. Figure 2.41 shows the network analysed for 17 weeks' duration; the weekly cost of each activity is also shown.

**Table 2.10** Optimising cost.

| Project durations | 15 | 17 | 18 | 19 | 23 | 24 |
|---|---|---|---|---|---|---|
| Direct costs (£) | 23 070 | 22 070 | 21 845 | 21 645 | 21 045 | 20 920 |
| Indirect costs (£) | 6 000 | 6 800 | 7 200 | 7 600 | 9 200 | 9 600 |
| Total costs (£) | 29 078 | 28 870 | 29 045 | 29 245 | 30 245 | 30 520 |

A graph can now be produced that shows the project direct cost curve, the indirect costs and the total costs, as shown in Fig. 2.42.

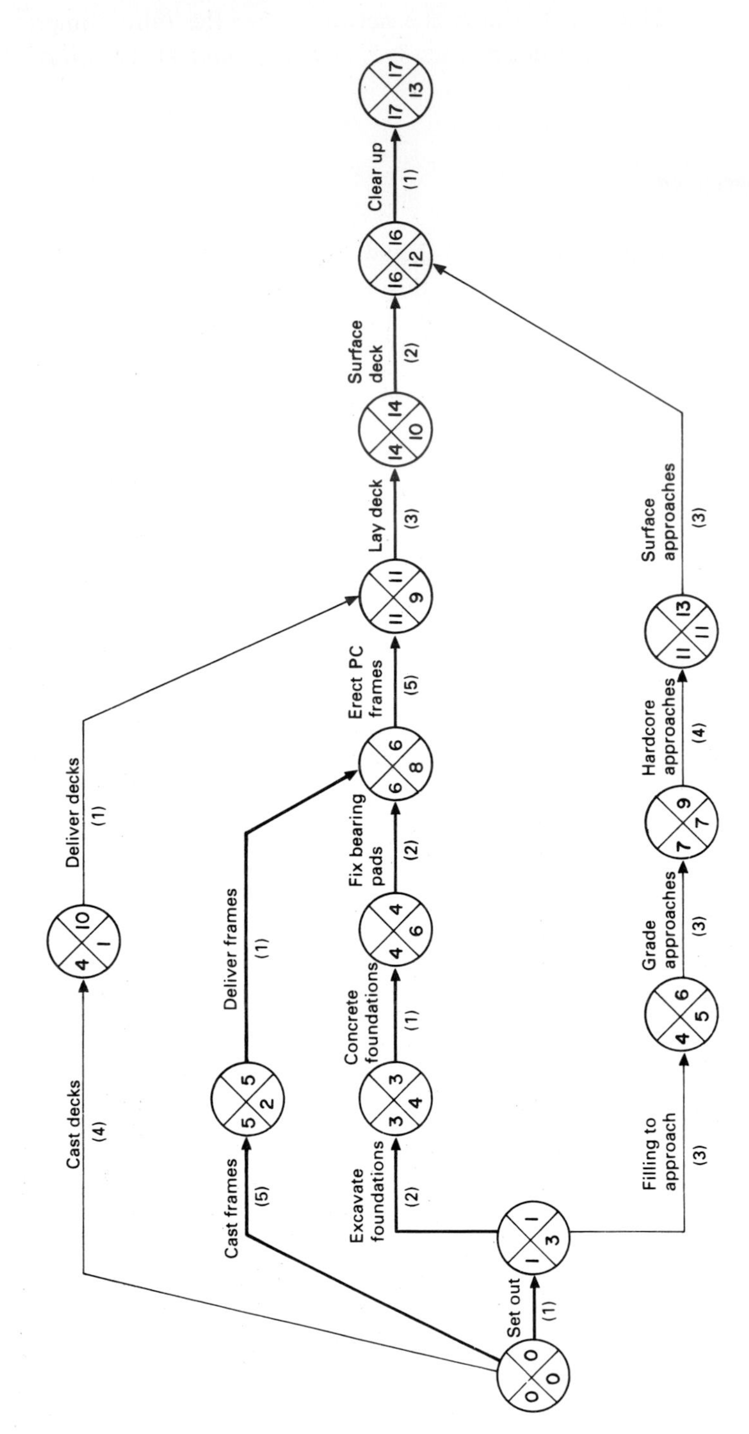


**Fig. 2.39** Bridge project: fourth compression.

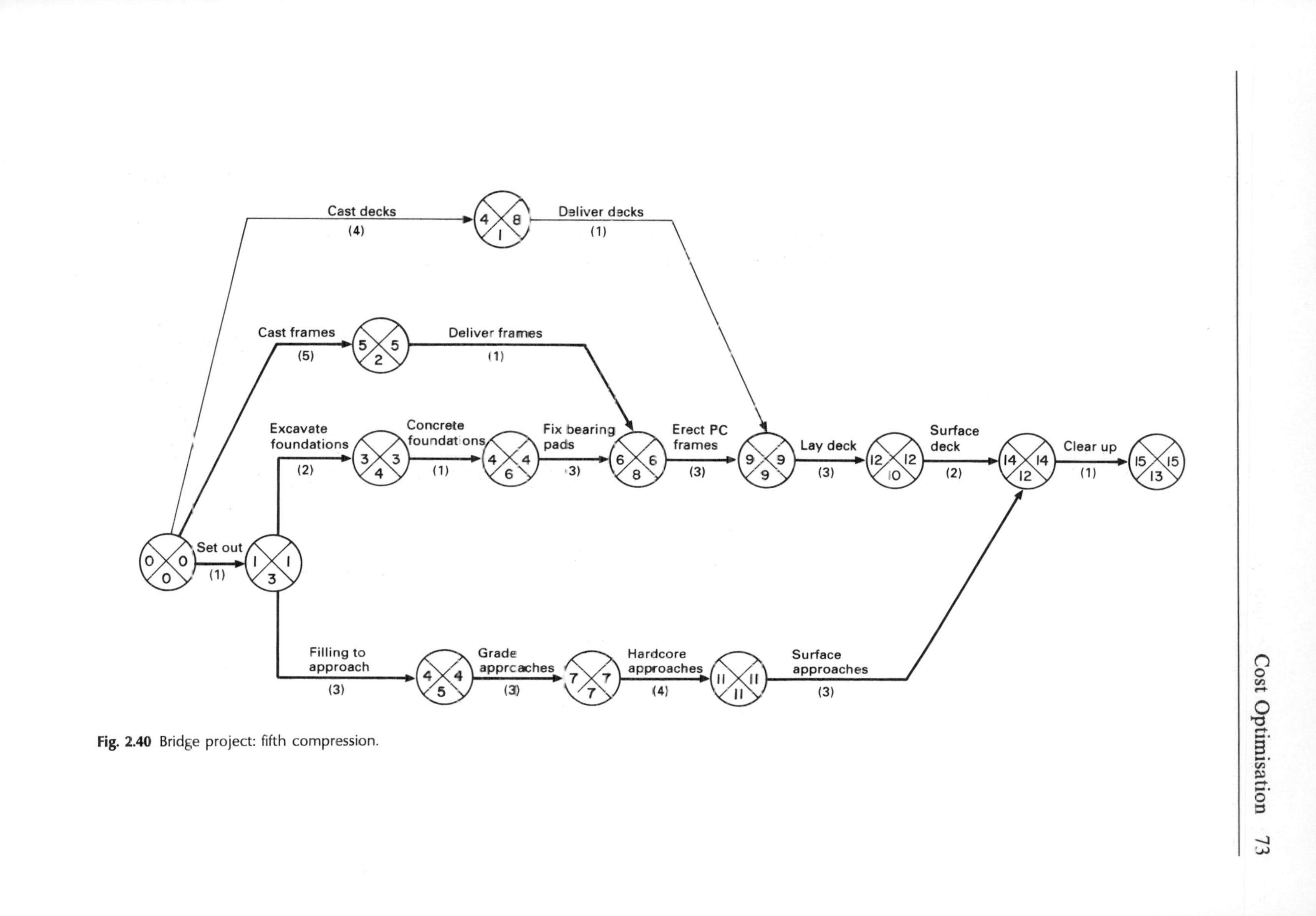


**Fig. 2.40** Bridge project: fifth compression.

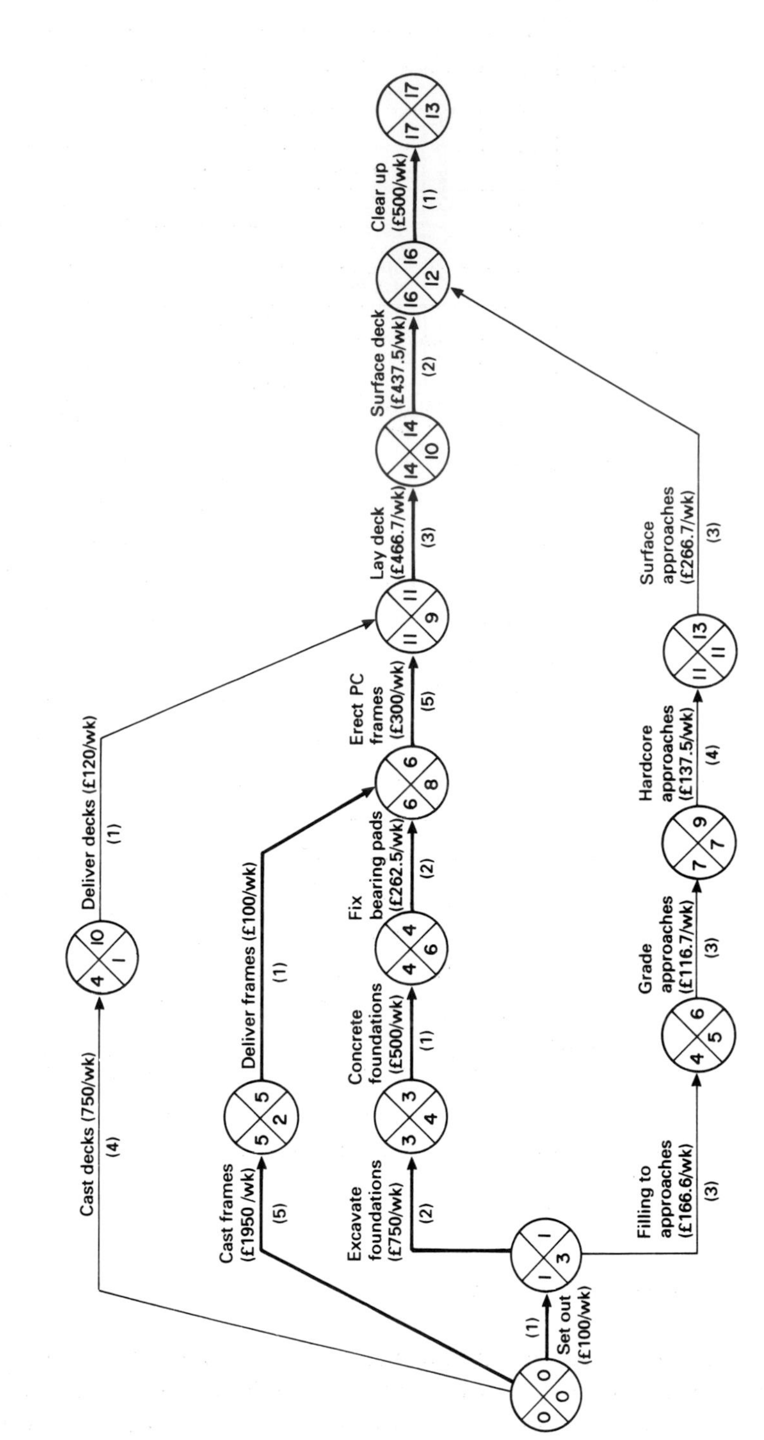


**Fig. 2.41** Bridge project: optimisation of cost.

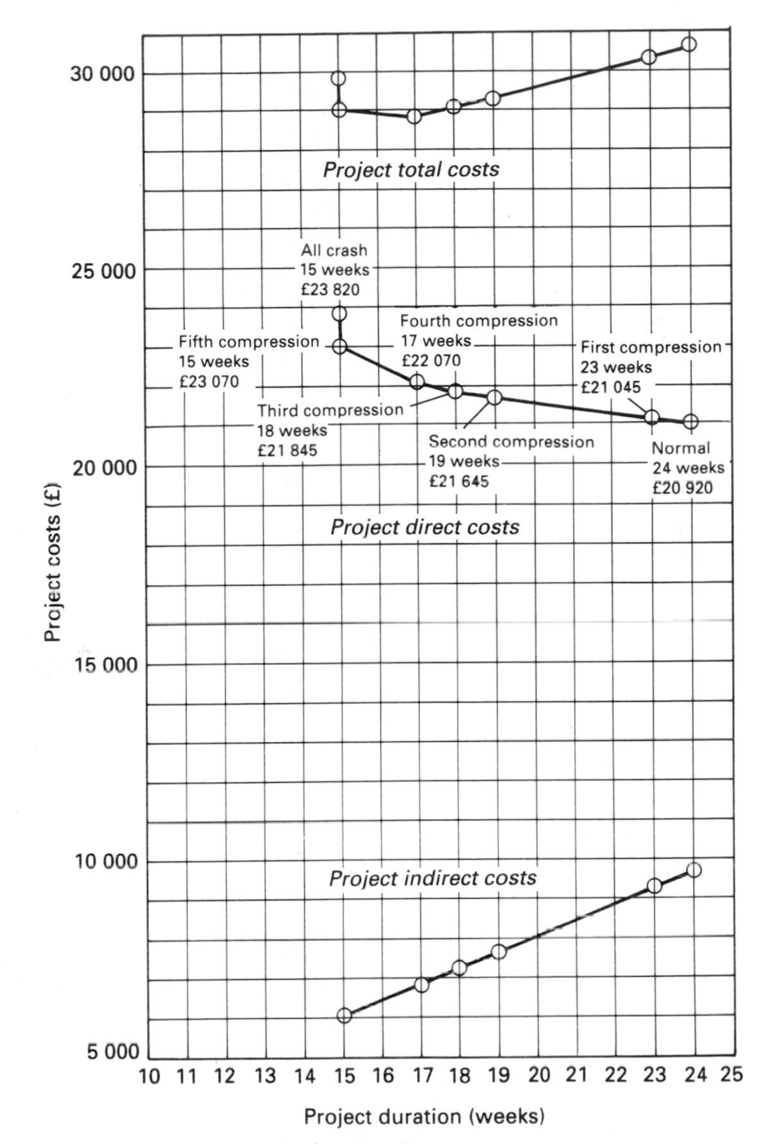


**Fig. 2.42** Bridge project: direct, indirect and total costs.

### *2.13.3 Optimal cost at minimum duration*

If project duration must be kept to a minimum then the approach could be varied and an 'all-crash' solution found first by crashing all the activities in the project, as shown in Fig. 2.43. The optimal least-time solution is then found by decompressing the non-critical activities,

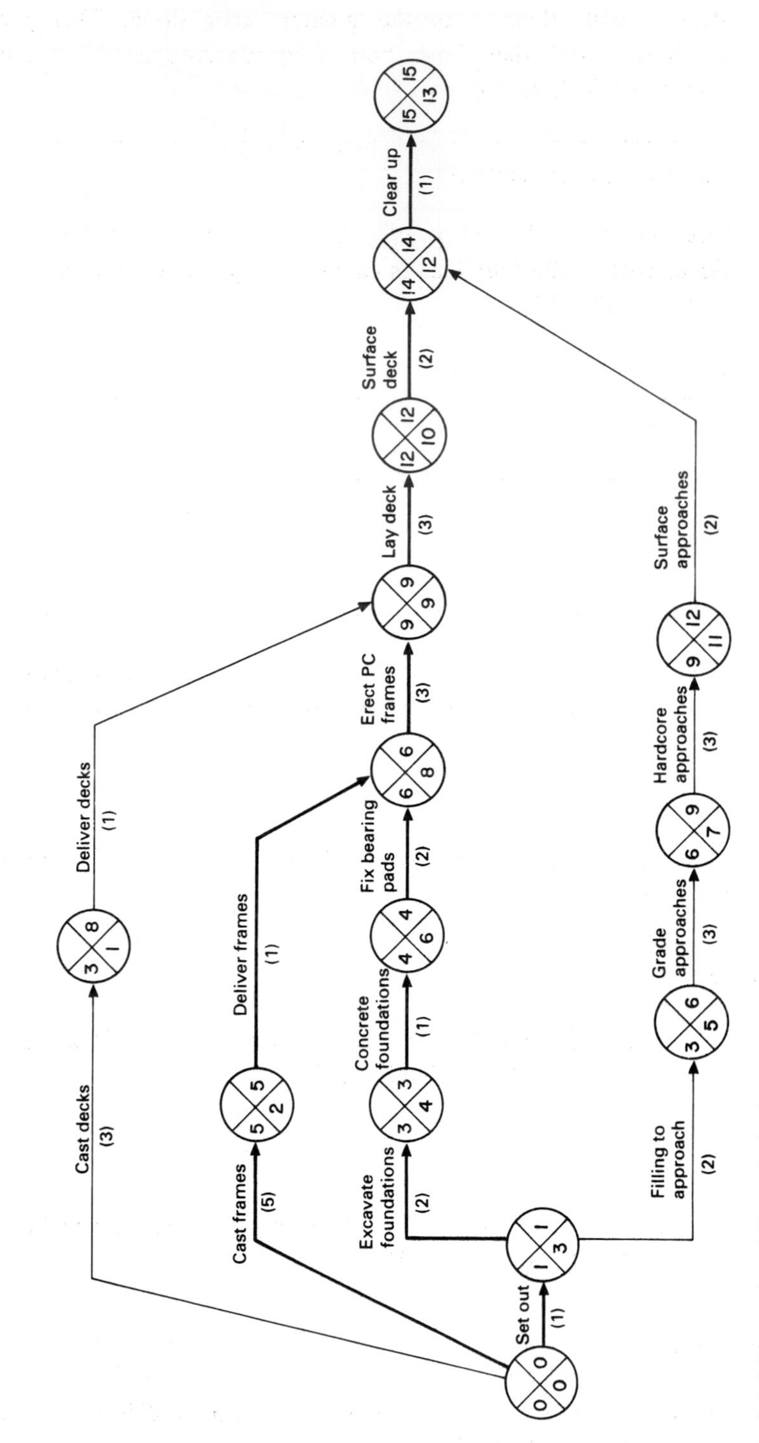


**Fig. 2.43** Bridge project: fully crashed solution.

starting with those with the greatest cost slope. This process is continued until they have been fully decompressed or until they become critical, as follows.

**Example 2.9: Optimal least-time solution**

The fully crashed solution is given in Fig. 2.43, and the 'all-crash' direct cost is £23 820. This is calculated by adding up all the crash costs in Table 2.7.

**Table 2.11** Non-critical activities.

| Activity | Maximum decompression | Cost slope |
|---|---|---|
| 0–1 | 1 | £300 |
| 11–12 | 1 | £300 |
| 3–5 | 1 | £100 |
| 7–11 | 1 | £ 50 |

The list of non-critical activities in order of greatest cost slope after omitting those that cannot be decompressed is shown in Table 2.11.

***First decompression***

Decompress activity 0–1 by 1 week at a saving of £300. The direct cost is now £23 820 – £300 – £23 520

***Second decompression***

Decompress activity 11–12 by 1 week at a saving of £300. The direct cost is now £23 520 – £300 = £23 220.

***Third decompression***

Decompress activity 3–5 by 1 week at a saving of £100. The direct cost is now £23 220 – £100 = £23 120.

***Fourth decompression***

Decompress activity 7–11 by 1 week at a saving of £50. The direct cost is now £23 120 – £50 = £23 070.

All available activities have now been fully decompressed, and the optimal cost at minimum duration is £23 070, which is of course the same result as that found in the fifth compression using the previous method. In this example, examination of the fully crashed network would have shown that all of the activities in the chains 0–1–9 and 3–5–7–11–12 could be fully decompressed, thus avoiding the step-by-step approach followed above. This is not always the case, however, as additional critical paths often emerge as activities are decompressed.

The 'all-crash' direct cost and all-crash total costs (£23 820 + £6000 = £29 820) are shown in Fig. 2.42.

### 2.13.4 *Other uses of the least-cost concept*

In order to make full use of the least-cost concept almost *all* the information is necessary at the tendering stage, and this is very unusual in practice. However, the least-cost concept can be used during the progress of the project. If critical activities are taking longer than planned, then *one* of the considerations in deciding how to gain back lost time is to make savings at the least cost. Naturally, saving time on activities with the least-cost slope would achieve this. Another use of the concept is to meet a stated project completion date. If, on first analysis, the project duration is longer than that set down in the conditions of contract, it can be reduced more economically by saving time on the activities with the least-cost slope.

### 2.13.5 *Decision rules*

Decision rules can be used to help select the best course of action, and one effective procedure is to increase the size of the gang for the trade that has the highest total time on the critical path. This will have an effect on many critical activities, and sometimes cuts down the project duration considerably. Very often gang sizes can be increased at little or no extra cost, particularly when the reduction in overheads due to the reduction in the project duration is taken into account.

### 2.13.6 *Complex compression*

When reducing project duration, the network diagram can be re-analysed at each stage. However, by careful observation, it is possible to perform a number of reductions in duration prior to total re-analysis.

The procedure for reducing the project duration is similar to that described in section 2.13.2, but includes in addition consideration of multiple cost slopes, alternativc mcthods of construction and taking account of discrete cost information.

**Example 2.10: Complex compression**

As the example on simple compression was based on an arrow diagram, this example will be based on a precedence diagram.

Figure 2.44 shows a network for the construction of a furniture showroom. The data for the activities which are compressible are shown in Fig. 2.45.

It can be seen from Fig. 2.45 that some of the activities do not have a simple cost slope, as follows:

- The cost slope for roof construction, plumbing and heating second fix and carpentry second fix increases when the reduction in duration goes beyond a certain point.
- The cost slope of 'Excavate foundations' varies depending on which method is chosen.
- The activities 'Erect frame' and 'Internal fittings' have two specific durations and associated costs with no interpolation between.

The procedure for finding the optimum solution is similar to the example in section 2.13.3 but the above factors have to be taken into account.

The all-normal solution is shown in Fig. 2.44. The project duration is 115 days, and the direct cost of labour and plant is £38 000. Indirect costs are to be charged at £140 per day. The priority of critical activities is shown in Table 2.12.

**Table 2.12** Priority of critical activities.

| Activity | Max. comp. | Cost | Notes |
|---|---|---|---|
| Alternate method for excavation | 4 | £200 | £50 per day equiv. cost slope |
| Plumbing and heating second fix | 2 | £60/day | |
| Roof construction | 2 | £120/day | |
| Internal fittings | 4 | £560 tot | £140 per day equiv. cost slope |
| Erect frame | 3 | £540 tot | £180 per day equiv. cost slope |
| Excavate foundations – Method B | 3 | £300/day | |

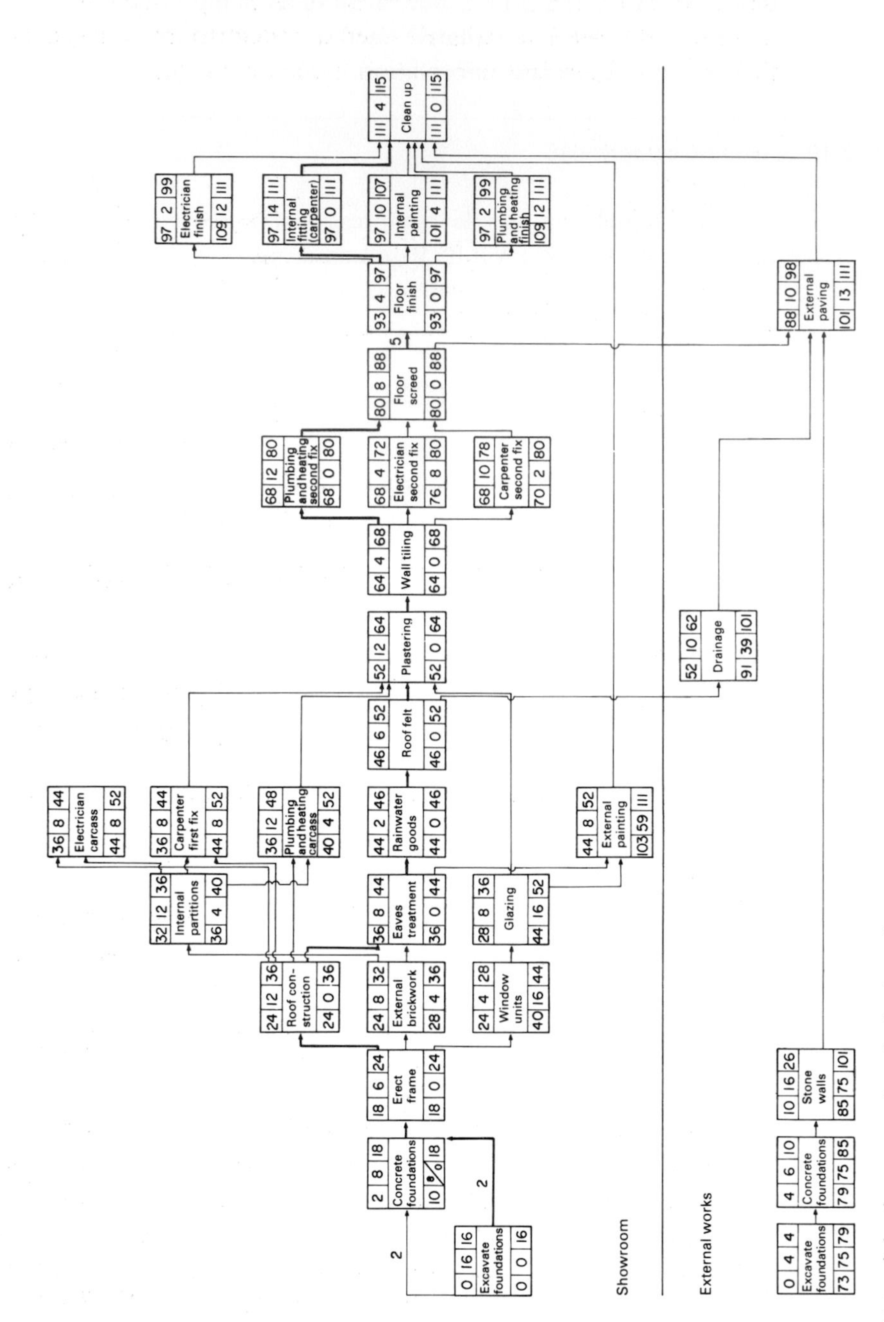


**Fig. 2.44** Network for the construction of a furniture showroom.

The equivalent cost slope has been given for some activities. Note that although internal fittings can be reduced by 7 days for £560, another path becomes critical after a reduction of 4 days, and therefore the equivalent cost slope is based on 4 days.

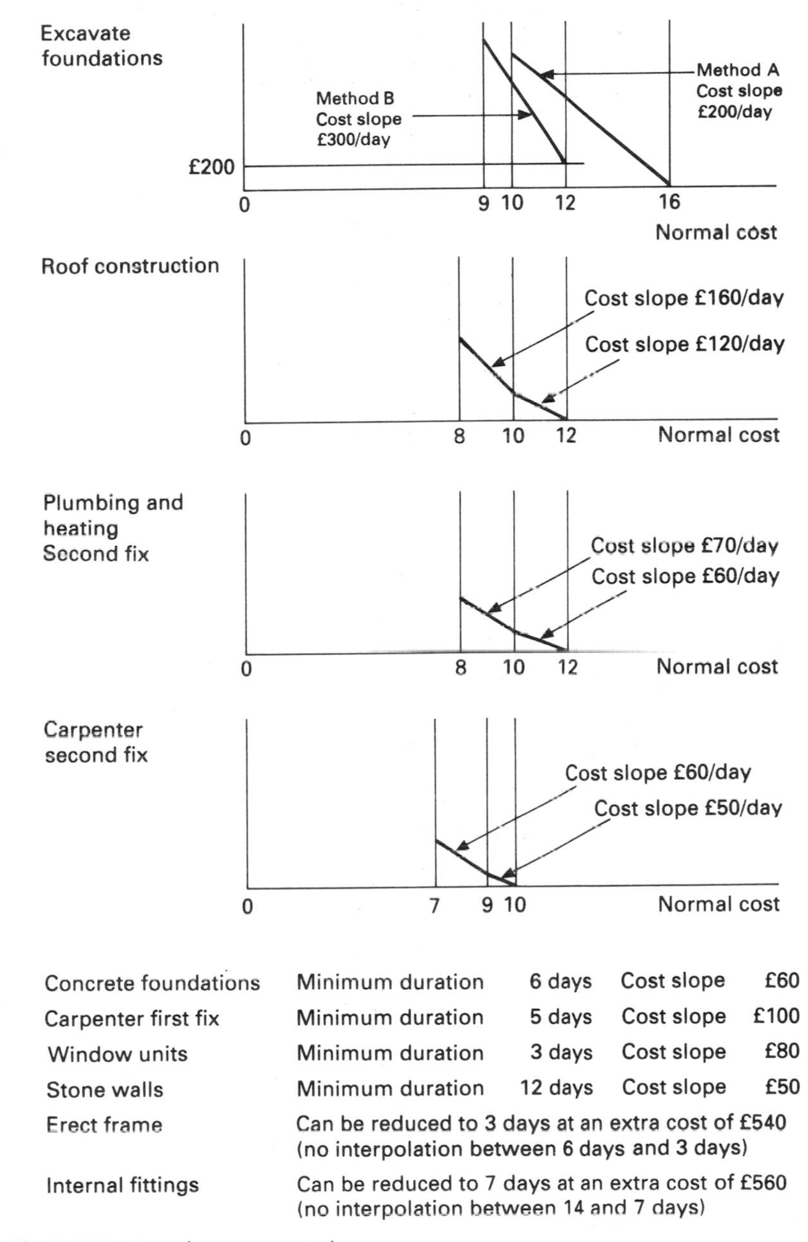


| | | | | |
|---|---|---|---|---|
| Concrete foundations | Minimum duration | 6 days | Cost slope | £60 |
| Carpenter first fix | Minimum duration | 5 days | Cost slope | £100 |
| Window units | Minimum duration | 3 days | Cost slope | £80 |
| Stone walls | Minimum duration | 12 days | Cost slope | £50 |
| Erect frame | Can be reduced to 3 days at an extra cost of £540 (no interpolation between 6 days and 3 days) | | | |
| Internal fittings | Can be reduced to 7 days at an extra cost of £560 (no interpolation between 14 and 7 days) | | | |

**Fig. 2.45** Furniture showroom: cost slopes.

### *First compression*

Compress 'Excavate foundations' by 4 days using method B at a cost of £200. The duration is now 111 days, and the direct cost is £38 000 + £200 = £38 200.

### *Second compression*

Compress 'Plumbing and heating second fix' by 2 days at a cost of 2 × £60 = £120. The duration is now 109 days, and the direct cost is £38 200 + £120 = £38 320.

Two paths are now critical in this area and the combined cost slopes are as shown in Table 2.13.

**Table 2.13** Combined cost slopes.

| Activity | Max. compression | | Cost per day |
|---|---|---|---|
| Carpentry second fix | 1 day | | £ 50 |
| Plumbing and heating second fix | 1 day | | £ 70 |
| | | Total | £120 |
| For further reduction: | | | |
| Carpentry second fix | 1 day | | £ 60 |
| Plumbing and heating second fix | 1 day | | £ 70 |
| | | Total | £130 |

### *Third compression*

Compress 'Roof construction' by 2 days, 'Carpentry second fix' by 1 day and 'Plumbing and heating second fix' by 1 day at a cost of 2 × £120 + £50 + £70 = £360.

The duration is now 106 days, and the direct cost is £38 320 + £360 = £38 680.

The greater cost slope of £160 per day now applies to roof construction.

### *Fourth compression*

Compress 'Carpentry second fix' and 'Plumbing and heating second fix' by a further 1 day at a cost of £130. The duration is now 105 days, and the direct cost is £38 680 + £130 = £38 810.

### *Fifth compression*

Compress 'Internal fittings' by 7 days (only 4 are effective) at a cost of £560. The duration is now 101 days, and the direct cost is £38 810 + £560 = £39 370.

### *Sixth compression*

Compress 'Roof construction' a further 2 days at a cost of 2 × £160 = £320. The duration is now 99 days, and the direct cost is £39 370 + £320 = £39 690.

### *Seventh compression*

Compress 'Erect frame' to 3 days at a total cost of £540. The duration is now 96 days, and the direct cost is £39 690 + £540 = £40 230.

### *Eighth compression*

Reduce 'Excavate foundations' a further 3 days at a cost of 3 × £300 = £900. The duration is now 93 days, and the direct cost is £40 230 + £900 = £41 130.

Figure 2.16 shows the network analysed for 93 days' duration.

### *Optimising the cost*

To determine the optimum cost of the project, indirect costs must be added as shown in Table 2.14. The optimum cost is £53 510, and the optimum project duration is 101 days.

*Note*: The cost is £53 510 for a project duration of 105 days, but the shortest duration has been selected as this gives more flexibility for controlling the project during progress of the work.

**Table 2.14** Optimising the cost.

| Project durations | 93 | 96 | 99 | 101 | 105 | 106 | 109 | 111 | 115 |
|---|---|---|---|---|---|---|---|---|---|
| Indirect costs (£) | 13 020 | 13 440 | 13 860 | 14 140 | 14 700 | 14 840 | 15 260 | 15 540 | 16 100 |
| Direct costs (£) | 41 130 | 40 230 | 39 690 | 39 370 | 38 810 | 38 680 | 38 320 | 38 200 | 38 000 |
| Total costs (£) | 54 150 | 53 670 | 53 550 | 53 510 | 53 510 | 53 520 | 53 580 | 53 740 | 54 100 |

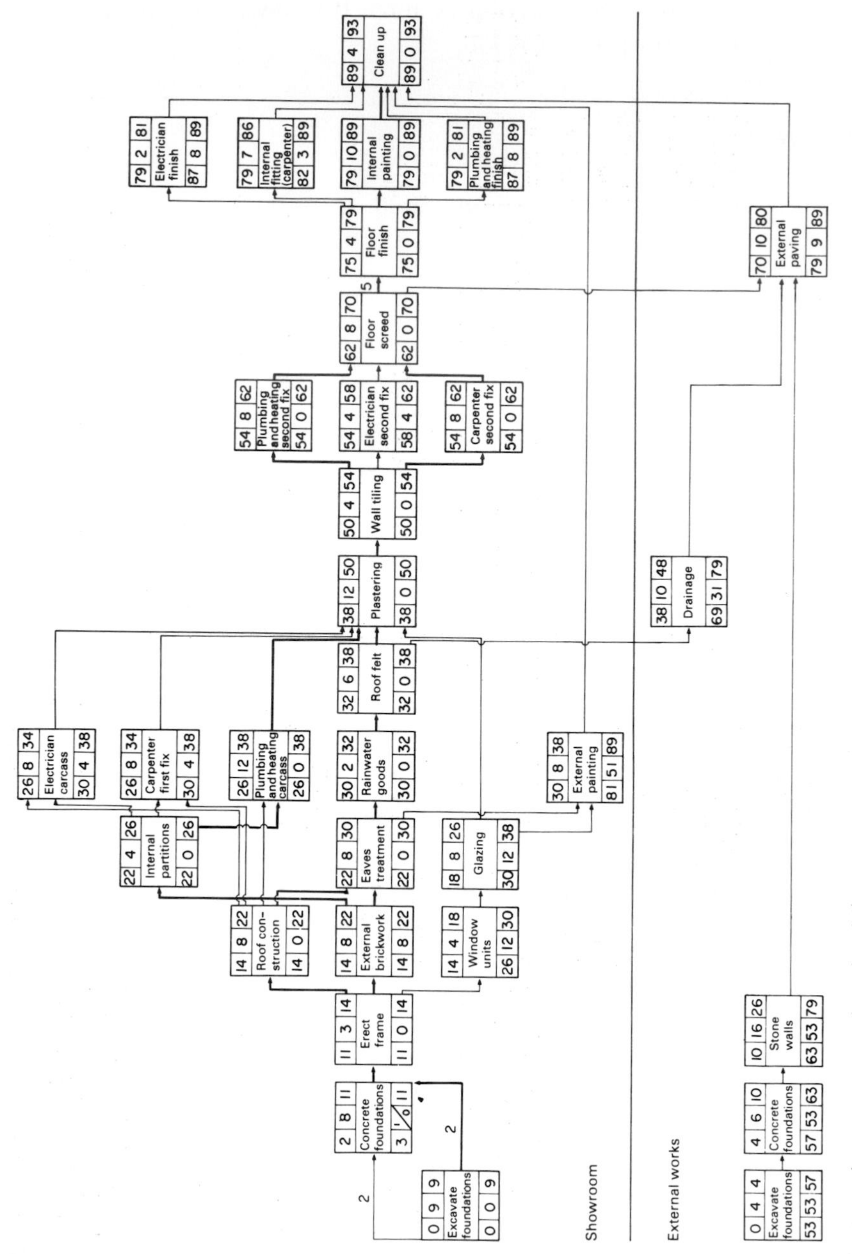


**Fig. 2.46** Furniture showroom: network analysed for 93 days' duration.

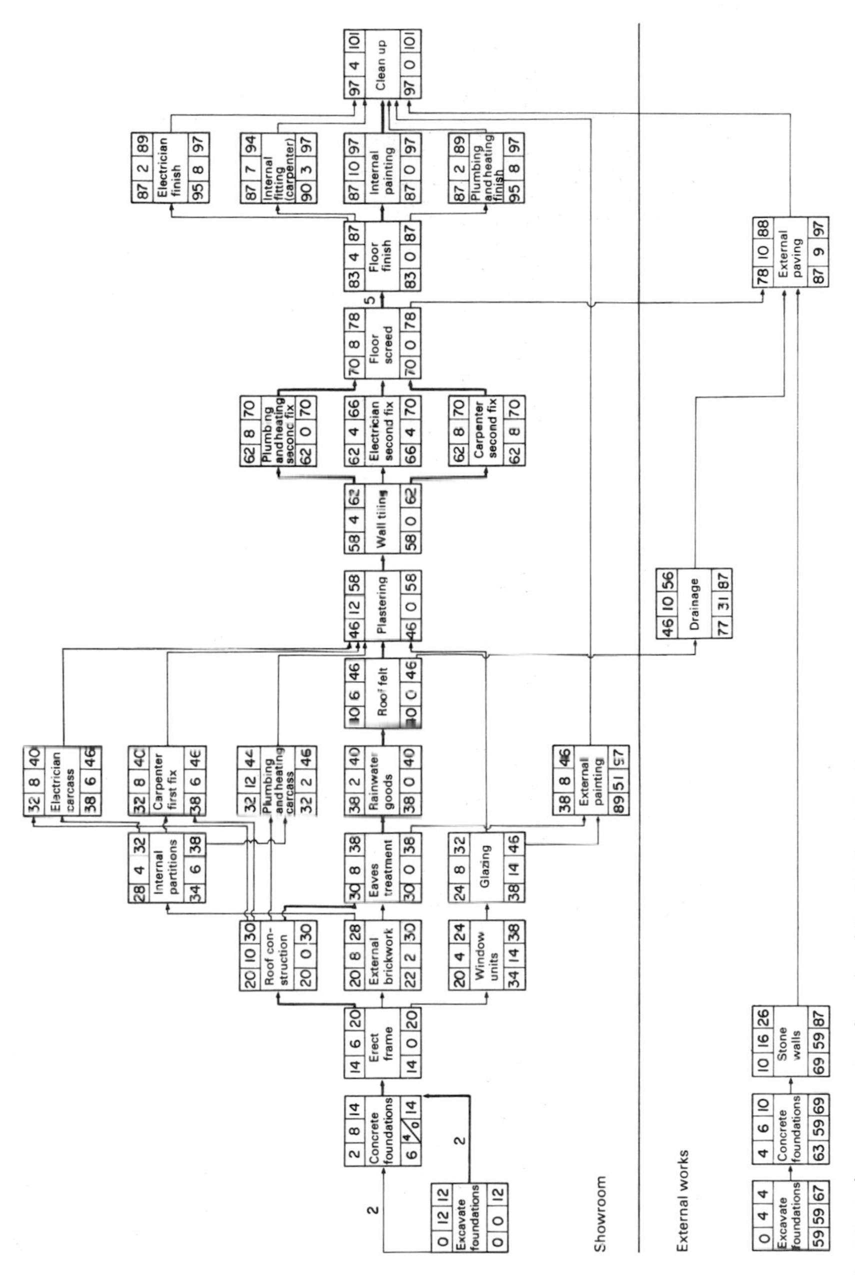

**Fig. 2.47** Furniture showroom: network analysed for 101 days' duration.

Figure 2.47 shows the network analysed for 101 days.

Graphs can now be produced that show the direct cost curve, the indirect cost curve and the total cost curve (see Figs 2.48 and 2.49).

***Optimum cost at minimum duration***

This can be found directly by following the procedure set out in example 2.9, but taking into account the additional factors set out at the beginning of this example.

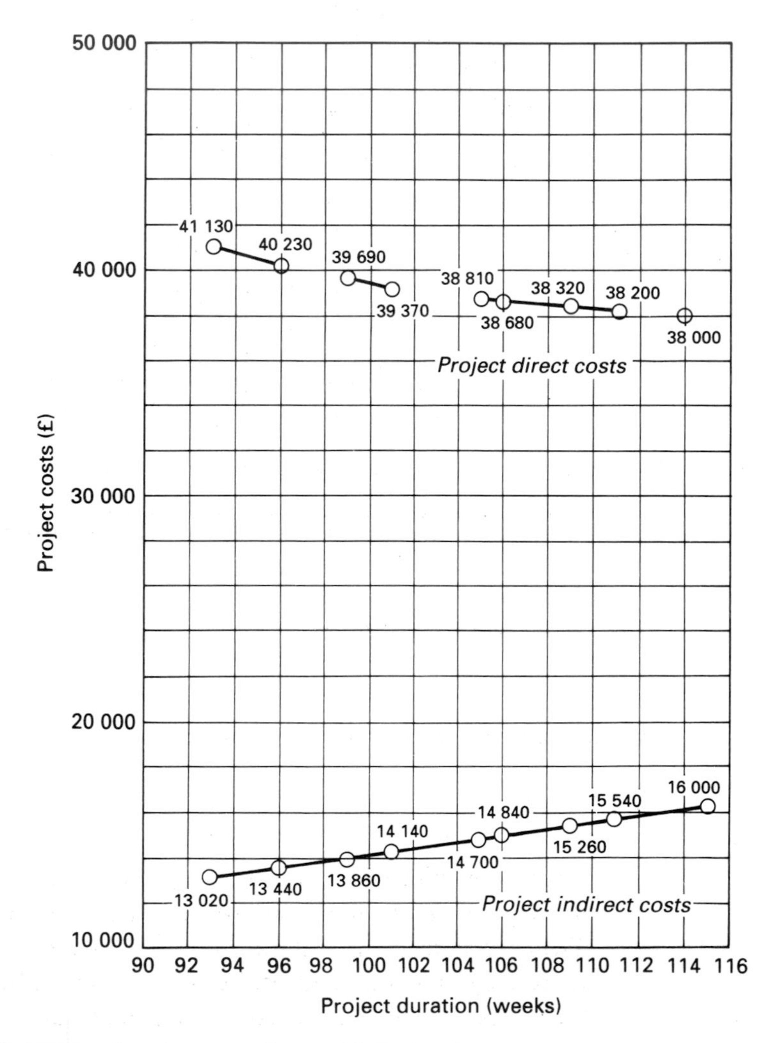


**Fig. 2.48** Furniture showroom: direct and indirect costs.

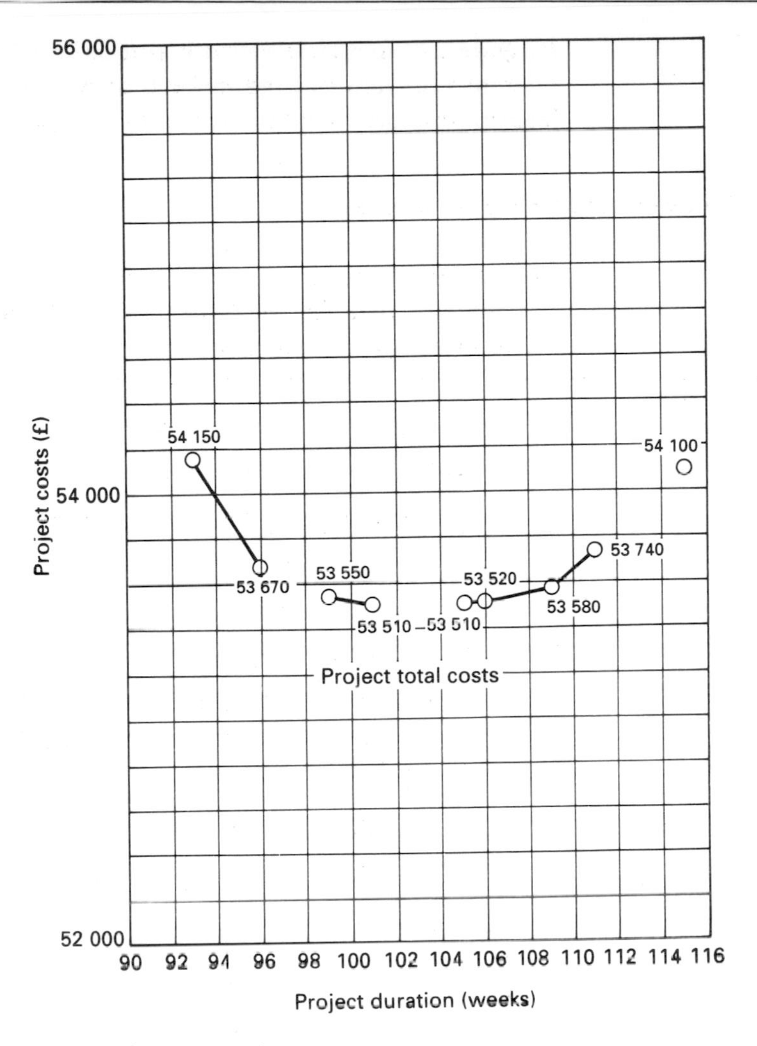


**Fig. 2.49** Furniture showroom: total costs.

## 2.14 Programme evaluation and review technique (PERT)

### *2.14.1 The use of PERT*

When using critical path analysis it is assumed that reasonably accurate information is available on the duration and cost of activities. In efficient companies this should be true for the majority of activities. There may be circumstances, however, when sufficient information is not available. This will be so when unique construction methods are being used, such as many of those encountered in the construction of the Sydney Opera House. It will also be so when networks are being used for projects involving research and development work, or when insufficient information is available. In these cases PERT can be used, as it takes account of uncertainty.

### 2.14.2 The effects of uncertainty

For activities that are not on the critical path, where a considerable amount of float is available, uncertainty may cause problems in scheduling labour, plant, material deliveries and subcontractor start times. For activities on the critical path, uncertainty will also result in the project duration being less clearly defined.

### 2.14.3 Allowing for uncertainty

To allow for uncertainty, three estimates are used for activity duration. These are:

- *Optimistic duration* ($d_o$): this is defined as the minimum duration if everything goes well (it is *not* the crashed duration).
- *Most likely duration* ($d_m$): this is based on analysis of work on previous projects, experience and judgement.
- *Pessimistic duration* ($d_p$): this is defined as the maximum time if everything goes wrong.

### 2.14.4 Activity probability distribution

The distribution can be skewed in either direction, and the range is roughly determined by $d_o$ and $d_p$ (Fig. 2.50). It is assumed that there is only a 1% chance of $d_o$ or $d_p$ being exceeded.

### 2.14.5 Calculating the expected mean duration and uncertainty

$$\text{expected mean duration,} \quad d_e = \frac{d_o + 4d_m + d_p}{6}$$

$$\text{standard deviation,} \quad \sigma d_e = \frac{d_p - d_o}{6}$$

$$\text{variance,} \quad \nu d_e = \left(\frac{d_p - d_o}{6}\right)^2$$

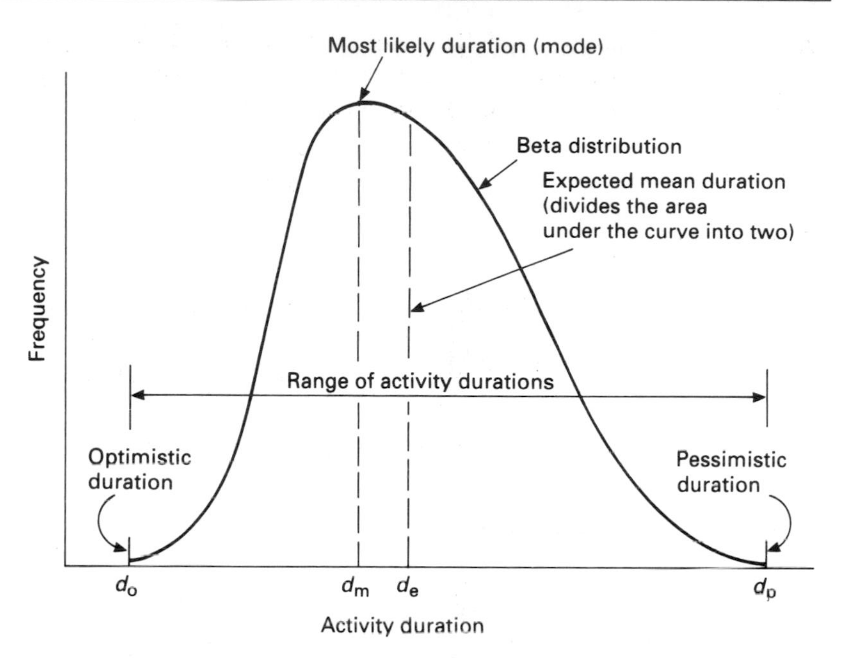


**Fig. 2.50** Activity probability distribution.

### 2.14.6 *Critical path calculations*

Calculations are carried out as set out in section 2.5, but the standard deviation and variance are also calculated.

### 2.14.7 *Event variance*

Event variance based on *earliest event times* is the sum of the variances up to the event being considered following the critical path from the start of the project. For non-critical events, it is the sum of the variances at each event, following the path that determined the earliest event time.

Event variance based on *latest event times* is the sum of the variances up to the event being considered following the critical path from the expected completion date. This variance gives the uncertainty remaining from the event being considered up to the end of the project measured along the critical path. For non-critical events it is the sum of the variances of each event, following the path that determined the latest event times.

### 2.14.8 *Multiple critical paths*

Where there is more than one critical path, the one with the maximum variance is used.

### 2.14.9 *Meeting specific event schedule times*

Event completion times are assumed to have a normal probability distribution with a mean value of $T_X$, the variance of $T_X$ being calculated as stated in section 2.14.7, and from this the standard deviation can be calculated. This assumption is strictly correct only for an infinite series, but is assumed to be accurate enough in practice for large projects. The event times in the example that follows are assumed to have normal distributions. To calculate the probability of achieving a specific event time, the area cut off by the specified time is calculated as a percentage of the total area beneath the curve (Fig. 2.51). Standard probability tables can be used for normal distribution functions, and a summarised table is given in Table 2.15. This table is accurate enough for construction projects.

The factor $z$ is calculated thus:

$$z = \frac{T_S - T_X}{\sigma T_X}$$

Using this value the probability of meeting the specific event scheduled time ($T_S$) can be read off from the table, interpolating as necessary.

### 2.14.10 *Determining a scheduled time based on a specific probability*

The scheduled time for a specific probability can be established by transposing the formula thus:

$$T_S = T_X + z\sigma T_X$$

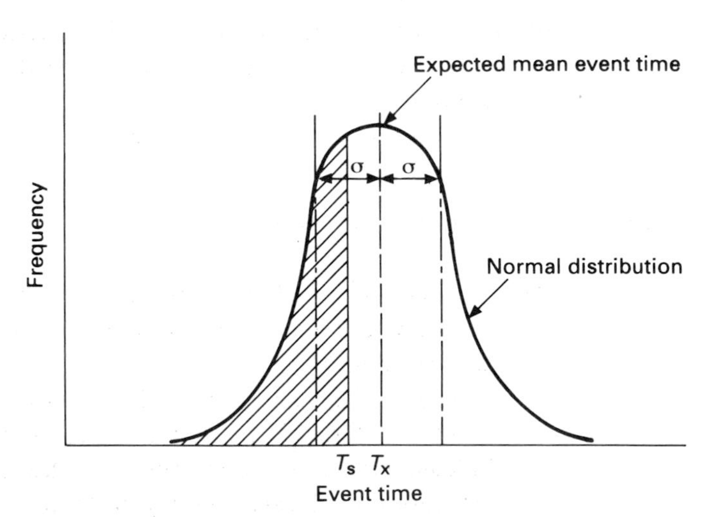


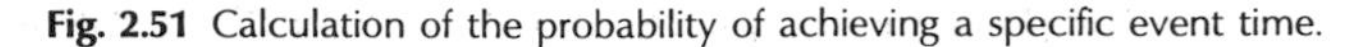
**Fig. 2.51** Calculation of the probability of achieving a specific event time.

**Table 2.15** Standard normal distribution.

| z | Probability | z | Probability | z | Probability | z | Probability |
|---|---|---|---|---|---|---|---|
| −2.8 | 0.0062 | −0.95 | 0.1711 | +0.05 | 0.5799 | +1.0 | 0.8413 |
| −2.2 | 0.0139 | −0.9 | 0.1841 | +0.1 | 0.5398 | +1.05 | 0.8531 |
| −2.0 | 0.0223 | −0.85 | 0.1977 | +0.15 | 0.5596 | +1.1 | 0.8643 |
| −1.9 | 0.0287 | −0.8 | 0.2119 | +0.20 | 0.5793 | +1.15 | 0.8749 |
| −1.8 | 0.0359 | −0.75 | 0.2266 | +0.25 | 0.5987 | +1.2 | 0.8849 |
| −1.7 | 0.0446 | −0.7 | 0.2412 | +0.3 | 0.6179 | +1.25 | 0.8944 |
| −1.6 | 0.0548 | −0.65 | 0.2578 | +0.35 | 0.6768 | +1.3 | 0.9032 |
| −1.5 | 0.0668 | −0.6 | 0.2743 | +0.4 | 0.6554 | +1.4 | 0.9192 |
| −1.4 | 0.0808 | −0.55 | 0.2912 | +0.45 | 0.6736 | +1.5 | 0.9332 |
| −1.3 | 0.0968 | −0.5 | 0.3085 | +0.5 | 0.6915 | +1.6 | 0.9452 |
| −1.25 | 0.1056 | −0.45 | 0.3264 | +0.55 | 0.7088 | +1.7 | 0.9554 |
| −1.2 | 0.1151 | −0.4 | 0.3446 | +0.6 | 0.7257 | +1.8 | 0.9641 |
| −1.15 | 0.1251 | −0.35 | 0.3632 | +0.65 | 0.7422 | +1.9 | 0.9713 |
| −1.1 | 0.1357 | −0.3 | 0.3821 | +0.7 | 0.7588 | +2.0 | 0.9777 |
| −1.05 | 0.1469 | −0.25 | 0.4013 | +0.75 | 0.7734 | +2.2 | 0.9861 |
| −1.0 | 0.1587 | −0.2 | 0.4207 | +0.8 | 0.7881 | +2.8 | 0.9938 |
| | | −0.15 | 0.4404 | +0.85 | 0.8023 | | |
| | | −0.1 | 0.4602 | +0.9 | 0.8159 | | |
| | | −0.05 | 0.4801 | +0.95 | 0.8289 | | |
| | | 0 | 0.5 | | | | |

**Example 2.11: Applying PERT to a construction project**

Figure 2.52 shows a building project, and Table 2.16 shows the three estimated activity durations for each activity, together with the calculated expected durations ($d_e$), standard deviation ($\sigma d_e$), and variance ($\nu d_e$).

The method of analysis is as follows.

***Event times and event variances (Fig. 2.52)***

(1) Calculate the earliest event times using the expected mean duration ($d_e$) by working through the diagram selecting the longest path.
(2) Calculate the latest event times, again using $d_e$, working backwards through the diagram selecting the longest path.
(3) Calculate the variance of each event from the first event using the variance of each activity ($\nu d_e$) by working through the diagram on the paths that determine the earliest event times.

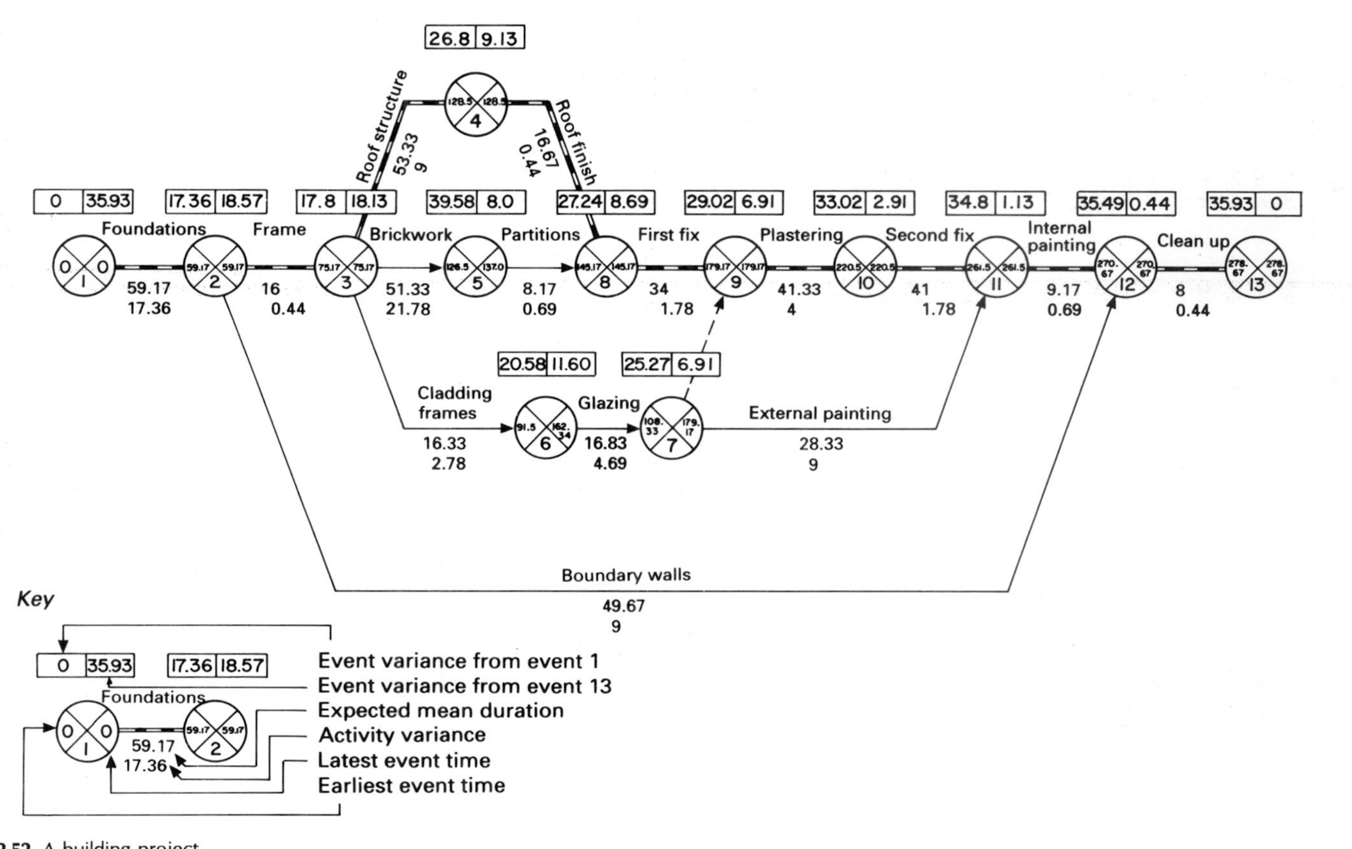


**Fig. 2.52** A building project.

**Table 2.16** Estimated activity durations for Fig. 2.52.

| Activity | $d_o$ | $d_m$ | $d_p$ | $d_e$ | $\sigma d_e$ | $\surd d_e$ |
|---|---|---|---|---|---|---|
| Foundations | 45 | 60 | 70 | 59.17 | 4.17 | 17.36 |
| Frame | 14 | 16 | 18 | 16.00 | 0.67 | 0.44 |
| Roof structure | 47 | 52 | 65 | 53.33 | 3.00 | 9.00 |
| Roof finish | 14 | 17 | 18 | 16.67 | 0.67 | 0.44 |
| Brickwork | 42 | 49 | 70 | 51.33 | 4.67 | 21.78 |
| Partitions | 6 | 8 | 11 | 8.17 | 0.83 | 0.69 |
| First fix | 30 | 34 | 38 | 34.00 | 1.33 | 1.78 |
| Plastering | 38 | 40 | 50 | 41.33 | 2.00 | 4.00 |
| Second fix | 37 | 41 | 45 | 41.00 | 1.33 | 1.78 |
| Internal painting | 7 | 9 | 12 | 9.17 | 0.83 | 0.69 |
| Cladding frames | 12 | 16 | 22 | 16.33 | 1.67 | 2.78 |
| Glazing | 12 | 16 | 25 | 16.83 | 2.17 | 4.69 |
| External painting | 22 | 27 | 40 | 28.33 | 3.00 | 9.00 |
| Brickwork to boundary walls | 42 | 49 | 60 | 49.67 | 3.00 | 9.00 |
| Clear up | 6 | 8 | 10 | 8.00 | 0.67 | 0.44 |

(4) Calculate the variance of each event from the last event, again using $\nu d_e$ and following the paths that determine the latest event times. It can be seen from Fig. 2.52 that the project duration is 278.67, and the variance of the last event is 35.93, giving a standard deviation of $\sqrt{35.93} = 5.99$. The probability of completing the project in 278.67 is 0.5 or 50% (as event times are assumed to be normal distributions).

### *Probability of achieving other project durations*

To calculate the probability of completing the project in 270, first calculate the factor $z$:

$$z = \frac{270 - 278.67}{5.99} = -1.45$$

From Table 2.15, probability is 0.07 or 7%.

### *Project duration for a specific probability*

To calculate the project duration if it is required to be 90% certain of completion on time, first look up the factor $z$ for 90% probability.

This equals 1.3:

$$T_s = T_x + z\sigma T_x$$

$$= 278.67 + 1.3 \times 5.99$$

$$= 286.46$$

To be 90% sure of completion on time:

project duration = 286.46

***Project duration determined by paths other than the critical path***

If a particular string of activities has very little float and a substantial standard deviation, this string could become critical. The probability of the string's becoming critical can be calculated.

The calculations become very complex when three or more strings could become critical, and will not therefore be discussed in this introduction to PERT.

In Fig. 2.52 string 3–5–8 may become critical. To calculate the probability of this, calculations start from the branching event (3).

The difference between two independent normal distributions results is another normal distribution whose variance is determined by adding the variances of the two component distributions. It is assumed that the two strings are approximately normal distributions for the sake of illustration.

Figure 2.53 shows the two distributions.

The variance from event 3 on 3–5–8 is 22.47 $\therefore \sigma = 4.74$
The variance from event 3 on 3–4–8 is 9.44 $\therefore \sigma = 3.07$

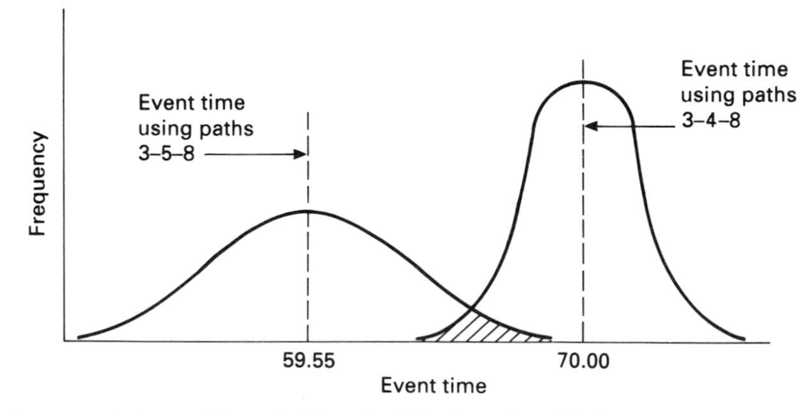


**Fig. 2.53** Calculation of the probability of activities becoming critical.

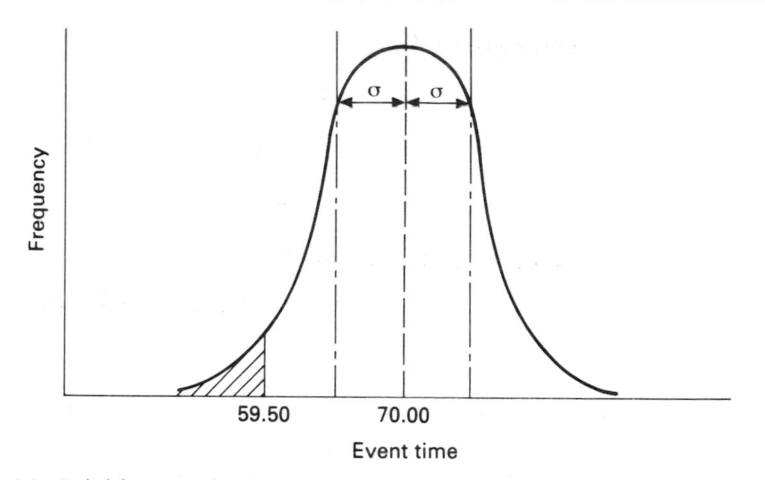


**Fig. 2.54** Probability curve for the time difference.

Sum of variance = 22.47 + 9.44 = 31.91

$$\sigma = 5.65$$

$$z = \frac{59.50 - 70.00}{5.65} = -1.86$$

probability = 0.03 or 3%

There is therefore a 3% probability that path 3–5–8 will affect the project duration.

Figure 2.54 shows the probability curve for the time difference.

Chapter 3

# Work Study

## 3.1 Introduction

The human aspect is a most important part of work study, both on site and at higher management level, and the subject consists just as much of human relations (i.e. problems of communication and eliciting co-operation) as it does of techniques. In this chapter only the application of work study techniques will be considered; for a more detailed consideration of the subject as a whole, other texts listed in the bibliography should be consulted.

### *3.1.1 Purpose of work study*

The purpose of work study is the provision of factual data to assist management in making decisions, and to enable them to utilise with the maximum of efficiency all available resources (i.e. labour, plant, materials and management) by applying a systematic approach to problems instead of using intuitive guesswork.

### *3.1.2 Breakdown of work study*

Work study has two main aspects, *method study* and *work measurement*, which are very closely related. For convenience they will be considered separately in this text, but their interdependence must be appreciated at all times.

## 3.2. Method study

The aim of method study is to provide information that will assist management in taking decisions related to the method it is proposed to use, by making a systematic analysis of a problem and developing alternative methods, thus determining the optimum layouts and the most effective use of resources.

There are six basic steps involved in carrying out a method study:

(1) Select the work to be studied and define the problem.
(2) Record the relevant facts using the recording techniques.
(3) Examine these facts critically and without bias to ascertain whether each element in the work is necessary, asking a series of questions and examining alternative solutions.
(4) Develop the best method from the alternatives and submit this to management for approval.
(5) Install the new method. Installation follows a decision by management to accept the method.
(6) Maintain the new method. This consists in checking that the new method is adhered to by regular inspection on site or by watching output records, which will indicate deviations from the standards set.

### 3.2.1 *Recording techniques*

There are three main groups of recording techniques available for setting down a problem. These are:

- Charts
  - Outline process charts
  - Flow process charts (man-type, material-type and equipment-type)
  - Two-handed process charts
  - Multiple activity charts
- Diagrams and models (two- and three-dimensional)
  - Flow diagrams
  - String diagrams
  - Cut-out templates (two-dimensional models)
  - Models (three-dimensional)
- Photographic
  - Photographs
  - Films

Templates – silhouettes of such things as plant, equipment, and storage areas – are very useful when considering layouts. It is also possible to obtain charting kits, which consist of translucent sheets and stick-on templates that may be used for die-line prints. This is an excellent method of recording when it is desired to compare one method with another.

*Flow process charts*

This technique can be useful in helping to solve problems of layout such as those in site workshop areas where the operations are likely to be repetitive.

All activities are shown by means of symbols, with a description against each. This is a very simple method of showing a sequence of work, and employs the following symbols:

○ Operation

□ Inspection

⇨ Transport

▽ Storage, e.g. stored for future use

D Temporary storage or delay

The movements of men, materials or equipment are followed through a process, and the symbols are used to indicate what is happening at the various stages. The distance travelled may be shown on the transport activities, and the time for each element can be given.

*Flow diagrams*

A flow diagram can be used in conjunction with a flow process chart to show where the activities take place. The same symbols that are employed on flow process charts are used, but in this case they are superimposed onto a drawing, and the descriptions are not necessary. All movements and distances can be clearly shown.

*String diagrams*

This technique is very useful in solving problems of movement. It is applied to repetitive situations, and is therefore most useful in working areas such as factories producing industrialised components, machine shops, precasting yards, and steel-bending areas on site. The diagram will show points of congestion and any excessive distances travelled.

The procedure for improving a layout is first to draw to scale a plan of the area under consideration with the work places or stacking areas etc., and all changes in direction denoted by pins. String or thread is then tied round the starting point and passed from pin to pin showing movement. Men, materials or machines can be denoted by different coloured string. If the string is then measured, any excessive distances travelled will be obvious and any points of congestion will be seen on the diagram.

Routes that are travelled regularly should be kept as short as possible, and alternative methods can be examined to obviate the faults of the first (templates can be very useful here).

### 3.2.2 *Critical examination*

The step 'Examine' is the key step in a method study, and consists of a detailed examination of every aspect of the work.

The purpose is to:

(1) establish the true facts surrounding the problem;
(2) establish the reasons for these facts and determine whether they are valid;
(3) on this foundation, to consider all the possible alternatives and hence the optimum solution.

Examination is carried out by a questioning technique, for which a critical examination sheet is very useful (Fig. 3.1). There should be liaison with the site management personnel who will carry out the work, and they should be encouraged to put their ideas forward.

Critical examination sheet

Description of operation:
Description of element:

| *Investigation* | *Primary question* | *Secondary question* | *Possible alternatives* | *Select alternatives* |
|---|---|---|---|---|
| Purpose | What is achieved? | Is it necessary? If so, why? | What else could be done? | What should be done? |
| Place | Where is it done? | Why there? | Where else could it be done? | Where should it be done? |
| Sequence | When is it done? | Why then? | When else could it be done? | When should it be done? |
| Person | Who does it? | Why him? | Who else could do it? | Who should do it? |
| Means | How is it done? | Why that way? | How else could it be done? | How should it be done? |

**Fig. 3.1** Critical examination sheet.

### 3.2.3 *Site layout problems*

When deciding on the relative positions of plant, working areas and storage areas on site, reference has to be made to the various activities to be performed: that is, the number and types of activity and the minimum transports and storages required. This basic information can be provided from materials flow process charts, which may be used in conjunction with a scale drawing of the site layout and paper templates. String diagrams can then be used either to evaluate or to represent visually the movement intensity for the alternative arrangements of the templates.

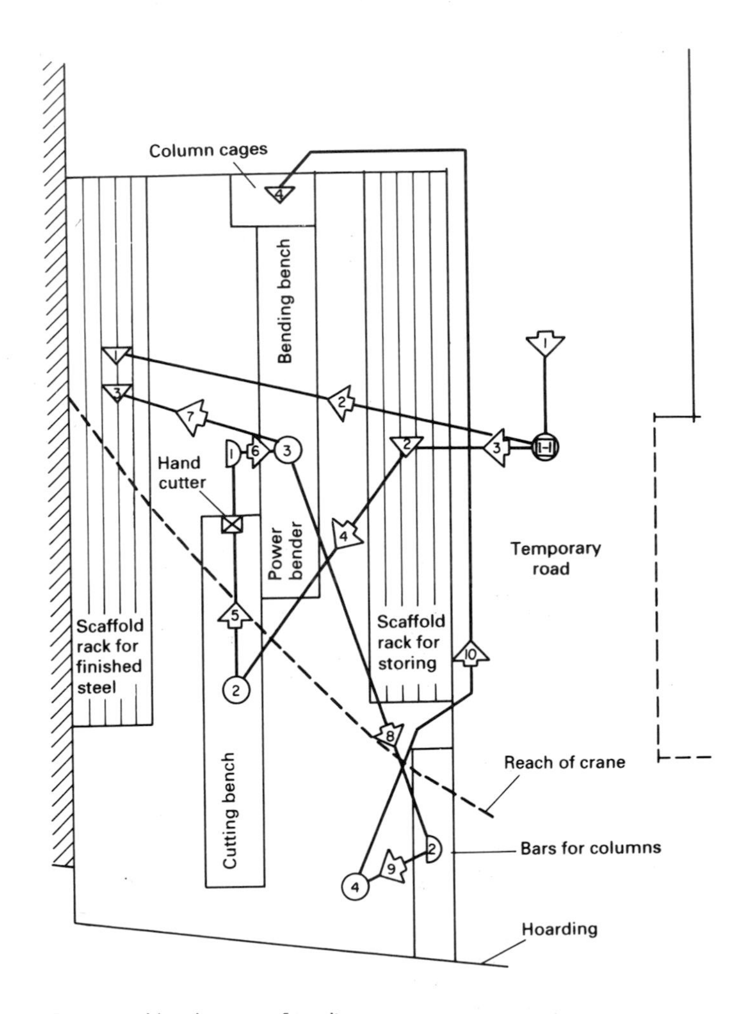


**Fig. 3.2** Cutting and bending area: flow diagram.

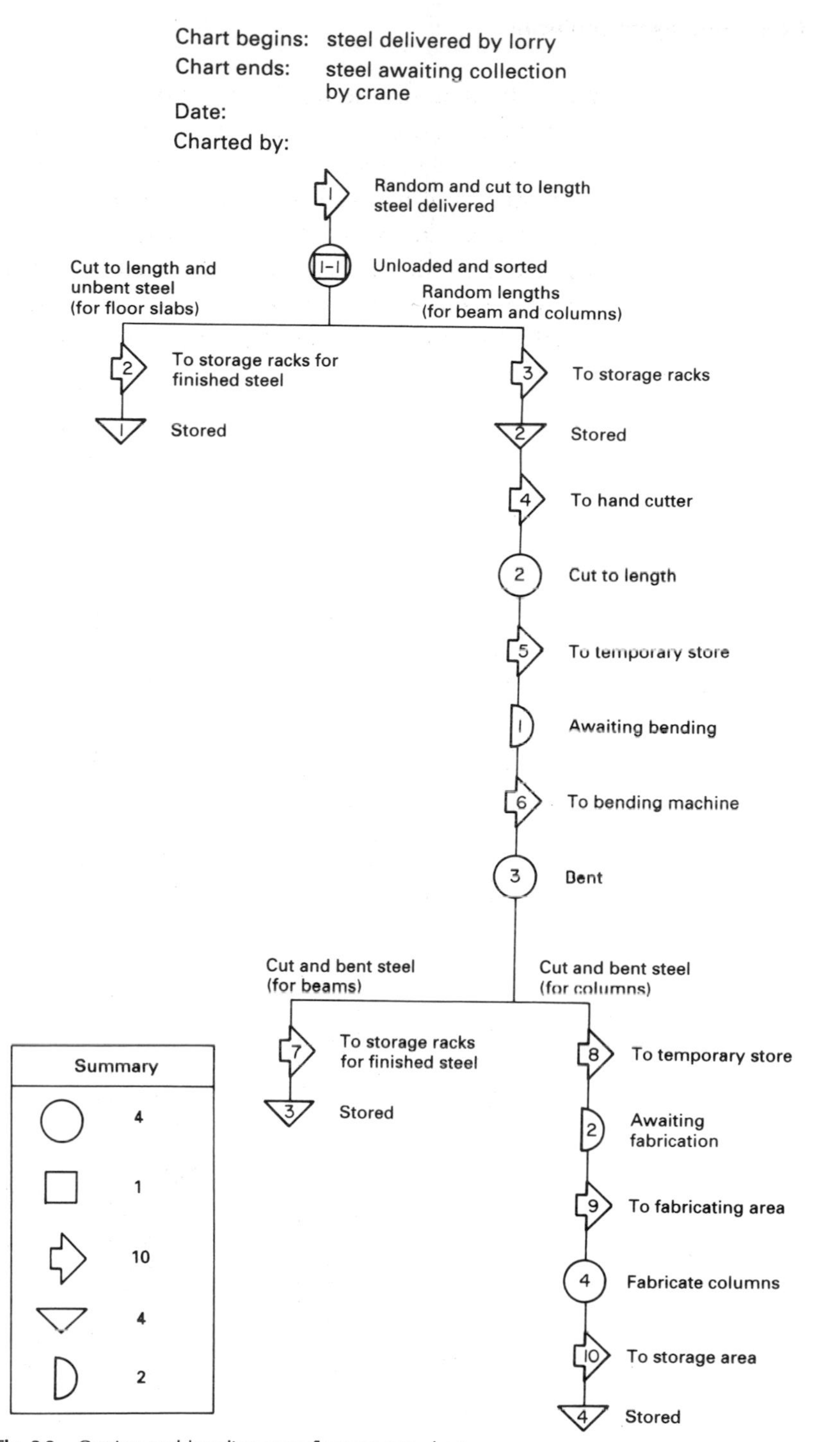


**Fig. 3.3** Cutting and bending area: flow process chart.

**Example 3.1: Steel-cutting and bending area**

As an example, flow charts, flow diagrams and string diagrams will be used to assist in considering the layout of a steel-bending area.

The space required for storing steel in the form in which it is delivered and after it has been prepared will be determined by the amount of steel involved and the timing of deliveries. A power bender will be required, but a hand cutter should be adequate for cutting purposes, as large-diameter bars and bars for the floor slab are being puchased cut to length.

The site manager has suggested the layout shown in Fig. 3.2, and a material-type flow diagram has been superimposed on it; a flow process chart has also been drawn up for this arrangement (Fig. 3.3).

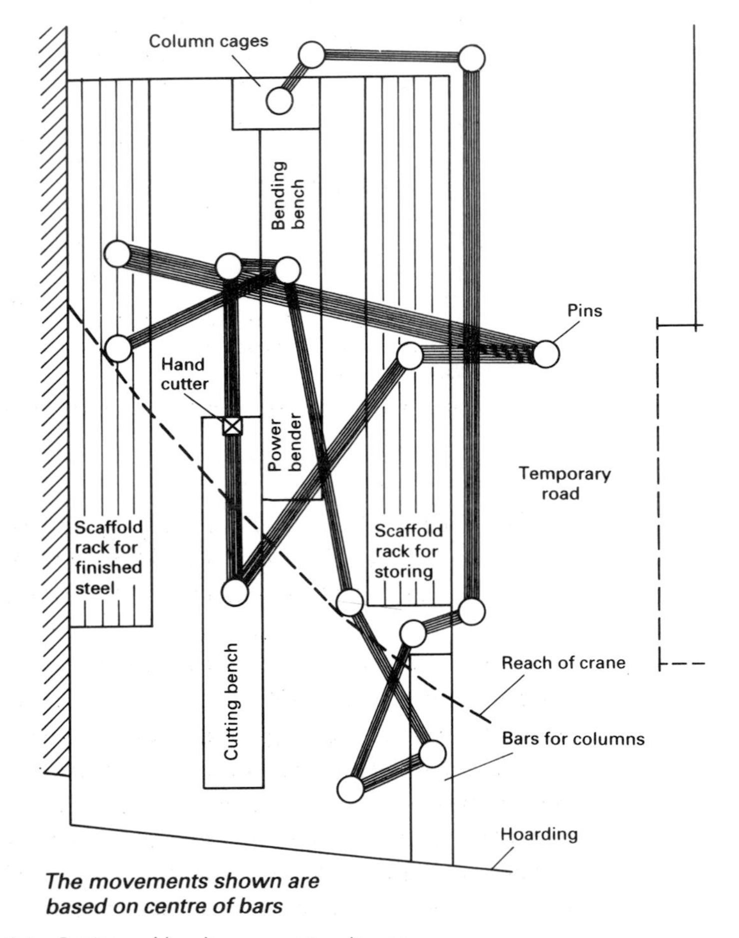

**Fig. 3.4** Cutting and bending area: string diagram.

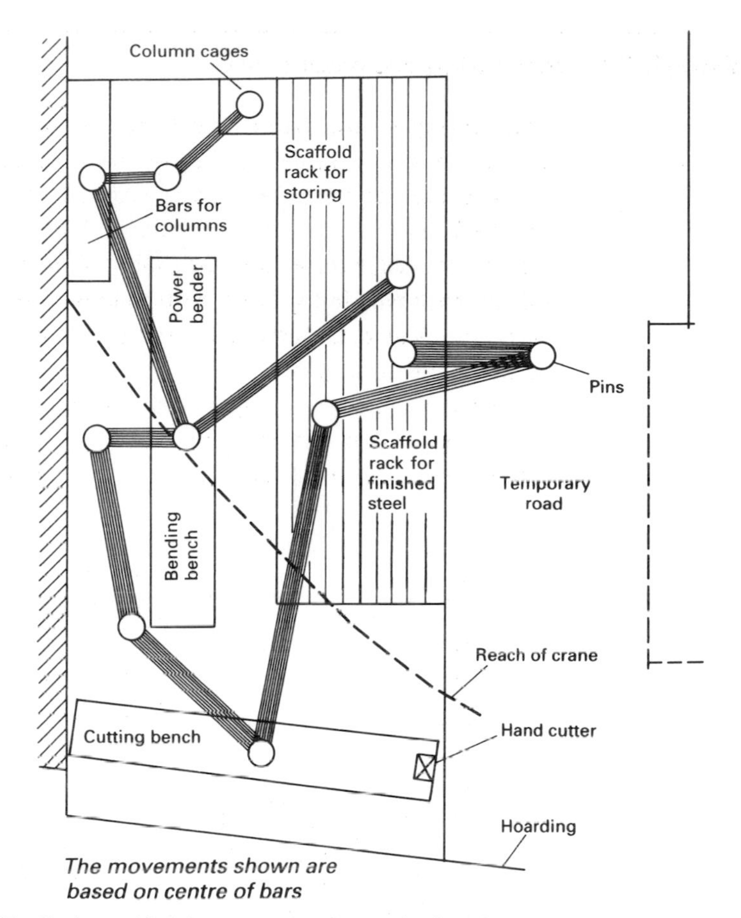


**Fig. 3.5** Cutting and bending area: string diagram for the selected alternative layout.

The chart and diagram are now critically examined to find improvements, and from the flow diagram (Fig. 3.2) it can be seen that a fair amount of movement is involved in transporting steel to and from the various positions where operations would take place. A string diagram is therefore produced to show the movement of materials for one day (Fig. 3.4), and it becomes evident that many long distances are travelled an excessive number of times.

Alternative layouts are examined to eliminate long transports on frequently used routes, and Fig. 3.5 shows the string diagram for the selected layout, with shortening of the routes travelled most frequently; the flow diagram for this layout is shown in Fig. 3.6. If required, different-coloured string can be used to show the movement of bars cut to length on delivery and those prepared on site.

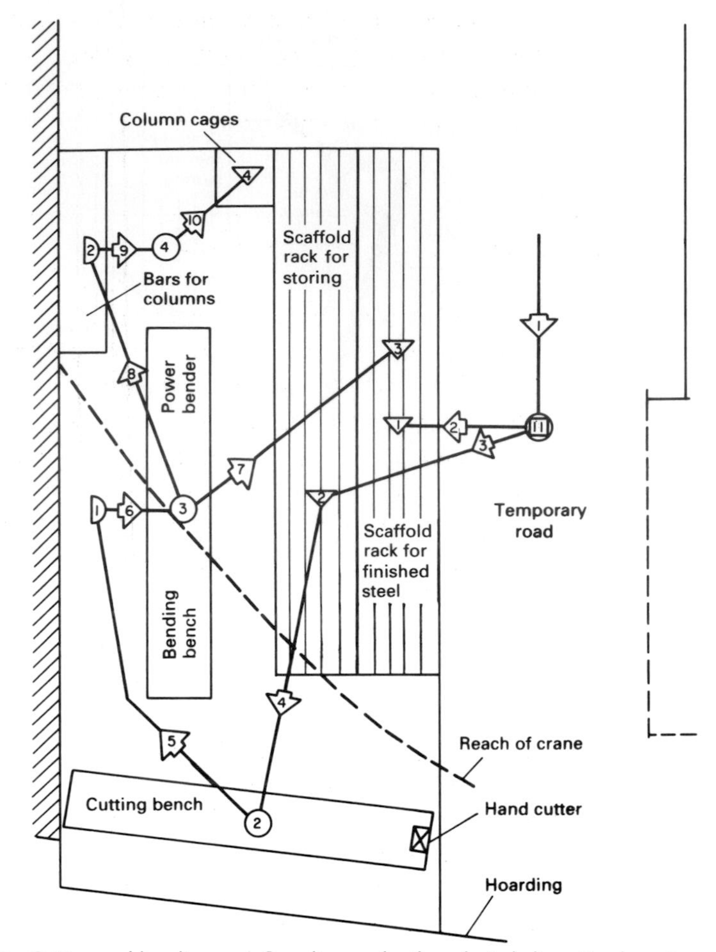


**Fig. 3.6** Cutting and bending area: flow diagram for the selected alternative layout.

Had the layout suggested by the site manager not been put forward, the first stage would have been to produce an outline process chart, followed by flow process charts, flow diagrams, and string diagrams.

### *3.2.4 Multiple activity charts*

This technique is used to help solve problems when a number of subjects (operatives, machines or equipment) are dependent on each other. It shows the occupied time (divided into elements of work if necessary) and unoccupied time for the subjects in both the present and the proposed methods. The subjects are recorded on a common time scale, and periods of unoccupied time can be readily seen. Alternatives may then be considered, the aim being to balance the work content for the subjects. It is necessary to plot more than one cycle of work, as the first taken may not be representative.

**Example 3.2: Laying concrete upper floor slabs**

Multiple activity charts are to be used to help increase efficiency in laying the upper concrete floor slabs in an amenity centre and office block. The operation is at present carried out by two men at the mixer (one driver and one loader), the tower crane transporting the concrete to the required position, and one labourer discharging the concrete from the skip and spreading and levelling it. It is assumd that a time study has been carried out and gives the information shown in Table 3.1 (the times stated are hypothetical).

***Method 1***

The multiple activity chart in Fig. 3.7 shows the operation as carried out at present. One man loads the hopper on the mixer with sand and coarse aggregate. The mixer driver checks the weight on a dial and discharges the cement from the silo above the hopper; he next elevates the hopper and discharges the contents into the mixer drum, feeds water into the drum, and lowers the hopper. When the concrete is mixed, the mixer driver places the skip in position and then discharges the concrete from the mixer into it; the crane then lifts the concrete in the skip and transports it to the desired position. Finally, a spreader discharges the skip and spreads the concrete, the crane returning with empty skip to the mixer.

The chart shows that the time cycle of 5.75 standard minutes is determind by the spreader, and this results in unoccupied time for other operatives and machines.

***Method 2***

The multiple activity chart in Fig. 3.8 shows an alternative method. Two spreaders are used here to discharge and spread concrete from

**Table 3.1** Time study, upper concrete floor slabs.

| Man/machine and element | Code | Description | Time in standard minutes |
|---|---|---|---|
| Loader | | | |
| Load hopper | L | Load mixer hopper with fine and coarse aggregate using hand scraper | 2.00 |
| Mixer driver | | | |
| Add cement | ct | Add cement to fine and coarse aggregate in hopper (from silo above mixer) | 0.25 |
| Mixer driver | | | |
| Raise hopper | R | Raise hopper up to opening in mixer drum | 0.25 |
| Mixer driver and mixer | | | |
| Mix concrete | M { T { | Tip contents into drum | 0.20 } |
| | | Add water | 0.05 } 2.00 |
| | | Mix concrete | 1.75 } |
| Mixer driver | | | |
| Lower hopper | H | Lower hopper from opening in mixer drum down to ground level | 0.25 |
| Mixer driver | | | |
| Discharge concrete | D | Position crane skip and discharge contents of mixer drum into it | 0.50 |
| Crane and driver | | | |
| Deliver concrete | DC | Lift concrete in skip, slew and deliver concrete to placing position | 1.50 |
| Spreader | | | |
| Discharge concrete | DL | Discharge concrete from crane skip | 0.25 |
| Spreader | | | |
| Spread concrete | SC | Spread and vibrate concrete between reinforcing steel and tamp | 5.50 |
| Crane and driver | | | |
| Return empty skip | RE | Return empty skip from placing position to concrete mixer | 1.50 |

the crane skip, so that the crane now determines the time cycle of 3.75 standard minutes – a marked increase in efficiency.

***Method 3***

The multiple activity chart in Fig. 3.9 shows another alternative. This procedure is similar to method 2 but the mixer driver also does the loading. The cycle time remains the same, and this results in a further increase in efficiency.

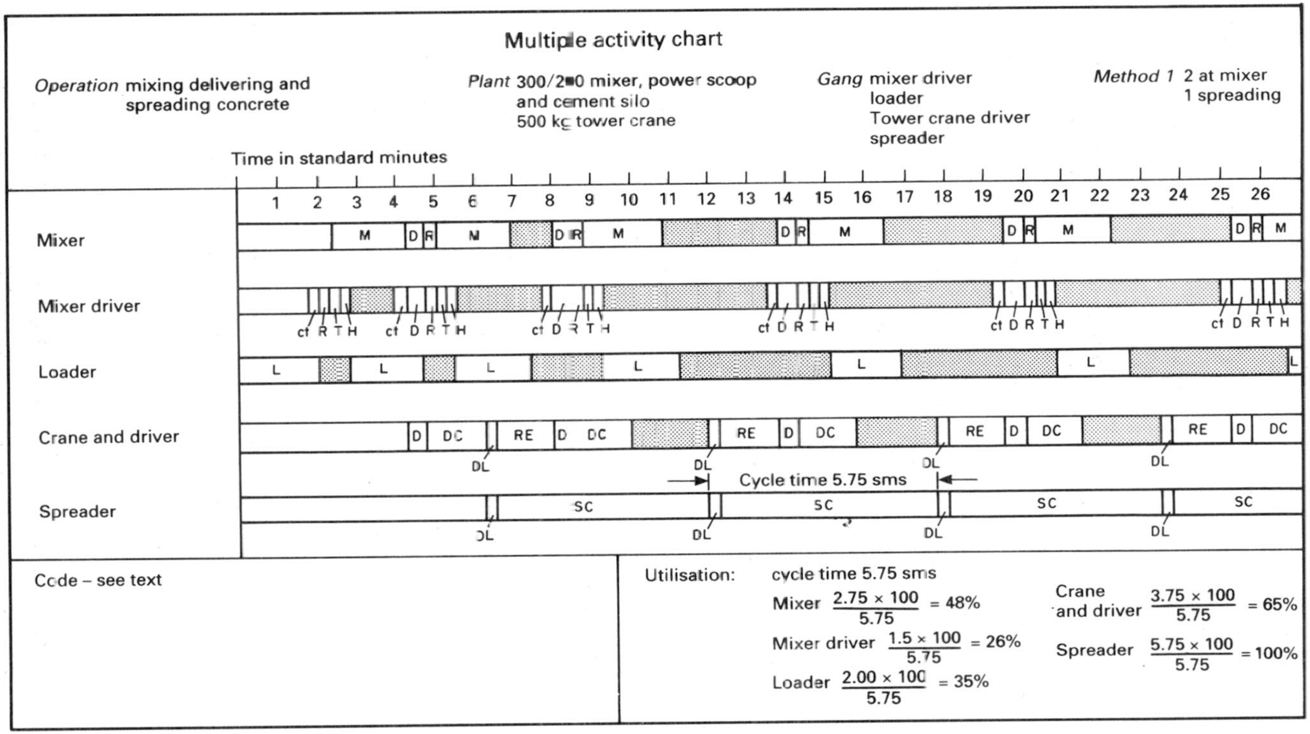

Multiple activity chart

*Operation* mixing delivering and spreading concrete

*Plant* 300/200 mixer, power scoop and cement silo 500 kg tower crane

*Gang* mixer driver, loader, Tower crane driver, spreader

*Method 1* 2 at mixer, 1 spreading

Time in standard minutes

1 2 3 4 5 6 7 8 9 10 11 12 13 14 15 16 17 18 19 20 21 22 23 24 25 26

Mixer: M D R M D R M D R M D R M D R M

Mixer driver: ct R T H ct D R T H ct D R T H ct D R T H ct D R T H ct D R T H

Loader: L L L L L L L

Crane and driver: D DC DL RE D DC DL RE D DC DL RE D DC DL RE D DC

Cycle time 5.75 sms

Spreader: DL SC DL SC DL SC DL SC

Code – see text

Utilisation: cycle time 5.75 sms

Mixer $\frac{2.75 \times 100}{5.75} = 48\%$

Mixer driver $\frac{1.5 \times 100}{5.75} = 26\%$

Loader $\frac{2.00 \times 100}{5.75} = 35\%$

Crane and driver $\frac{3.75 \times 100}{5.75} = 65\%$

Spreader $\frac{5.75 \times 100}{5.75} = 100\%$

**Fig. 3.7** Multiple activity chart for laying concrete upper floor slabs.

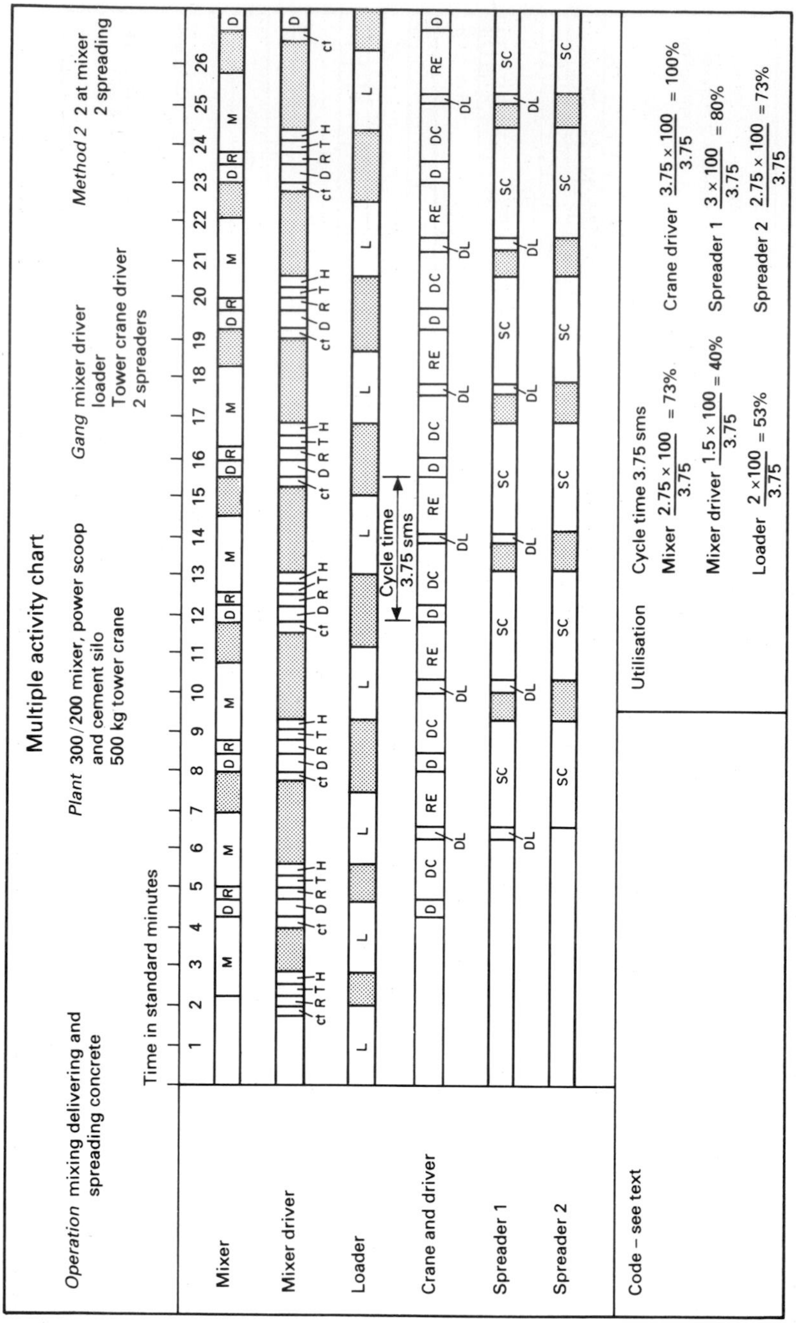

**Multiple activity chart**

*Operation* mixing delivering and spreading concrete

*Plant* 300/200 mixer, power scoop and cement silo
500 kg tower crane

*Gang* mixer driver
loader
Tower crane driver
2 spreaders

*Method 2* 2 at mixer
2 spreading

Time in standard minutes

1 2 3 4 5 6 7 8 9 10 11 12 13 14 15 16 17 18 19 20 21 22 23 24 25 26

Mixer: M D R M D R M D R M D R M D R M D R M D R M D

Mixer driver: ct R T H ct D R T H ct D R T H ct D R T H ct D R T H ct D R T H ct D R T H ct D

Loader: L L L L L L L L L

Cycle time 3.75 sms

Crane and driver: D DC DL RE D DC DL RE D DC DL RE D DC DL RE D DC DL RE D DC DL RE D

Spreader 1: DL SC DL SC DL SC DL SC DL SC DL SC

Spreader 2: SC SC SC SC SC SC

Code – see text

Utilisation

Cycle time 3.75 sms

Mixer $\frac{2.75 \times 100}{3.75} = 73\%$

Mixer driver $\frac{1.5 \times 100}{3.75} = 40\%$

Loader $\frac{2 \times 100}{3.75} = 53\%$

Crane driver $\frac{3.75 \times 100}{3.75} = 100\%$

Spreader 1 $\frac{3 \times 100}{3.75} = 80\%$

Spreader 2 $\frac{2.75 \times 100}{3.75} = 73\%$

**Fig. 3.8** Multiple activity chart for alternative operation: note the increase in efficiency.

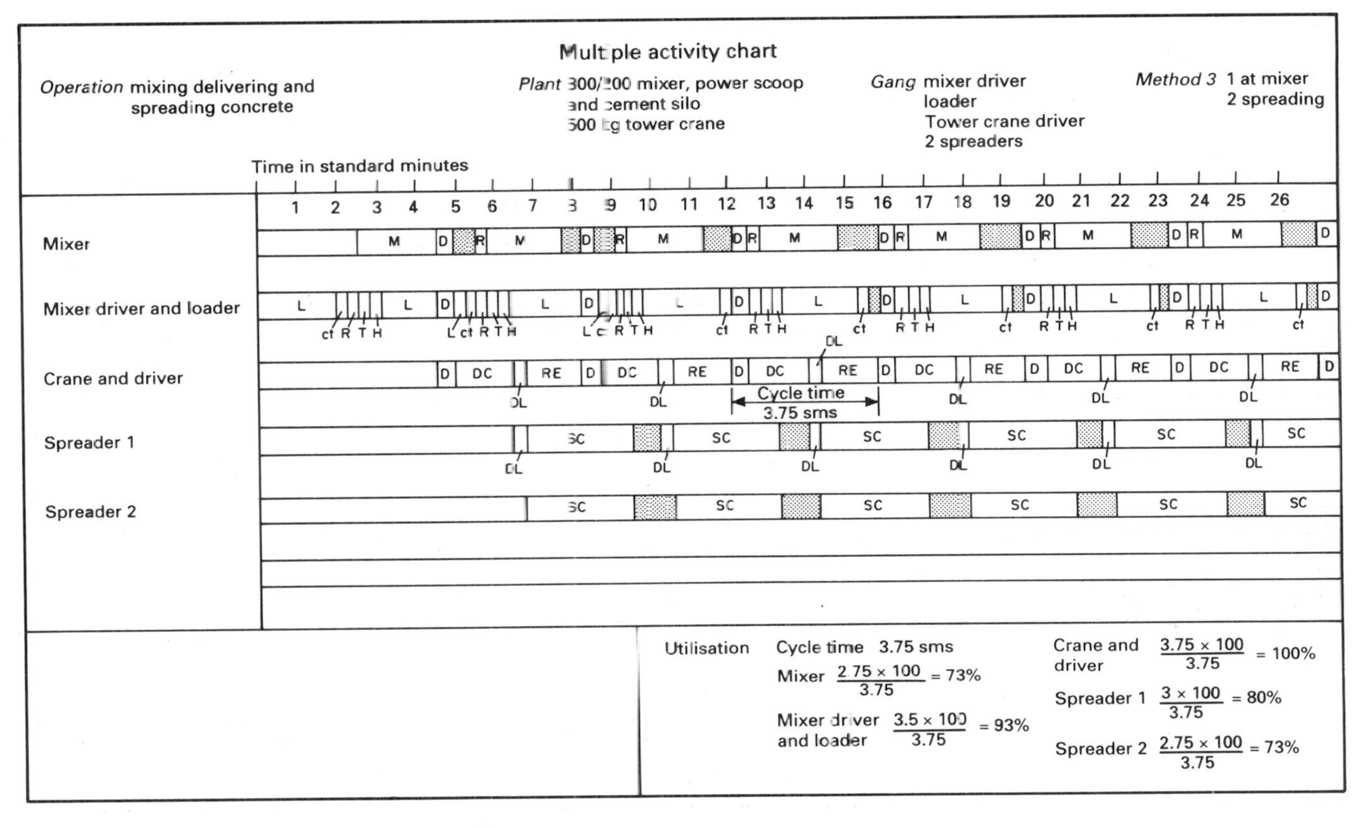


**Fig. 3.9** A second alternative, with a further increase in efficiency.

***Other methods***

Additional alternatives could be tried, and one possibility is to use two crane skips, which might cut down some of the time at present wasted while the crane waits for the skip to be filled at the mixer. There would of course be extra time involved in attaching and detaching the skip.

## 3.3 Work measurement

The aim of work measurement is to determine the time that it takes for a qualified worker to carry out a specific job at a defined level of performance, and to eliminate ineffective elements of work. It seeks to provide the standard times for jobs, and thus supplies basic, essential data for management.

### 3.3.1 *Measurement techniques common in the construction industry*

The work measurement techniques common in the construction industry are:

- ❑ time study
- ❑ activity sampling
- ❑ synthesis from standard and synthetic data
- ❑ analytical estimating.

### 3.3.2 *Time study*

The stages involved in carrying out a time study are:

(1) selecting the work to be measured;
(2) analysing and breaking the work down into elements;
(3) rating and timing each element;
(4) extending the observed time to basic time;
(5) selecting basic times, allocating allowances and building up the final standard.

Timing and rating are obviously carried out in the field, and consequently in the example given it is assumed that the times and ratings have already been obtained.

The methods of timing commonly used in the construction industry are cumulative timing and flyback timing.

*Cumulative timing* is the more common, as it is better for observing a number of operatives in a gang and requires only an accurate wrist-watch. The cumulative time is recorded after each element.

*Flyback timing* is carried out with a flyback stopwatch, the observer recording the time for each element as work proceeds. The watch has a flyback button on it that returns the hands to zero when pressed; on releasing the button the watch recommences timing.

To check the accuracy, the start and finish times are taken, and the difference between them is compared with the total of the readings.

**Example 3.3: Working up the standard time for an operation**

To determine a standard time, a number of time studies of the operation will have to be completed; it is assumed that this has been done using cumulative timing.

In the example the following terms are used:

- *Element*: a distinct part of an operation that can be easily defined.
- *Rating*: the method used to take into account variations in working pace.
- *Cumulative time*: the time recorded on the study sheet.
- *Observed time*: the time taken to perform an element of work.
- *Standard rating*: the average rate at which a qualified man will work given sufficient motivation and instruction.
- *Basic time*: the time required for carrying out an element of work at standard rating.
- *Relaxation allowance*: the time given to allow a worker to recover from the effects of fatigue and to attend to personal needs.
- *Contingency allowance*: the time given to allow for occurrences throughout the day, e.g. a painter cleaning brushes.

***Method used for calculating standard time***

(1) Calculate the observed time for each element by subtracting the previous cumulative time from the cumulative time for the element in question. Check that the total observed time equals the last cumulative time recorded (Table 3.2).

(2) Extend the observed times to basic times: that is, show all times at the standard rating by taking the observed rating into account. The formula for this calculation (Table 3.3) is:

$$\frac{\text{observed time} \times \text{observed rating}}{\text{standard rating}}$$

(3) Calculate the average of the basic times over the series of studies (Table 3.4).
(4) Add the appropriate relaxation allowances and contingency allowances (Table 3.5), remembering that unoccupied time allowance and interference allowances must also sometimes be made.

This procedure will give the standard time for each element; to calculate the standard time for an operation, add the elements in the operation together.

*Note*: This example is very much simplified, and the times are hypothetical.

**Table 3.2** Operation: concreting upper floor slabs.

| Element | Rating | Cumulative time (min) | Observed time (min) |
|---|---|---|---|
| Check time | | 2.75 | 2.75 |
| A. Discharge crane skip | 85 | 2.99 | 0.24 |
| B. Spread concrete | 75 | 4.89 | 1.90 |
| | 80 | | |
| | 70 | | |
| C. Vibrate concrete | 100 | 5.33 | 0.44 |
| D. Level and tamp concrete | 95 | | |
| | 90 | | |
| | 85 | | |
| | 90 | 7.63 | 2.30 |
| | | | 7.63 |

**Table 3.3** Extend observed times to basic times.

| Element | Rating | Cumulative time (min) | Observed time (min) | Basic time (min) |
|---|---|---|---|---|
| Check time | | 2.75 | 2.75 | |
| A. Discharge crane skip | 85 | 2.99 | 0.24 | $\frac{0.24 \times 85}{100} = 0.20$ |
| B. Spread concrete | 75 | 4.89 | 1.90 | $\frac{1.90 \times 75}{100} = 1.43$ |
| | 80 | | | |
| | 70 | | | |
| C. Vibrate concrete | 100 | 5.33 | 0.44 | $\frac{0.44 \times 100}{100} = 0.44$ |
| D. Level and tamp concrete | 95 | 7.63 | 2.30 | $\frac{2.30 \times 90}{100} = 2.07$ |
| | 90 | | 7.63 | |
| | 85 | | | |
| | 90 | | | |

**Table 3.4** Calculate average of basic times.

| Element | Basic time (min) | Total basic times (min) | Frequency | Average (min) |
|---|---|---|---|---|
| A. Discharge crane skip | 0.20, 0.18, 0.19, 0.18, 0.21, 0.20, 0.18, 0.19, 0.19, 0.18 | 1.90 | 10 | 0.19 |
| B. Spread concrete | 1.43, 1.27, 1.31, 1.49, 1.45, 1.45, 1.72, 1.48, 1.80, 1.60 | 15.00 | 10 | 1.50 |
| C. Vibrate concrete | 0.44, 0.46, 0.44, 0.42, 0.44, 0.45, 0.45, 0.44, 0.43, 0.43 | 4.40 | 10 | 0.44 |
| D. Level and tamp concrete | 2.07, 2.13, 2.35, 2.35, 2.23, 2.27, 2.24, 2.00, 2.16, 2.20 | 22.00 | 10 | 2.20 |

Assume that the total relaxation allowance and contingency allowance are as shown in Table 3.5. The standard time for the operation is $0.25 + 2.03 + 0.58 + 2.97 = 5.83$ standard minutes.

**Table 3.5** Total relaxation allowance and contingency allowance.

| Element | Basic time (min) | Total relaxation allowance (%) | Contingency allowance (%) | Total allowances (%) | Standard time (min) |
|---|---|---|---|---|---|
| A. Discharge crane skip | 0.19 | 25 | 5 | 30 | $\frac{0.19 \times 130}{100} = 0.25$ |
| B. Spread concrete | 1.50 | 30 | 5 | 35 | $\frac{1.50 \times 135}{100} = 2.03$ |
| C. Vibrate concrete | 0.44 | 26 | 5 | 31 | $\frac{0.44 \times 131}{100} = 0.58$ |
| D. Level and tamp concrete | 2.20 | 30 | 5 | 35 | $\frac{2.20 \times 135}{100} = 2.97$ |

## 3.4 Activity sampling

Activity sampling is also known as *snap observation studies* or *random observation studies*. It is a very useful technique in the construction industry, and is used to determine the activity levels of machines and operatives. Using this method a number of subjects can be observed concurrently.

### 3.4.1 *Basis of the method*

In this procedure the principles of the statistical method of random samples are used. Snap readings are taken at intervals, and the percentage of readings taken per element will give a result very near to the actual percentage on each, provided the sample is big enough. From this it is possible to calculate the time taken on each element if necessary.

One observer can record the activities of a whole gang, and readings can be almost continuous. The timing may be by either the fixed interval method (easier to use on site) or the random interval method, which is better for work of a cyclical nature.

It is essential that a recording is made the instant a subject is observed.

### 3.4.2 *Uses of activity sampling*

On site, activity sampling can be used:

- to assess unoccupied time as a basis for analysing cause;
- to find the percentage of time spent on each element of work in an operation by each operative and/or machine;
- to find the percentage utilisation of machines or operatives as a basis for cutting down unoccupied time (rated activity sampling can be used for this, which is an extension of the method).

***Example 3.4: Activity sampling***

The following example (Table 3.6) shows the application of activity sampling to determine the proportions of productive and non-productive time for three excavators owned by a contractor.

A total of 999 observations were made on the three excavators, and at each observation the machines were recorded as being productive, suffering from major delays, or suffering from minor delays.

Weather delays, opening up working areas, and machine repairs were defined as a major delay, while insufficient lorries, spotting lorries, trimming and clearing up, moving the machine, routine maintenance, and operator delays were held to constitute minor delays.

If it is felt that the delays are excessive, a further analysis can be carried out to ascertain the actual cause of delay under the more detailed headings listed above. As stated previously, the accuracy of the result is determined to some extent by the size of the sample, and formulae are available that will give:

- the numbr of observations to produce a given accuracy, and
- the limit of accuracy obtained from the observations made.

These formulae are:

$$\text{number of observations required} = \frac{4P(100 - P)}{L^2}$$

where $P$ is the expected percentage occurrence of the activity and $L$ is the limit of accuracy required (percentage $\pm$)

$$\text{limit of accuracy} = 2\sqrt{\left(\frac{P(100 - P)}{N}\right)}$$

where $N$ is the number of observations and $P$ is the percentage of observations recorded for a particular activity.

There is a 95% certainty that the value of $P$ will be somewhere between $P + L$ and $P - L$.

These formulae will now be applied to the example to ascertain:

- the number of observations necessary to give a limit of accuracy of $\pm 3\%$ on minor delays, and
- the accuracy achieved with the number of observations taken.

Before starting the calculations it is necessary to estimate what the percentage on the activity will be, either from a short preliminary study or from other information. In this example, an estimate of 33% has been made.

(a) Number of observations required

$$= \frac{4P(100 - P)}{L^2}$$

$$= \frac{4 \times 33(100 - 33)}{3^2}$$

$$= \frac{4 \times 33 \times 67}{9}$$

$$= 983 \text{ observations}$$

999 observations were actually taken.

(b) Limit of accuracy of result obtained (Table 3.6)

$$= 2\sqrt{\left(\frac{P(100 - P)}{N}\right)}$$

$$= 2\sqrt{\left(\frac{21 \times 79}{999}\right)}$$

$$= 2\sqrt{1.6}$$

$$= 2 \times 1.26$$

$$= \pm 2.52\%$$

The formulae can also be applied to major delays and productive time if required.

**Table 3.6** Proportions of productive and non-productive times.

| Number of tours of site | | Excavator 1 | Excavator 2 | Excavator 3 | |
|---|---|---|---|---|---|
| 1 | | P | SD | P | |
| 2 | | SD | P | LD | |
| 3 | | LD | LD | P | |
| etc | | etc | etc | etc | |
| 333 | | P | SD | LD | |
| | | | | | Totals |
| Totals for | P | 133 | 137 | 140 | 410 |
| 999 | SD | 73 | 70 | 67 | 210 |
| observations | LD | 127 | 126 | 126 | 379 |

P = productive time; SD = minor delays; LD = major delays

Calculations:

$$\text{productive time } \frac{410 \times 100}{999} = 40\%$$

$$\text{minor delays } \frac{210 \times 100}{999} = 21\%$$

$$\text{minor delays } \frac{379 \times 100}{999} = 38\%$$

**Example 3.5: Carrying out an activity sampling study**

Assume that the activities of a carpenter over a 2 day period are as shown in Fig. 3.10 (obviously this information is not available when the study is carried out because the work has not yet been started). Activity sampling is to be used to determine the percentage of time spent on studying drawings, working on beam formwork, working on slab formwork, unoccupied and absent from the job. It has been decided to take 30 observations in total, initially. A table of random numbers could be used as shown in Table 3.7 to determine the timing of each observation. Assume we start the study with random No 01, row 3 column 2 and read off along the rows. The numbers can now be converted to time as shown in Table 3.8. By applying these times to Fig. 3.10, the percentage of observations on each activity can be calculated and the limit of accuracy obtained as described in example 3.4 (see Table 3.9). Clearly, the accuracy of the percentages on each activity would probably be unacceptable, and futher calculations would be carried out to determine the number of additional observations necessary to reach the accuracy required.

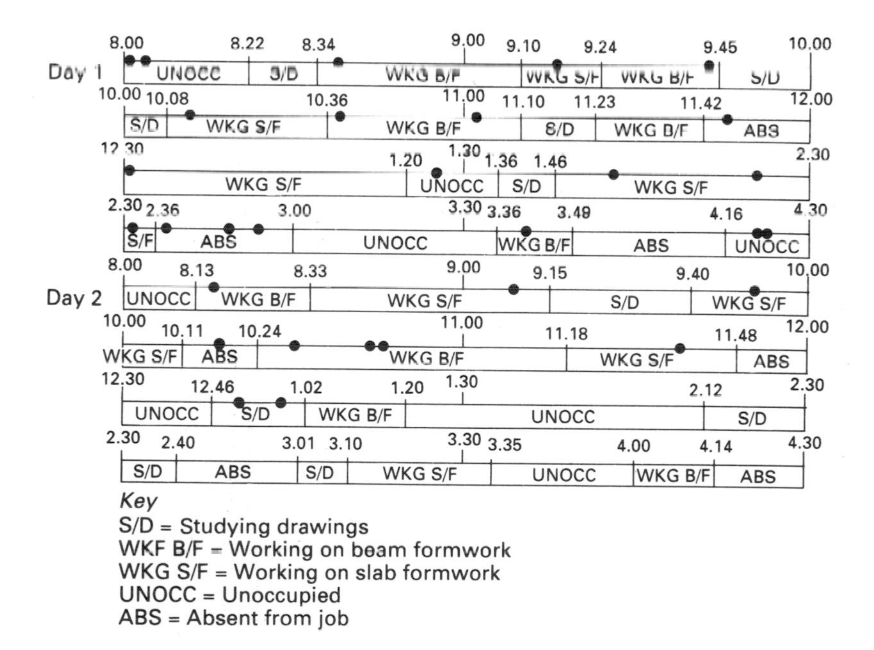


**Fig. 3.10** Activities of a carpenter over a 2 day period.

**Table 3.7** Random numbers.

| | | | | | | | | | | | |
|---|---|---|---|---|---|---|---|---|---|---|---|
| 02 | 49 | 59 | 26 | 39 | 45 | 12 | 32 | 18 | 56 | 10 | 03 |
| 22 | 02 | 43 | 19 | 23 | 55 | 39 | 05 | 09 | 17 | 44 | 52 |
| 00 | 01 | 03 | 34 | 38 | 27 | 29 | 26 | 24 | 44 | 15 | 54 |
| 31 | 26 | 10 | 06 | 11 | 05 | 47 | 41 | 01 | 23 | 53 | 42 |
| 26 | 13 | 16 | 01 | 51 | 43 | 07 | 42 | 21 | 05 | 24 | 57 |
| 34 | 56 | 36 | 31 | 41 | 17 | 44 | 50 | 25 | 53 | 02 | 03 |
| 45 | 07 | 16 | 18 | 42 | 51 | 33 | 55 | 43 | 20 | 46 | 54 |
| 23 | 09 | 21 | 01 | 04 | 23 | 08 | 04 | 23 | 52 | 38 | 01 |
| 17 | 43 | 07 | 28 | 19 | 54 | 40 | 40 | 36 | 54 | 07 | 58 |
| 14 | 04 | 50 | 10 | 11 | 25 | 36 | 00 | 39 | 37 | 59 | 59 |
| 34 | 16 | 05 | 29 | 25 | 28 | 30 | 32 | 35 | 58 | 02 | 51 |
| 35 | 08 | 52 | 47 | 52 | 10 | 05 | 36 | 02 | 48 | 17 | 31 |

**Table 3.8** Random numbers converted to time.

| Random number | Time |
|---|---|
| 1 | 8.01 |
| 3 | 8.04 |
| 34 | 8.38 |
| 38 | 9.16 |
| 27 | 9.43 |
| 29 | 10.12 |
| 26 | 10.38 |
| 24 | 11.02 |
| 44 | 11.46 |
| 15 | 12.31 |
| 54 | 1.25 |
| 31 | 1.56 |
| 26 | 2.22 |
| 10 | 2.32 |
| 06 | 2.38 |
| 11 | 2.49 |
| 05 | 2.54 |
| 47 | 3.41 |
| 41 | 4.22 |
| 01 | 4.23 |
| 23 | 8.16 |
| 53 | 9.09 |
| 42 | 9.51 |
| 26 | 10.17 |
| 13 | 10.30 |
| 16 | 10.46 |
| 01 | 10.47 |
| 51 | 11.38 |
| 43 | 12.51 |
| 07 | 12.58 |
| 748 | |

**Table 3.9** Limits of accuracy.

| Activity | Tally | Total | % | Limit of accuracy $L = 2\sqrt{\left(\frac{P(100-P)}{N}\right)}$ |
|---|---|---|---|---|
| (1) Study drawings | \|\| | 2 | 6.67 | 9.12% |
| (2) Working on beam formwork | ~~\|\|\|\|~~ \|\|\|\| | 9 | 30.00 | 16.74% |
| (3) Working on slab formwork | ~~\|\|\|\|~~ \|\|\|\| | 9 | 30.00 | 16.74% |
| (4) Unoccupied | ~~\|\|\|\|~~ | 5 | 16.67 | 13.60% |
| (5) Absent | ~~\|\|\|\|~~ | 5 | 16.67 | 13.60% |
| Total | | 30 | 100.01 | |

## 3.5 Synthesis and synthetic data

The purpose of synthetic data is to enable time values to be compiled for jobs where direct measurement is unnecessary or impracticable. Many operations are very similar to ones that have been carried out before, perhaps with variation due to physical dimensions (examples are: sawing different lengths of timber and paving different areas). These various jobs have many common elements, and a suitably referenced stock of data relating to these elements can be built up over a period of time and used for quickly compiling the work content of various jobs.

## 3.6 Analytical estimating

To calculate the work content of non-repetitive jobs, the work value can be compiled by analytical estimating, using whatever information is available from past time studies or standard data and estimating the time for the remaining elements.

## 3.7 Financial incentives

### *3.7.1 Aims of financial incentive schemes*

The aim of a sound financial incentive scheme is to increase productivity by increasing the efficiency of the individual, and enable him to increase his earnings.

A good scheme will achieve these objectives, but a poor one will cause nothing but trouble. Incentive schemes should be very closely related to short-term planning and cost control. Careful supervision is necessary to ensure that quality is maintained.

### 3.7.2 *Features of a sound scheme*

The main features of a sound financial incentive scheme are as follows.

- The amount of bonus paid to the operatives should be a direct proportion of the time saved, and there should be no limit to the amount that can be earned.
- Targets should be issued for all operations before work commences. The extent and nature of the operation should also be clearly understood. The length of time for an operation should ideally be kept to about two days.
- Targets should be set for small gangs (there are exceptions to this in repetitive work similar to flow production when group bonuses may be more appropriate).
- Targets should not be altered during an operation without the agreement of both parties.
- The system used for calculating bonus earnings should be easily understandable by all persons concerned.
- Operatives must not be penalised for lost time that is caused by delays outside their control.
- Bonus earnings should be paid weekly.
- Arrangements should be made for dealing with time lost due to reasons outside the operatives' control.

### 3.7.3 *Basis for setting targets*

*The estimate*

For large projects where the estimate is based on a bill of quantities the rates used are not really suitable for direct use as bonus targets. The figures used in the bill are average: for example, brickwork operations will vary in difficulty and therefore speed of completion, depending on their situation. Brickwork in small panels takes longer per unit area to build than a large expanse of uninterrupted brickwork. This is often not clear from the traditional bill of quantities. In addition, it is a laborious job to collect together bill items that are relevant to a particular operation.

For small projects where the builder has to prepare the estimate from drawings it is easier to extract the labour content of operations and set targets based on this. By this means a check can easily be made on profits. The labour content in the estimate should of course be based on data from past projects.

*Data from past projects*

This is probably the most reliable information reasonably available to most firms where work measurement has not yet been used. The data will be built up over a period of years, and can be used as a basis for estimating and as a basis for setting bonus targets. Great care is necessary in analysing these data as it must be appreciated that the level of productivity – that is, the *rating* – will vary considerably. The conditions under which operations are carried out should also be analysed.

*Work measurement*

This is by far the best method but, as stated above, is not at present available to the vast majority of firms. The targets based on work measurement are more accurate, and the method to be used is taken into account. Targets based on work measurement are less likely to be questioned by the operatives, but where the targets are questioned a check should be made before making adjustments.

*Other methods*

Experience, bargaining and other methods are used but are not very reliable generally and will not be considered here.

### 3.7.4 Expected earnings when using bonus schemes

The general level of earnings may well be fixed by competition from local firms, and this may determine the basis for setting targets.

It is generally felt, however, that an operative should be paid $\frac{1}{3}$ of his basic weekly earnings as a bonus when he works at a rating of 100, and the following discussion is based on this.

### 3.7.5 Rate of payment for time saved

Operatives are paid differing rates of payment for time saved depending on circumstances. This will affect the targets set in order to allow the $\frac{1}{3}$ bonus at 100 rating.

Examples of rates of payment that could be used are:

- payment of full rate for each hour saved (100% system);
- payment of 75% of hourly rate for each hour saved (75% system);
- payment of 50% of hourly rate for each hour saved (50% system);
- payment of 25% of hourly rate for each hour saved (25% system).

When the payments made for time saved are less than the full hourly rate the systems are known as *geared systems*.

### 3.7.6 Effects of different rates of payment on earnings and direct costs

*Effects on earnings*

Assume an operation takes $x$ hours to perform at 100 rating and at this $\frac{1}{3}$ bonus will be earned.

- Payment of all time saved (100% system):

$$\text{target} = x + \frac{x}{3} = 1\tfrac{1}{3}x$$

- Payment of 75% of time saved (75% system):

$$\text{target} = x + \frac{4/3}{3}x = 1\tfrac{4}{9}x$$

- Payment of 50% of time saved (50% system):

$$\text{target} = x + \tfrac{2}{3}x = 1\tfrac{2}{3}x$$

- Payment of 25% of time saved (25% system):

$$\text{target} = x + \tfrac{4}{3}x = 2\tfrac{1}{3}x$$

Rating at which bonus will start to be earned:

- 100% system: target is $\frac{4}{3}$ of operations time at 100 rating. Therefore bonus starts at $\frac{3}{4} \times 100 = 75$ rating
- 75% system: target is $\frac{13}{9}$ of operations time at 100 rating. Therefore bonus starts at $\frac{9}{13} \times 100 = 69.2$ rating
- 50% system: target is $\frac{5}{3}$ of operations time at 100 rating. Therefore bonus starts at $\frac{3}{5} \times 100 = 60$ rating
- 25% system: target is $\frac{7}{3}$ of operations time at 100 rating. Therefore bonus starts at $\frac{3}{7} \times 100 = 42.9$ rating

*Effects on direct costs*

It is assumed that the labour content of estimated prices is based on outputs that approximate to a rating of 75. The following comparisons show the effects of different systems of payment on actual direct costs.

- 100% system

At $37\frac{1}{2}$ rating:

$$\text{cost} = \frac{75}{37\frac{1}{2}} \times 100 = 200\%$$

At 40 rating:

$$\text{cost} = \frac{75}{40} \times 100 = 187\tfrac{1}{2}\%$$

At 60 rating:

$$\text{cost} = \frac{75}{60} \times 100 = 125\%$$

At 75 rating and above cost is constant at 100%.

- 75% system – target $1\frac{4}{9}x$
Cost as above until bonus is earned at 42.9 rating.

At 75 rating:

$$\text{time required} = \tfrac{4}{3}x$$

$$\text{time saved} = 1\tfrac{4}{9}x - \tfrac{4}{3}x = \tfrac{1}{9}x$$

$$\text{bonus} = \tfrac{3}{4} \times \tfrac{1}{9}x = \tfrac{1}{12}x$$

$$\text{cost} = \frac{\frac{4}{3}x + \frac{1}{12}x}{\frac{4}{3}x} = 106\tfrac{1}{4}\%$$

At 100 rating:

$$\text{time required} = x$$

$$\text{time saved} = 1\tfrac{4}{9}x - x = \tfrac{4}{9}x$$

$$\text{bonus} = \tfrac{1}{3}x$$

$$\text{cost} = \frac{x + \frac{1}{3}x}{\frac{4}{3}x} = 100\%$$

At 125 rating:

$$\text{time required} = \tfrac{4}{5}x$$

$$\text{time saved} = 1\tfrac{4}{9}x - \tfrac{4}{5}x = \tfrac{29}{45}x$$

$$\text{bonus} = \tfrac{29}{60}x$$

$$\text{cost} = \frac{\frac{4}{5}x + \frac{29}{60}x}{\frac{4}{3}x} = 96\tfrac{1}{4}\%$$

At 150 rating:

$$\text{time required} = \tfrac{2}{3}x$$

$$\text{time saved} = 1\tfrac{4}{9}x - \tfrac{2}{3}x = \tfrac{7}{9}x$$

$$\text{bonus} = \tfrac{7}{12}x$$

$$\text{cost} = \frac{\tfrac{2}{3}x + \tfrac{7}{12}x}{\tfrac{4}{3}x} = 93\tfrac{3}{4}\%$$

At 175 rating:

$$\text{time required} = \tfrac{4}{7}x$$

$$\text{time saved} = 1\tfrac{4}{9}x - \tfrac{4}{7}x = \tfrac{55}{63}x$$

$$\text{bonus} = \tfrac{55}{84}x$$

$$\text{cost} = \frac{\tfrac{4}{7}x + \tfrac{55}{84}x}{\tfrac{4}{3}x} = 92\%$$

❑ 50% system – target $= 1\tfrac{2}{3}x$
Cost as above until bonus is earned at 60 rating.

At 75 rating:

$$\text{time required} = \tfrac{100}{75}x = 1\tfrac{1}{3}x$$

$$\text{target} = 1\tfrac{2}{3}x$$

$$\text{time saved} = 1\tfrac{2}{3}x - 1\tfrac{1}{3}x = \tfrac{1}{3}x$$

$$\text{bonus} = \tfrac{1}{6}x$$

$$\text{cost} = \frac{1\tfrac{1}{3}x + \tfrac{1}{6}x}{1\tfrac{1}{3}x} = 112\tfrac{1}{2}\%$$

At 100 rating:

$$\text{time required} = x$$

$$\text{time saved} = 1\tfrac{2}{3}x - x = \tfrac{2}{3}x$$

$$\text{bonus} = \tfrac{1}{3}x$$

$$\text{cost} = \frac{x + \tfrac{1}{3}x}{1\tfrac{1}{3}x} = 100\%$$

At 125 rating:

$$\text{time required} = \tfrac{100}{125}x = \tfrac{4}{5}x$$

$$\text{time saved} = 1\tfrac{2}{3}x - \tfrac{4}{5}x = \tfrac{13}{15}x$$

$$\text{bonus} = \tfrac{13}{30}x$$

$$\text{cost} = \frac{\tfrac{4}{5}x + \tfrac{13}{30}x}{1\tfrac{1}{3}x} = 92\tfrac{1}{2}\%$$

At 150 rating:

$$\text{time required} = \tfrac{100}{150}x = \tfrac{2}{3}x$$

$$\text{time saved} = 1\tfrac{2}{3}x - \tfrac{2}{3}x = x$$

$$\text{bonus} = \tfrac{1}{2}x$$

$$\text{cost} = \frac{\tfrac{2}{3}x + \tfrac{1}{2}x}{1\tfrac{1}{3}x} = 87\tfrac{1}{2}\%$$

At 175 rating:

$$\text{time required} = \tfrac{100}{175}x = \tfrac{4}{7}x$$

$$\text{time saved} = 1\tfrac{2}{3}x - \tfrac{4}{7}x = \tfrac{23}{21}x$$

$$\text{bonus} = \tfrac{23}{42}x$$

$$\text{cost} = \frac{\tfrac{4}{7}x + \tfrac{23}{42}x}{1\tfrac{1}{3}x} = 84\%$$

- 25% system – target $= 2\tfrac{1}{3}x$
  Cost as above until bonus is earned at 42.9 rating
  At 50 rating:

$$\text{time required} = \tfrac{100}{50}x = 2x$$

$$\text{time saved} = 2\tfrac{1}{3}x - 2x = \tfrac{1}{3}x$$

$$\text{bonus} = \tfrac{1}{12}x$$

$$\text{cost} = \frac{2x + \tfrac{1}{12}x}{1\tfrac{1}{3}x} = 156\tfrac{1}{4}\%$$

At 75 rating:

$$\text{time required} = \tfrac{100}{75}x = \tfrac{4}{3}x$$

$$\text{time saved} = 2\tfrac{1}{3}x - 1\tfrac{1}{3} = x$$

$$\text{bonus} = \tfrac{1}{4}x$$

$$\text{cost} = \frac{\tfrac{4}{3}x + \tfrac{1}{4}x}{1\tfrac{1}{3}x} = 118\tfrac{3}{4}\%$$

At 100 rating:

$$\text{time required} = x$$

$$\text{time saved} = 2\tfrac{1}{3}x - x = 1\tfrac{1}{3}x$$

$$\text{bonus} = \tfrac{1}{3}x$$

$$\text{cost} = \frac{x + \tfrac{1}{3}x}{1\tfrac{1}{3}x} = 100\%$$

At 125 rating:

$$\text{time required} = \tfrac{100}{125}x = \tfrac{4}{5}x$$

$$\text{time saved} = 2\tfrac{1}{3}x - \tfrac{4}{5}x = 1\tfrac{8}{15}x$$

$$\text{bonus} = \tfrac{23}{60}x$$

$$\text{cost} = \frac{\tfrac{4}{5}x + \tfrac{23}{60}x}{1\tfrac{1}{3}x} = 88\tfrac{3}{4}\%$$

At 150 rating:

$$\text{time required} = \tfrac{100}{150}x = \tfrac{2}{3}x$$

$$\text{time saved} = 2\tfrac{1}{3}x - \tfrac{2}{3}x = 1\tfrac{2}{3}x$$

$$\text{bonus} = \tfrac{5}{12}x$$

$$\text{cost} = \frac{\tfrac{2}{3}x + \tfrac{5}{12}x}{1\tfrac{1}{3}x} = 81\tfrac{1}{4}\%$$

At 175 rating:

$$\text{time required} = \tfrac{100}{175}x - \tfrac{4}{7}x$$

$$\text{time saved} = 2\tfrac{1}{3}x - \tfrac{4}{7}x = \tfrac{37}{21}x$$

$$\text{bonus} = \tfrac{37}{84}x$$

$$\text{cost} = \frac{\tfrac{4}{7}x + \tfrac{37}{84}x}{1\tfrac{1}{3}x} = 76\%$$

### 3.7.7 General comments

*The 100% system*

As shown in Fig. 3.11, the cost is constant when the rating exceeds 75. The operatives earn a greater bonus when the rating exceeds 100 when this system is used.

*Geared systems*

Direct costs are higher than the 100% system below 100 rating down to the level at which bonus earnings start. Direct costs break even at a rating of 100.

### 3.7.8 Effects of indirect costs on overall costs

In practice, savings for the employer are greater than those resulting from direct costs. The all-in-labour rate used in the estimate includes

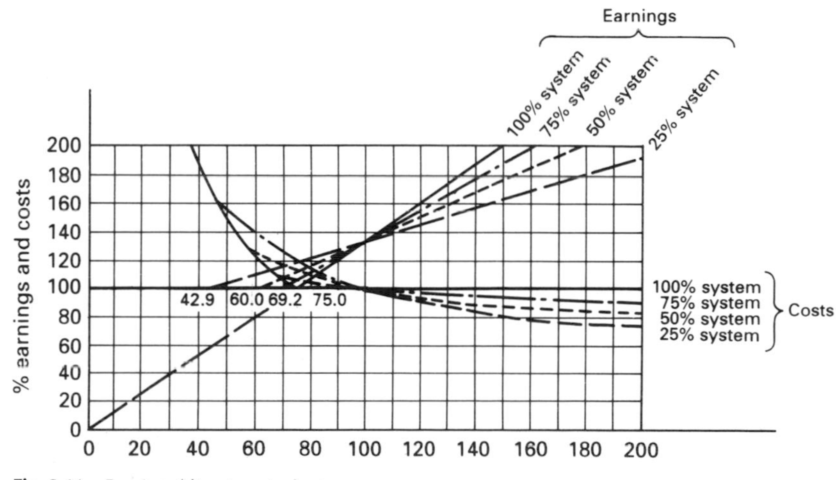


**Fig. 3.11** Earnings/direct cost chart.

indirect costs which cover sick pay allowances, public holiday pay, annual holiday pay and death benefit, National Insurance, CITB levy, severance pay allowance, employers' liability insurance, inclement weather and other NWR allowances.

These items vary in cost but at present add approximately one third to the basic weekly wage. For illustration purposes it is assumed that the basic hourly rate is £5.25 and the all-in rate for estimating purposes is £7.00.

The operation used in the examples that follow is assumed to take 45 man-hours at 100 rating. The 100%, 50% and 25% systems only will be considered in this section.

**Example 3.6: Effects on cost when the estimate is based on a rating of 75**

***Earnings and cost calculations***

The results shown in Fig. 3.12 and Tables 3.10–3.12 are calculated as follows.

$$\text{time taken } (T_t) = \frac{\text{time at 100 rating} \times 100}{\text{rating required}}$$

$$\text{time saved } (T_s) = \text{target} - \text{time taken}$$

$$\text{Bonus hours } (B_h) = \text{time saved} \times \text{\% payback}$$

$$\text{bonus pay} = \text{bonus hours} \times £5.25$$

$$\text{percentage cost} = \frac{100(T_t \times £7.00 + B_p)}{60 \times £7.00}$$

$$\text{percentage earnings} = \frac{100(T_t \times £5.25 + B_p)}{T_t \times £5.25}$$

***General comments***

*100% system or direct system (Fig. 3.12 and Table 3.10)*
The rate of increase in earnings rises rapidly above 75 rating.

The 100% system is suitable where the targets are set very accurately: that is, in factory conditions, particularly when work measurement is used as a basis for the targets. As shown in Fig. 3.12, savings are made when the rating exceeds 75, and these savings increase as the rating increases.

For site work it is very difficult to set really accurate targets, particularly in view of the fact that the vast majority of firms do not use work measurement as a basis for setting targets.

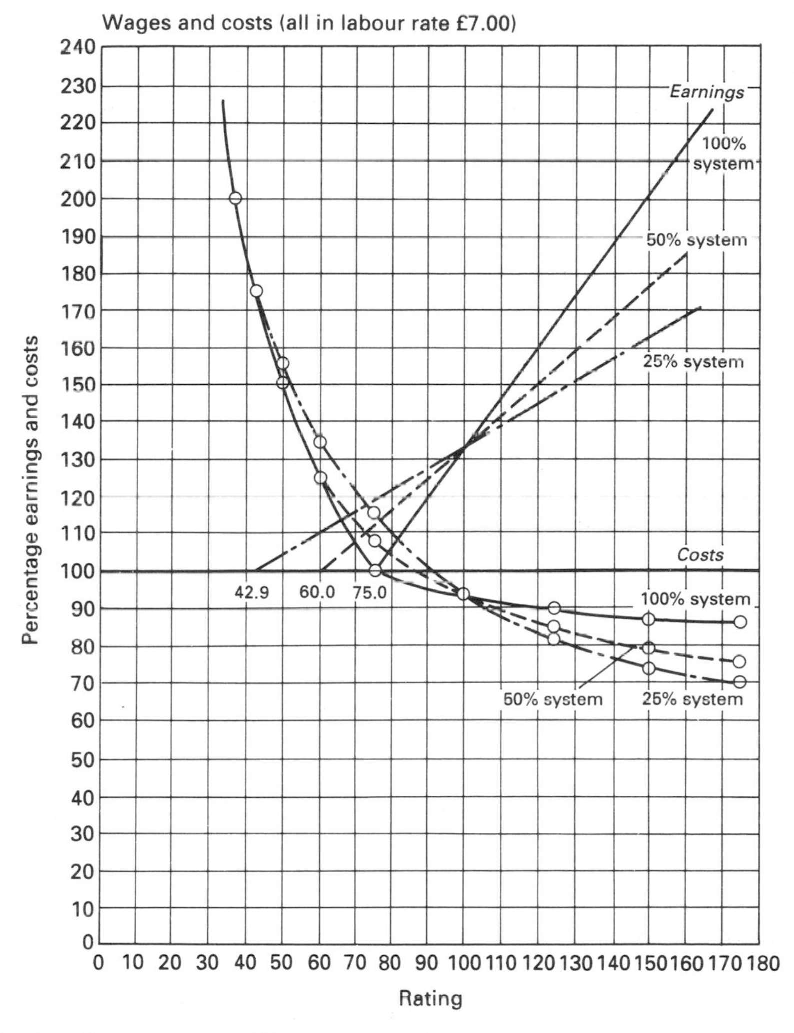


**Fig. 3.12** Wages and costs: all-in labour rate £7.00.

**Table 3.10** Direct system (pay back all time saved). Target = 45 + 15 = 60 hours.

| | Rating | | | | | | | |
|---|---|---|---|---|---|---|---|---|
| | $37\frac{1}{2}$ | 50 | 60 | 75 | 100 | 125 | 150 | 175 |
| Time taken | 120 | 90 | 75 | 60 | 45 | 36 | 30 | 25.71 |
| Time saved | Nil | Nil | Nil | Nil | 15 | 24 | 30 | 34.29 |
| Bonus hours | Nil | Nil | Nil | Nil | 15 | 24 | 30 | 34.29 |
| Bonus pay (£) | Nil | Nil | Nil | Nil | 78.75 | 126 | 157.50 | 180.02 |
| Percentage cost | 200 | 150 | 125 | 100 | 94 | 90 | 88 | 86 |
| Percentage earnings | 100 | 100 | 100 | 100 | 133 | 167 | 200 | 233 |

If tight targets are set, the operatives want them adjusted. If the targets are too easy, the operatives earn high wages, and the company can lose money even at ratings above 75. When productivity is below 75, rating costs are high, but when productivity is high, gains for the company are lower than in geared systems.

**Table 3.11** Geared system (pay back 50% of time saved). Target = 45 + 30 = 75 hours.

| | Rating | | | | | | | |
|---|---|---|---|---|---|---|---|---|
| | $37\frac{1}{2}$ | 50 | 60 | 75 | 100 | 125 | 150 | 175 |
| Time taken | 120 | 90 | 75 | 60 | 45 | 36 | 30 | 25.71 |
| Time saved | Nil | Nil | Nil | 15 | 30 | 39 | 45 | 49.29 |
| Bonus hours | Nil | Nil | Nil | 7.5 | 15 | 19.5 | 22.5 | 24.65 |
| Bonus pay (£) | Nil | Nil | Nil | 39.38 | 78.75 | 102.38 | 118.13 | 129.39 |
| Percentage cost | 200 | 150 | 125 | 109 | 94 | 84 | 78 | 74 |
| Percentage earnings | 100 | 100 | 100 | 113 | 133 | 154 | 175 | 196 |

**Table 3.12** Geared system (pay back 25% of time saved). Target = 45 + 60 = 105 hours.

| | Rating | | | | | | | |
|---|---|---|---|---|---|---|---|---|
| | $37\frac{1}{2}$ | 50 | 60 | 75 | 100 | 125 | 150 | 175 |
| Time taken | 120 | 90 | 75 | 60 | 45 | 36 | 30 | 25.71 |
| Time saved | Nil | 15 | 30 | 45 | 60 | 69 | 75 | 79.29 |
| Bonus hours | Nil | 3.75 | 7.5 | 11.25 | 15 | 17.25 | 18.75 | 19.82 |
| Bonus pay (£) | Nil | 19.69 | 39.38 | 59.06 | 78.75 | 90.56 | 98.44 | 104.07 |
| Percentage cost | 200 | 155 | 134 | 114 | 94 | 82 | 73 | 68 |
| Percentage earnings | 100 | 104 | 110 | 119 | 133 | 148 | 163 | 177 |

*50% system (Fig. 3.12 and Table 3.11)*
Below 60, rating the costs are identical to those for the 100% system. Between 60 and 100, rating costs are higher than the 100% system, and at ratings exceeding 100 they are less. At a rate of 88.2, the costs are equal to those included in the unit rates.

The 50% system is the one most used in the construction industry.

*25% system (Fig. 3.12 and Table 3.12)*
Below 42.9 rating, the costs are identical to those for the 100% system. Between 42.9 and 100 rating, costs are higher than the 100% system, and at ratings exceeding 100 they are less. At a rating of 90.7, costs are equal to those included in the unit rates.

This system is not often used as there appears to be little incentive for operatives to work at high ratings when they only receive 25% of the time saved. It is also very costly at low ratings. Two advantages are:

(i) Inaccurate targets have less effect on bonus earnings because the rating at which bonus starts to be earned is very low.
(ii) The bonus earned does not tend to vary as much each week.

At least one national contractor uses a system similar to this one.

**Example 3.7: Effects on cost when the estimate is based on a rating of 100**

If it is assumed that the operatives work at 100 rating, the 'all-in' rate will include one third bonus. In the examples that follow, the all-in rate is therefore

$$\text{bonus} = \tfrac{1}{3} \text{ of } £5.25 = £1.75$$

$$\text{all-in rate based on 75 rating} = £7.00 \text{ (see section 3.7.8)}$$

$$\text{all-in rate based on 100 rating} = £8.75$$

The labour element in the unit rates will be less than that included when the rates are based on 75 rating. The percentage costs will therefore be higher.

**Table 3.13** Direct system (pay back all time saved).

| | Rating | | | | | | | |
|---|---|---|---|---|---|---|---|---|
| | $37\frac{1}{2}$ | 50 | 60 | 75 | 100 | 125 | 150 | 175 |
| Percentage cost | 214 | 160 | 134 | 107 | 100 | 96 | 93 | 91 |

**Table 3.14** Geared system (pay back 50% of time saved).

| | Rating | | | | | | | |
|---|---|---|---|---|---|---|---|---|
| | $37\frac{1}{2}$ | 50 | 60 | 75 | 100 | 125 | 150 | 175 |
| Percentage cost | 214 | 160 | 133 | 117 | 100 | 90 | 83 | 79 |

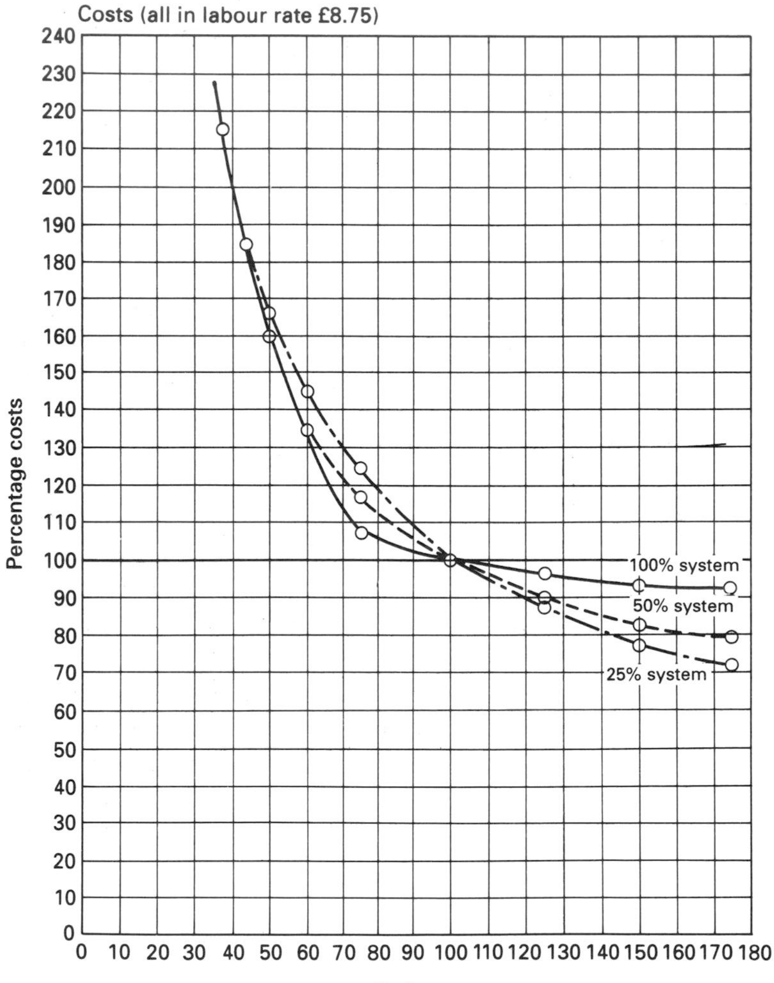


**Fig. 3.13** Costs: all-in labour rate £8.75.

**Table 3.15** Geared system (pay back 25% of time saved).

| | Rating | | | | | | | |
|---|---|---|---|---|---|---|---|---|
| | $37\frac{1}{2}$ | 50 | 60 | 75 | 100 | 125 | 150 | 175 |
| Percentage cost | 214 | 165 | 143 | 122 | 100 | 87 | 78 | 72 |

***Cost calculations***

The results shown in Tables 3.13–3.15 and Fig. 3.13 are calculated as follows:

$$\text{percentage cost} = \frac{100\,(\text{time taken} \times £7.00 + \text{bonus pay})}{45 \times £8.75}$$

Percentage earnings are identical to those calculated previously and are not therefore shown on Fig. 3.13.

***General comments***

As shown in Fig. 3.13, savings are made when the rating exceeds 100 for all the systems, and these savings increase gradually as the rating increases. At low ratings, costs are considerably higher than in the previous method.

### 3.7.9 *Conclusions*

Considerable increases in productivity and wages can result from a good scheme. Savings in excess of those illustrated are achieved in practice, because any reduction in project duration results in savings in overheads. Good supervision is necessary to ensure that quality is maintained, and careful control of material usage is essential. Care must also be taken to ensure that safety is not overlooked.

Chapter 4

# Budgetary and Cost Control

## 4.1 Budgets and budgetary control

### *4.1.1 Introduction*

Budgets have been used for planning the income and expenditure of a country's government for many years. With the growing complexity of the building industry it is essential that firms also plan their policy well into the future, and it is through budgets that such plans can be converted into the quantitative and monetary terms that are a company's fundamental objectives.

### *4.1.2 Policy budgets in the construction industry*

By compiling budgets a firm can arrive at a selling or tendering price for its projects, having considered what resources are available to carry out the work. The firm will need to know, for example, how much capital is necessary. Having acquired this information it can then decide whether to expand or contract any particular aspect of the business, bearing in mind competition and demand. From past results it may also be possible to determine a change in demand that could give an indication of possible new expansion plans.

### *4.1.3 Project budgets*

It is important for any contractor to know what capital is going to be needed for a project and when it will be required. In order to ascertain this, the contractor will have to draw up a programme for the project and from this find the rate of expenditure and rate of income, based on a particular time period. The difference between these two will give the amount of capital required in this time period (examples are given later in this chapter).

If required, budgets can be prepared for the labour to be used (based on the labour graph), for materials and plant (based on schedules and the programme), and for the site overheads (based on the programme).

### 4.1.4 *Achievement of budgets*

Once a budget has been established, the next objective is to achieve the target set. This is attained through a system of budgetary control. Like any other control technique, budgetary control is a continuous comparison of the actual achievement with that planned, which highlights any deviations from the plan so that action can be taken either to bring things back on course or to change the plan.

### 4.1.5 *Examples of budgets and budgetary control*

Two examples will be given in the following pages. The first is an example of repetitive construction. This example is very much simplified and is put forward for illustrative purposes only. A more detailed example of non-repetitive construction is given in example 4.2, in which the cash flow for the project is also ascertained.

**Example 4.1: Repetitive construction**

See Fig. 4.1. It is assumed that a contract has been entered into with the client, and that payment will be made in stages for each unit, such as excavation, concrete foundations, brickwork, roof carcass, and roof finishes. Payment will be received one week after completion of a stage, and the contract makes no provision for the retention money.

It has been ascertained that the total cost of completing these shells is £136 550, which is built up in the following manner:

| | |
|---|---|
| Excavation | £1 650 = £165 each block |
| Concrete foundations | £31 900 = £3190 each block |
| Brickwork | £63 400 = £6340 each block |
| Roof construction | £13 200 = £1320 each block |
| Roof finish | £26 400 = £2640 each block |
| | £136 550 |

By allocating these costs to the programme of work it can be seen that the distribution of cost will be that shown in Table 4.1.

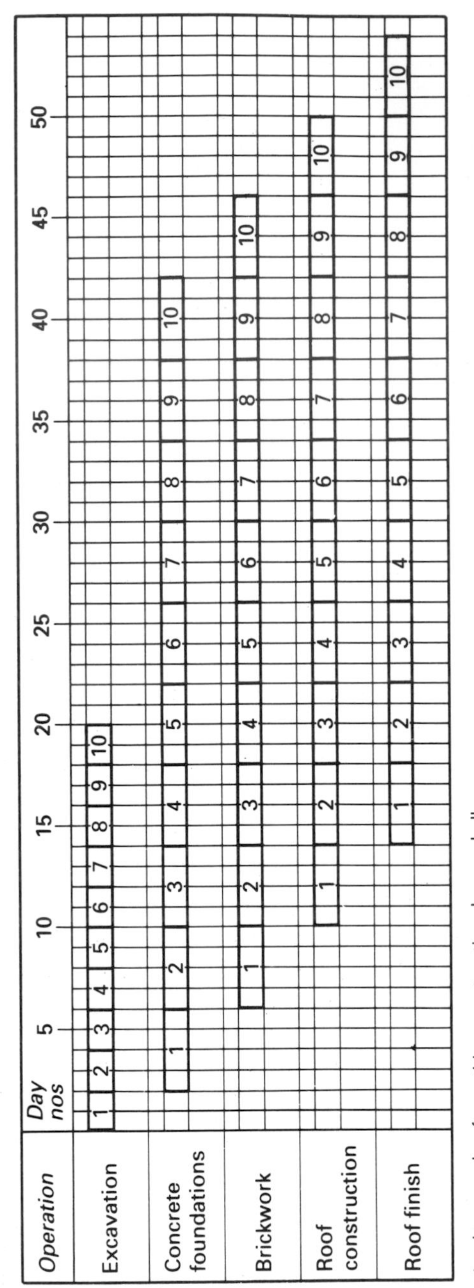

**Fig. 4.1** A simple example of repetitive construction: house shells.

**Table 4.1** Distribution of cost.

| | Weekly expenditure (£) | Cumulative expenditure (£) |
|---|---|---|
| Week 1 | 2 805.0 | 2 805.0 |
| Week 2 | 10 740.0 | 13 545.0 |
| Week 3 | 14 635.0 | 28 180.0 |
| Week 4 | 17 275.0 | 45 455.0 |
| Week 5 | 16 862.5 | 62 317.5 |
| Week 6 | 16 862.5 | 79 180.0 |
| Week 7 | 16 862.5 | 96 042.5 |
| Week 8 | 16 862.5 | 112 905.0 |
| Week 9 | 14 470.0 | 127 375.0 |
| Week 10 | 6 535.0 | 133 910.0 |
| Week 11 | 52 640.0 | 136 550.0 |

*Note*: In practice these figures would be taken to the nearest £10, but here precise calculations have been made so that the reader can follow the working.

The contract sum £157 100, and the value of each stage (as ascertained from the priced bill of quantities) is as follows:

| | |
|---|---|
| Excavation | £1 900 = £190 each block |
| Concrete foundations | £36 700 = £3670 each block |
| Brickwork | £72 900 = £7290 each block |
| Roof construction | £15 200 = £1520 each block |
| Roof finish | £30 400 = £3040 each block |
| | £157 100 |

By allocating these values to the programme of work it can be seen that the distribution of the value will be as shown in Table 4.2 (only completed stages will be measured at the end of the week).

**Table 4.2** Distribution of value.

| | Weekly expenditure (£) | Cumulative expenditure (£) |
|---|---|---|
| Week 1 | 0 | 0 |
| Week 2 | 380 | 380 |
| Week 3 | 15 200 | 15 580 |
| Week 4 | 12 860 | 28 440 |
| Week 5 | 16 090 | 44 530 |
| Week 6 | 15 520 | 60 050 |
| Week 7 | 31 040 | 91 090 |

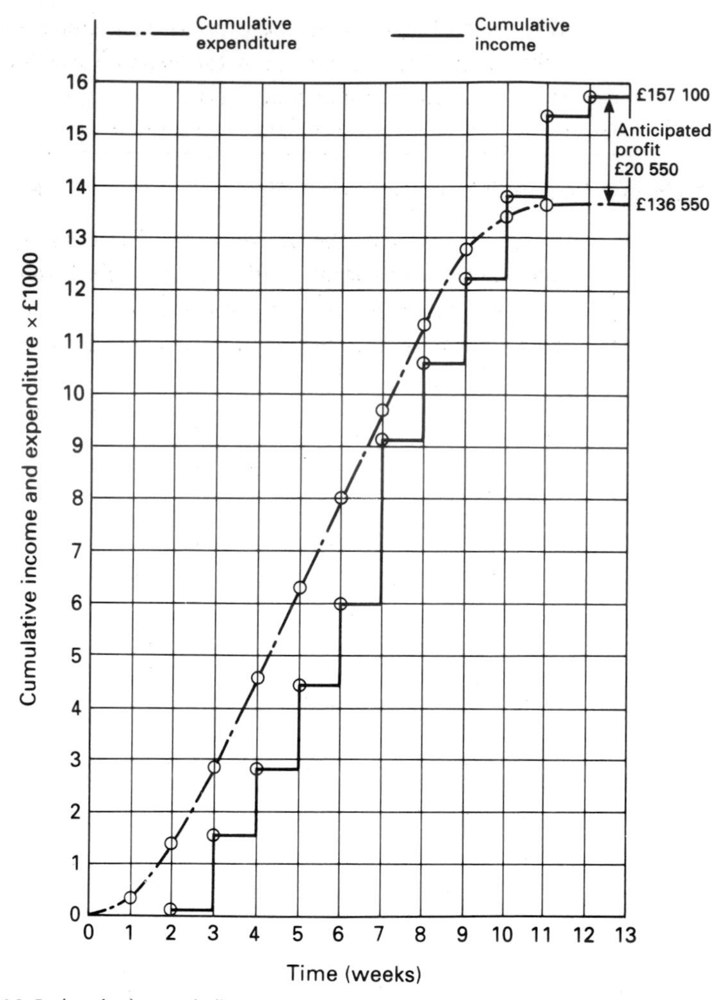


**Fig. 4.2** Budget for house shells.

**Table 4.2** *Continued.*

| | Weekly expenditure (£) | Cumulative expenditure (£) |
|---|---|---|
| Week 8 | 15 520 | 106 610 |
| Week 9 | 15 520 | 122 130 |
| Week 10 | 15 520 | 137 650 |
| Week 11 | 16 410 | 154 060 |
| Week 12 | 43 040 | 157 100 |
| | 157 100 | |

Both sets of figures can be plotted on a chart field (Fig. 4.2) and by subtracting the income from the expenditure the capital required at that particular time can be found.

**Example 4.2: Project budget and cash flow**

Figure 4.3 shows the programme for a small office building. All operations are programmed at their earliest times, but the amount of float available in each operation is indicated by a dotted line. The expenditure on and income from each operation has been ascertained, and is shown in Table 4.3.

**Table 4.3** Expenditure and income.

| Operation | Planned expenditure | Anticipated income |
|---|---|---|
| Set up site | 210 | 240 |
| Setting out | 60 | 80 |
| Excavate to RL | 1 000 | 1 200 |
| Piling | 3 600 | 4 800 |
| Drains | 300 | 360 |
| Pad foundations | 630 | 720 |
| Strip foundations | 500 | 600 |
| In situ conc frame | 22 000 | 33 000 |
| GF slab | 800 | 900 |
| Brick infill panels | 8 700 | 9 750 |
| PC stairs and floors | 2 400 | 2 700 |
| Hardwood frames | 1 000 | 1 200 |
| PC roof slab | 2 000 | 2 500 |
| Block parts GF | 1 750 | 2 100 |
| Glazing | 240 | 300 |
| External doors | 1 600 | 1 800 |
| Roof covering | 1 000 | 1 200 |
| External plumbing | 180 | 210 |
| Block parts first floor | 2 100 | 2 450 |
| Plumbing first fix | 900 | 1 050 |
| Electrician first fix | 400 | 480 |
| External painting | 240 | 300 |
| External works | 7 564 | 9 000 |
| Joiner first fix | 800 | 880 |
| Plasterer | 1 800 | 2 160 |
| Floor screed | 600 | 700 |
| Quarry tiles | 100 | 140 |
| Plumbing second fix | 1 000 | 1 200 |
| Joiner second fix | 1 000 | 1 200 |
| Internal painting | 420 | 504 |
| Vinyl tiling | 240 | 300 |
| Electrician second fix | 360 | 540 |
| Clean and handover | 50 | 60 |
| Prelims | 13 175 | 13 950 |

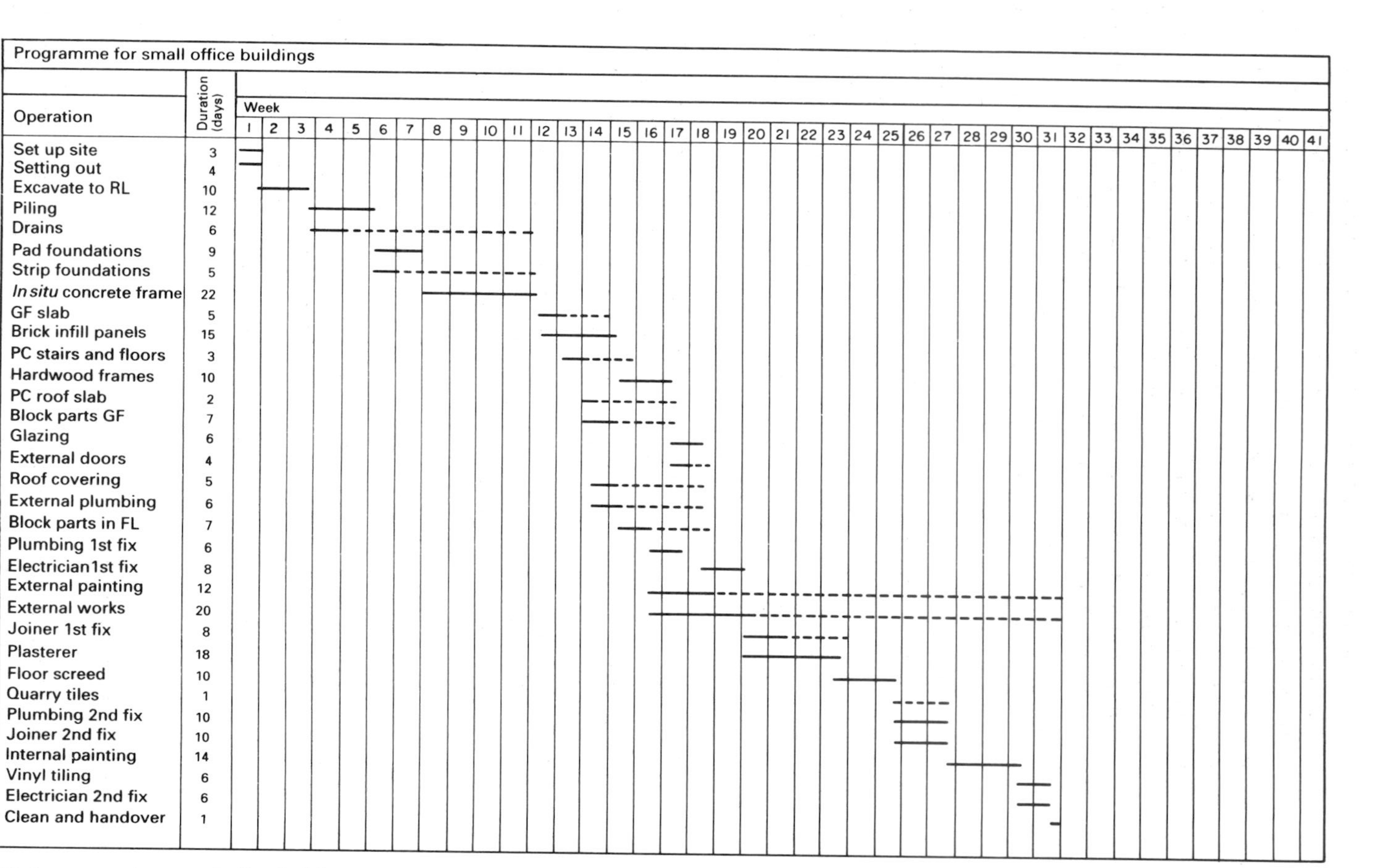


**Fig. 4.3** Programme for a small office building.

It is assumed that a contract has been entered into with the client, and that payment for the work carried out will be made one month after the end of the month in which the work is carried out. The retention will be 5% throughout. Half the retention will be released one month after the completion. By allocating the planned expenditure and anticipated income to the master programme (see Fig. 4.4 operations at earliest times), it is possible to arrive at the expenditure and income for each month. These figures can then be processed further to arrive at the cumulative amounts for expenditure and income, having made the necessary adjustments for retention and payment delays. The figures for income and expenditure have been recorded graphically in Fig. 4.5.

The cash flow for the project with all operations taking place at their earliest times can now be ascertained by plotting the monthly planned expenditure and anticipated income, as shown in Fig. 4.6.

With all operations starting at their earliest time, it can be seen that the maximum amount of working capital required is £37 781 at the end of month 4, and that the project becomes self-financing at the end of month 6.

For the purpose of comparison, the process can be repeated to ascertain the effect on the budget and cash flow of carrying out all operations at their latest times. Figure 4.7 shows the new allocation of planned expenditure and anticipated income with the appropriate processing at the foot of the chart. Figure 4.8 shows the new graphical presentation of the planned expenditure and anticipated income, and Fig. 4.9 shows the new cash flow for the project. Notice that the maximum amount of working capital is now reduced to £33 117 at the end of month 4, but the project does not become self-financing until the end of month 8.

The acid test as to which of these conditions is the most acceptable financially is to ascertain which of the extremes is the least expensive in terms of interest charges. There are of course an infinite number of alternative solutions in between the extremes of earliest and latest times that can be used to give the desired cash flow.

**Example 4.3: Multi-project cash flow**

The same technique can be used to ascertain the cash flow for all projects being undertaken by a company. Table 4.4 shows the planned expenditure and anticipated income for four projects. The cash flows

| Programme for small office building – expenditure and income at earliest times | | | | | | | | | | | | | | | | | |
|---|---|---|---|---|---|---|---|---|---|---|---|---|---|---|---|---|---|
| Operation | Duration (days) | Planned expenditure | Anticipated income | Weeks 1 | 2 | 3 | 4 | 5 | 6 | 7 | 8 | 9 | 10 | 11 | 12 | 13 | 14 |
| Set up site | 3 | 210 | 240 | 210<br>240 | | | | | | | | | | | | | |
| Setting out | 4 | 60 | 80 | 60<br>80 | | | | | | | | | | | | | |
| Excavate to reduced level | 10 | 1000 | 1200 | 100<br>120 | 500<br>600 | 400<br>480 | | | | | | | | | | | |
| Piling | 12 | 3600 | 4800 | | | 300<br>400 | 1500<br>2000 | 1500<br>2000 | 300<br>400 | | | | | | | | |
| Drains | 6 | 300 | 360 | | | 50<br>60 | 250<br>300 | | | | | | | | | | |
| Pad foundations | 9 | 630 | 720 | | | | | | 280<br>320 | 350<br>400 | | | | | | | |
| Strip foundations | 5 | 500 | 600 | | | | | | 400<br>480 | 100<br>120 | | | | | | | |
| *In situ* concrete frames | 22 | 22000 | 33000 | | | | | | | | 5000<br>7500 | 5000<br>7500 | 5000<br>7500 | 5000<br>7500 | 2000<br>3000 | | |
| Ground floor slab | 5 | 800 | 900 | | | | | | | | | | | | 480<br>540 | 320<br>360 | |
| Brick infill panels | 15 | 8700 | 9750 | | | | | | | | | | | | 1740<br>1950 | 2900<br>3250 | 2900<br>3250 |
| Precast stairs and floors | 3 | 2400 | 2700 | | | | | | | | | | | | | 2400<br>2700 | |
| Hardwood frames | 10 | 1000 | 1200 | | | | | | | | | | | | | | |
| Precast roof slab | 2 | 2000 | 2500 | | | | | | | | | | | | | | 2000<br>2500 |
| Block partitions of GF | 7 | 1750 | 2100 | | | | | | | | | | | | | | 1250<br>1500 |
| Glazing | 6 | 240 | 300 | | | | | | | | | | | | | | |
| External doors | 4 | 1600 | 1800 | | | | | | | | | | | | | | |
| Roof covering | 5 | 1000 | 1200 | | | | | | | | | | | | | | 600<br>720 |
| External plumbing | 6 | 180 | 210 | | | | | | | | | | | | | | 90<br>105 |
| Block partitions 1st FL | 7 | 2100 | 2450 | | | | | | | | | | | | | | |
| Plumbing 1st fix | 5 | 900 | 1050 | | | | | | | | | | | | | | |
| Electrician 1st fix | 8 | 400 | 480 | | | | | | | | | | | | | | |
| External painting | 12 | 240 | 300 | | | | | | | | | | | | | | |
| External works | 20 | 7564 | 9000 | | | | | | | | | | | | | | |
| Joiner 1st fix | 8 | 800 | 880 | | | | | | | | | | | | | | |
| Plasterer | 18 | 1800 | 2160 | | | | | | | | | | | | | | |
| Floor screed | 10 | 600 | 700 | | | | | | | | | | | | | | |
| Quarry tiles | 1 | 100 | 140 | | | | | | | | | | | | | | |
| Plumbing 2nd fix | 10 | 1000 | 1200 | | | | | | | | | | | | | | |
| Joiner 2nd fix | 10 | 1000 | 1200 | | | | | | | | | | | | | | |
| Internal painting | 14 | 420 | 504 | | | | | | | | | | | | | | |
| Vinyl tiling | 6 | 240 | 300 | | | | | | | | | | | | | | |
| Electrician 2nd fix | 6 | 360 | 540 | | | | | | | | | | | | | | |
| Clean and handover | 1 | 50 | 60 | | | | | | | | | | | | | | |
| Preliminaries | | 13175 | 13950 | 425<br>450 | 425<br>450 | 425<br>450 | 425<br>450 | 425<br>450 | 425<br>450 | 425<br>450 | 425<br>450 | 425<br>450 | 425<br>450 | 425<br>450 | 425<br>450 | 425<br>450 | 425<br>450 |
| Planned expenditure | | 78719 | | 795 | 925 | 1175 | 2175 | 1925 | 1405 | 875 | 5425 | 5425 | 5425 | 5425 | 4645 | 6045 | 7265 |
| Anticipated income | | | 98574 | 890 | 1050 | 1390 | 2750 | 2450 | 1650 | 970 | 7950 | 7950 | 7950 | 7950 | 5940 | 6760 | 8525 |
| Planned expenditure (monthly) | | | | | | | 5070 | | | | 9630 | | | | 20920 | | |
| Anticipated income (monthly) | | | | | | | 6080 | | | | 13020 | | | | 29790 | | |
| 5% retention on anticipated income | | | | | | | 304 | | | | 651 | | | | 1490 | | |
| Income less retention time adjusted | | | | | | | | | | | 5776 | | | | 12369 | | |
| Cumulative expenditure | | | | | | | 5070 | | | | 14700 | | | | 35620 | | |
| Cumulative income time adjusted | | | | | | | | | | | 5776 | | | | 18145 | | |
| Difference between expenditure and income | | | | | | | −5070 | | | | −14700 | | | | −29844 | | |

**Fig. 4.4** Office building: expenditure and income at earliest times.

| 15 | 16 | 17 | 18 | 19 | 20 | 21 | 22 | 23 | 24 | 25 | 26 | 27 | 28 | 29 | 30 | 31 | 32 | 33 | 34 | 35 | 36 | 37 | 38 |
|---|---|---|---|---|---|---|---|---|---|---|---|---|---|---|---|---|---|---|---|---|---|---|---|
| 1160<br>1300<br>300<br>360<br>500<br>600<br>400<br>480<br>90<br>105<br>900<br>1050 | 500<br>600<br>1200<br>1400<br>300<br>350<br>40<br>50<br>756<br>900 | 200<br>240<br>120<br>150<br>1200<br>1350<br>600<br>700<br>100<br>125<br>1891<br>2250 | 120<br>150<br>400<br>450<br>100<br>120<br>100<br>125<br>1891<br>2250 | 250<br>300<br>1891<br>2250 | 50<br>60<br>1135<br>1350<br>400<br>440<br>400<br>480 | 400<br>440<br>500<br>600 | 500<br>600 | 400<br>480<br>60<br>70 | 300<br>350 | 240<br>280<br>100<br>140<br>100<br>120<br>100<br>120 | 500<br>600<br>500<br>600 | 400<br>480<br>400<br>480<br>30<br>36 | 150<br>180 | 150<br>180 | 90<br>108<br>80<br>100<br>120<br>180 | 160<br>200<br>240<br>360<br>50<br>60 | | | | | | | |
| 425 | 425 | 425 | 425 | 425 | 425 | 425 | 425 | 425 | 425 | 425 | 425 | 425 | 425 | 425 | 425 | 425 | | | | | | | |
| 450 | 450 | 450 | 450 | 450 | 450 | 450 | 450 | 450 | 450 | 450 | 450 | 450 | 450 | 450 | 450 | 450 | | | | | | | |
| 3775 | 3221 | 4536 | 3036 | 2566 | 2410 | 1325 | 925 | 885 | 725 | 965 | 1425 | 1255 | 575 | 575 | 715 | 875 | | | | | | | |
| 4345 | 3750 | 5265 | 3545 | 3000 | 2780 | 1490 | 1050 | 1000 | 800 | 1110 | 1650 | 1446 | 630 | 630 | 838 | 1070 | | | | | | | |
| | 20306 | | | | 12548 | | | | 3860 | | | | 4220 | | | 2165 | | | | | | | |
| | 23380 | | | | 14590 | | | | 4340 | | | | 4836 | | | 2538 | | | | | | | |
| | 1169 | | | | 730 | | | | 217 | | | | 242 | | | 127 | | | | | | | |
| | 28300 | | | | 22211 | | | | 13860 | | | | 4123 | | | | 4594 | | | | 2411 | | |
| | 55926 | | | | 68474 | | | | 72334 | | | | 76554 | | | 78719 | | | | | | | |
| | 46445 | | | | 68656 | | | | 82516 | | | | 86639 | | | | 91233 | | | | 93644<br>2465 | 2465 return | |
| | - 37781 | | | | - 22029 | | | | - 3678 | | | | + 5962 | | | | + 7920 | | | | + 12514<br>+ 17390 | month 15 | |

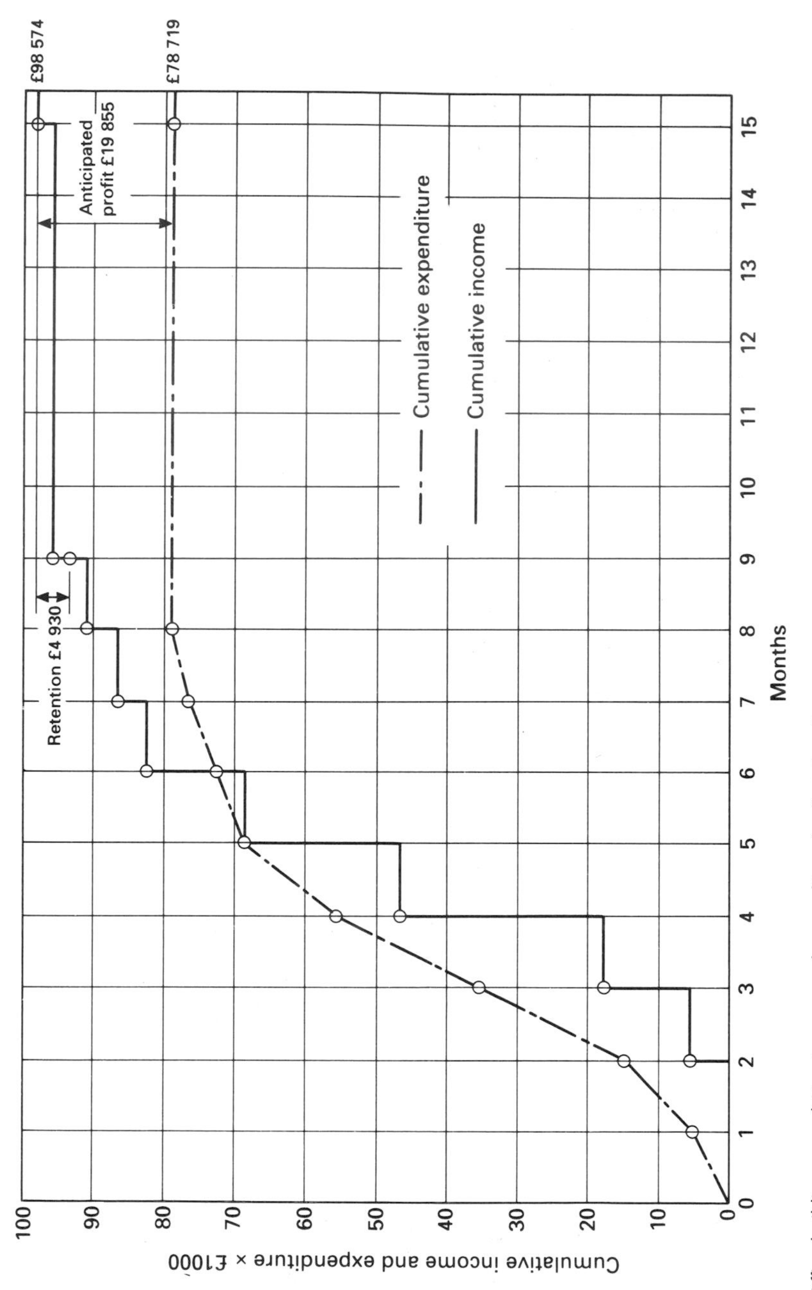

**Fig. 4.5** Office building: cumulative income and expenditure (earliest times).

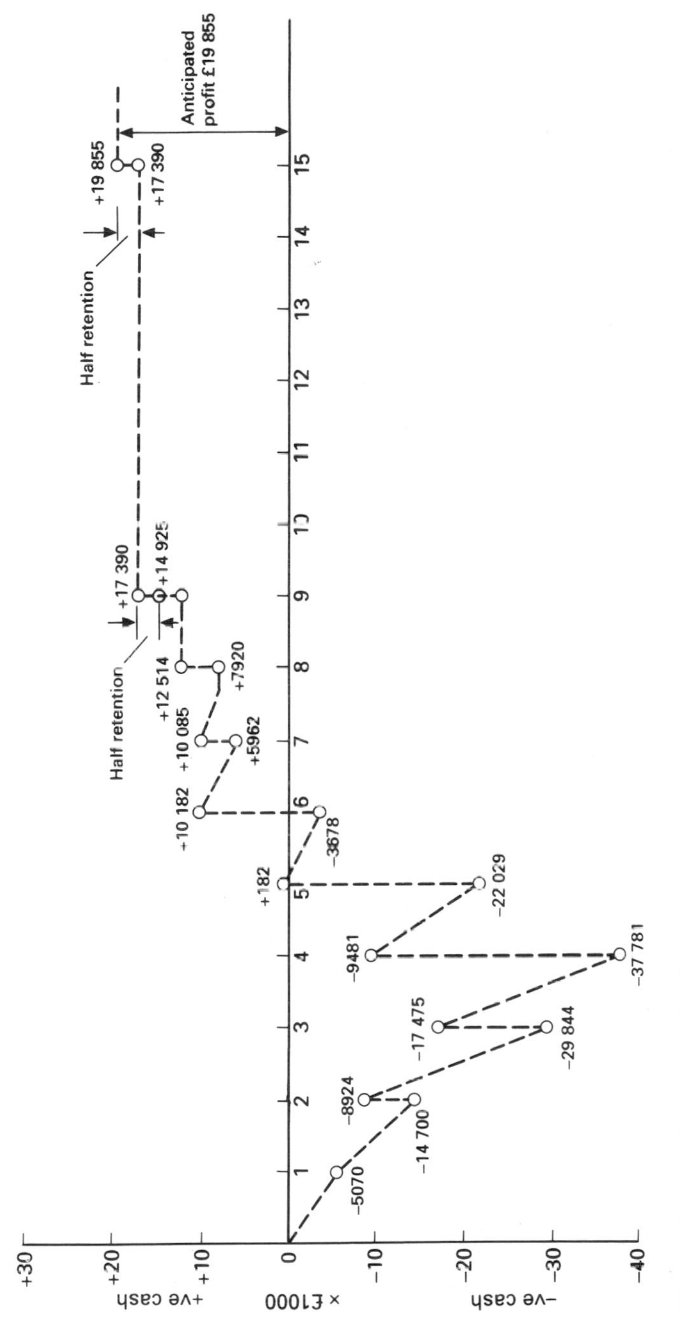


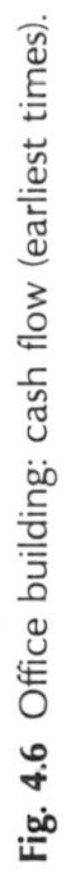
**Fig. 4.6** Office building: cash flow (earliest times).

Programme for small office building – expenditure and income at latest time

| Operation | Duration (days) | Planned expenditure | Anticipated income | Weeks 1 | 2 | 3 | 4 | 5 | 6 | 7 | 8 | 9 | 10 | 11 | 12 | 13 | 14 |
|---|---|---|---|---|---|---|---|---|---|---|---|---|---|---|---|---|---|
| Set up site | 3 | 210 | 240 | 210 / 240 | | | | | | | | | | | | | |
| Setting out | 4 | 60 | 80 | 60 / 80 | | | | | | | | | | | | | |
| Excavate to reduced level | 10 | 1000 | 1200 | 100 / 120 | 500 / 600 | 400 / 480 | | | | | | | | | | | |
| Piling | 12 | 3600 | 4800 | | | 300 / 400 | 1500 / 2000 | 1500 / 2000 | 300 / 400 | | | | | | | | |
| Drains | 6 | 300 | 360 | | | | | | | | | | | 250 / 300 | 50 / 60 | | |
| Pad foundations | 9 | 630 | 720 | | | | | | 280 / 320 | 350 / 400 | | | | | | | |
| Strip foundations | 5 | 500 | 600 | | | | | | | | | | | 400 / 480 | 100 / 120 | | |
| *In situ* concrete frames | 22 | 22000 | 33000 | | | | | | | | 5000 / 7500 | 5000 / 7500 | 5000 / 7500 | 5000 / 7500 | 2000 / 3000 | | |
| Ground floor slab | 5 | 800 | 900 | | | | | | | | | | | | | | 800 / 900 |
| Brick infill panels | 15 | 8700 | 9750 | | | | | | | | | | | | 1740 / 1950 | 2900 / 3250 | 2900 / 3250 |
| Precast stairs and floors | 3 | 2400 | 2700 | | | | | | | | | | | | | | |
| Hardwood frames | 10 | 1000 | 1200 | | | | | | | | | | | | | | |
| Precast roof slab | 2 | 2000 | 2500 | | | | | | | | | | | | | | |
| Block partitions of GF | 7 | 1750 | 2100 | | | | | | | | | | | | | | |
| Glazing | 6 | 240 | 300 | | | | | | | | | | | | | | |
| External doors | 4 | 1600 | 1800 | | | | | | | | | | | | | | |
| Roof covering | 5 | 1000 | 1200 | | | | | | | | | | | | | | |
| External plumbing | 6 | 180 | 210 | | | | | | | | | | | | | | |
| Block partitions 1st FL | 7 | 2100 | 2450 | | | | | | | | | | | | | | |
| Plumbing 1st fix | 5 | 900 | 1050 | | | | | | | | | | | | | | |
| Electrician 1st fix | 8 | 400 | 480 | | | | | | | | | | | | | | |
| External painting | 12 | 240 | 300 | | | | | | | | | | | | | | |
| External works | 20 | 7564 | 9000 | | | | | | | | | | | | | | |
| Joiner 1st fix | 8 | 800 | 880 | | | | | | | | | | | | | | |
| Plasterer | 18 | 1800 | 2160 | | | | | | | | | | | | | | |
| Floor screed | 10 | 600 | 700 | | | | | | | | | | | | | | |
| Quarry tiles | 1 | 100 | 140 | | | | | | | | | | | | | | |
| Plumbing 2nd fix | 10 | 1000 | 1200 | | | | | | | | | | | | | | |
| Joiner 2nd fix | 10 | 1000 | 1200 | | | | | | | | | | | | | | |
| Internal painting | 14 | 420 | 504 | | | | | | | | | | | | | | |
| Vinyl tiling | 6 | 240 | 300 | | | | | | | | | | | | | | |
| Electrician 2nd fix | 6 | 360 | 540 | | | | | | | | | | | | | | |
| Clean and handover | 1 | 50 | 60 | | | | | | | | | | | | | | |
| Preliminaries | | 13175 | 13950 | 425 / 450 | 425 / 450 | 425 / 450 | 425 / 450 | 425 / 450 | 425 / 450 | 425 / 450 | 425 / 450 | 425 / 450 | 425 / 450 | 425 / 450 | 425 / 450 | 425 / 450 | 425 / 450 |
| Planned expenditure | | 78719 | | 795 | 925 | 1125 | 1925 | 1925 | 1005 | 775 | 5425 | 5425 | 5425 | 6075 | 4315 | 3325 | 4125 |
| Anticipated income | | | 98574 | 890 | 1050 | 1330 | 2450 | 2450 | 1170 | 850 | 7950 | 7950 | 7950 | 8730 | 5580 | 3700 | 4600 |
| Planned expenditure (monthly) | | | | | | | 4770 | | | | 9130 | | | | 21240 | | |
| Anticipated income (monthly) | | | | | | | 5720 | | | | 12420 | | | | 30210 | | |
| 5% retention on anticipated income | | | | | | | 286 | | | | 621 | | | | 1511 | | |
| Income less retention time adjusted | | | | | | | | | | | 5434 | | | | 11789 | | |
| Cumulative expenditure | | | | | | | 4770 | | | | 13900 | | | | 35140 | | |
| Cumulative income time adjusted | | | | | | | | | | | 5434 | | | | 17233 | | |
| Difference between expenditure and income | | | | | | | −4770 | | | | −13900 | | | | −29706 | | |

**Fig. 4.7** Office building: expenditure and income at latest times.

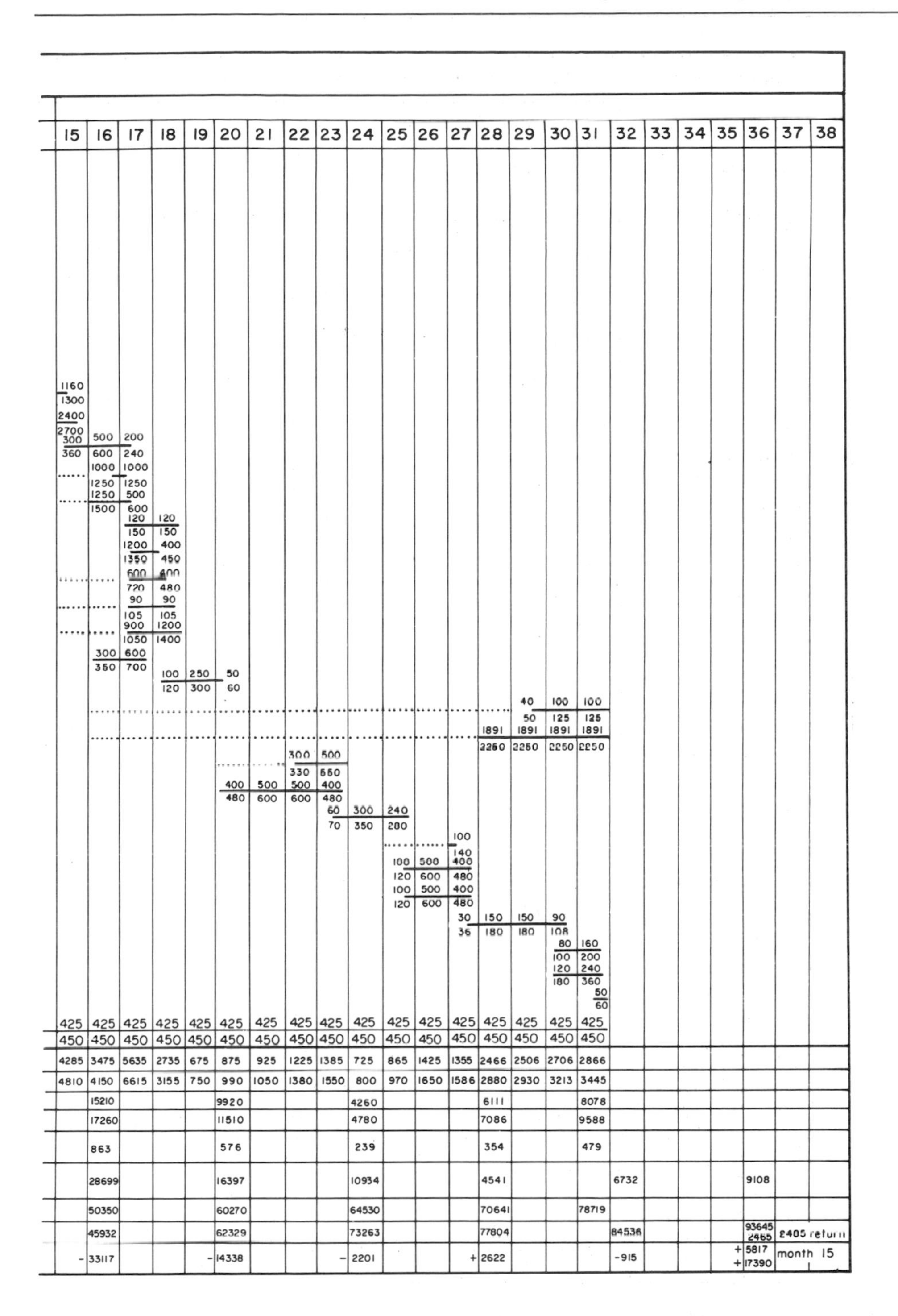
15 16 17 18 19 20 21 22 23 24 25 26 27 28 29 30 31 32 33 34 35 36 37 38
425 425 425 425 425 425 425 425 425 425 425 425 425 425 425 425 425
450 450 450 450 450 450 450 450 450 450 450 450 450 450 450 450 450
4285 3475 5635 2735 675 875 925 1225 1385 725 865 1425 1355 2466 2506 2706 2866
4810 4150 6615 3155 750 990 1050 1380 1550 800 970 1650 1586 2880 2930 3213 3445
15210 9920 4260 6111 8078
17260 11510 4780 7086 9588
863 576 239 354 479
28699 16397 10934 4541 6732 9108
50350 60270 64530 70641 78719
45932 62329 73263 77804 84536 93645 2465 £405 return
- 33117 - 14338 - 2201 + 2622 -915 + 5817 + 17390 month 15

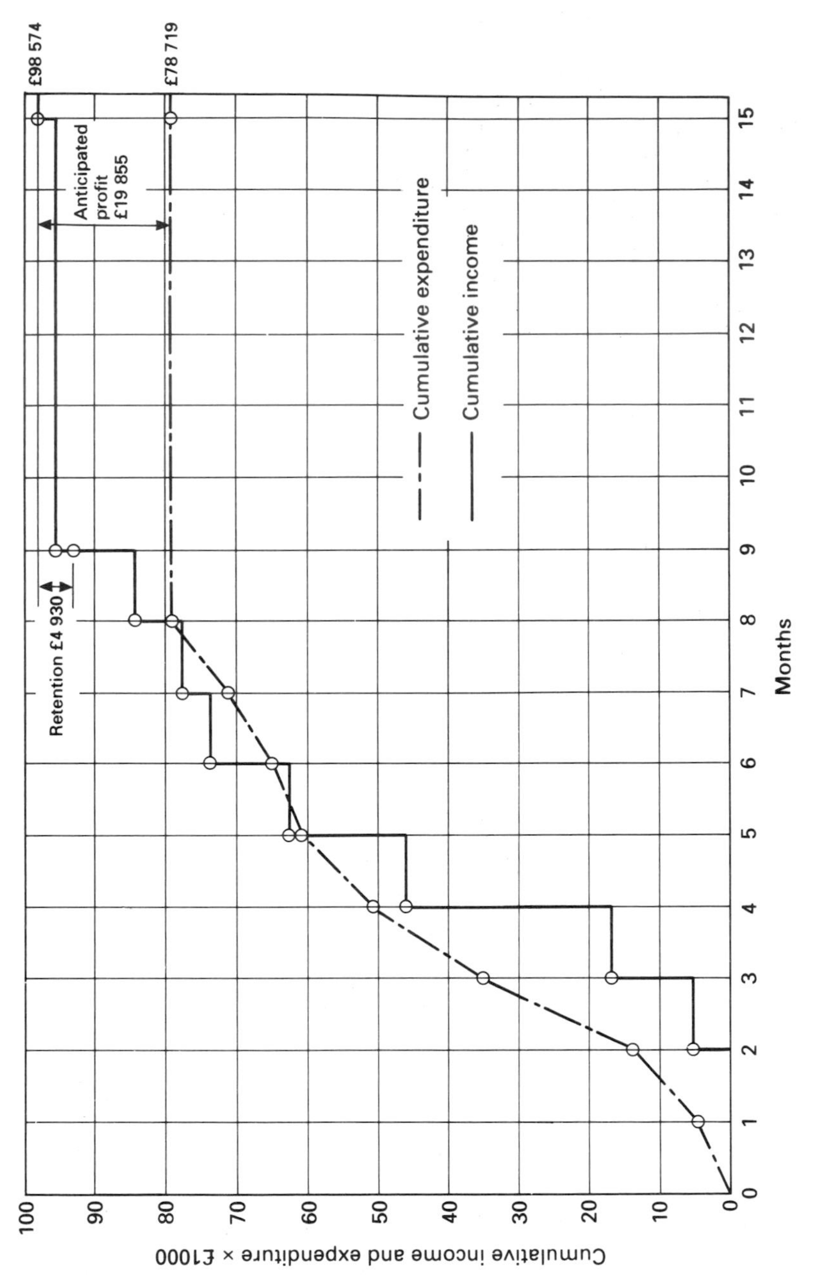

**Fig. 4.8** Office building: cumulative income and expenditure (latest times).

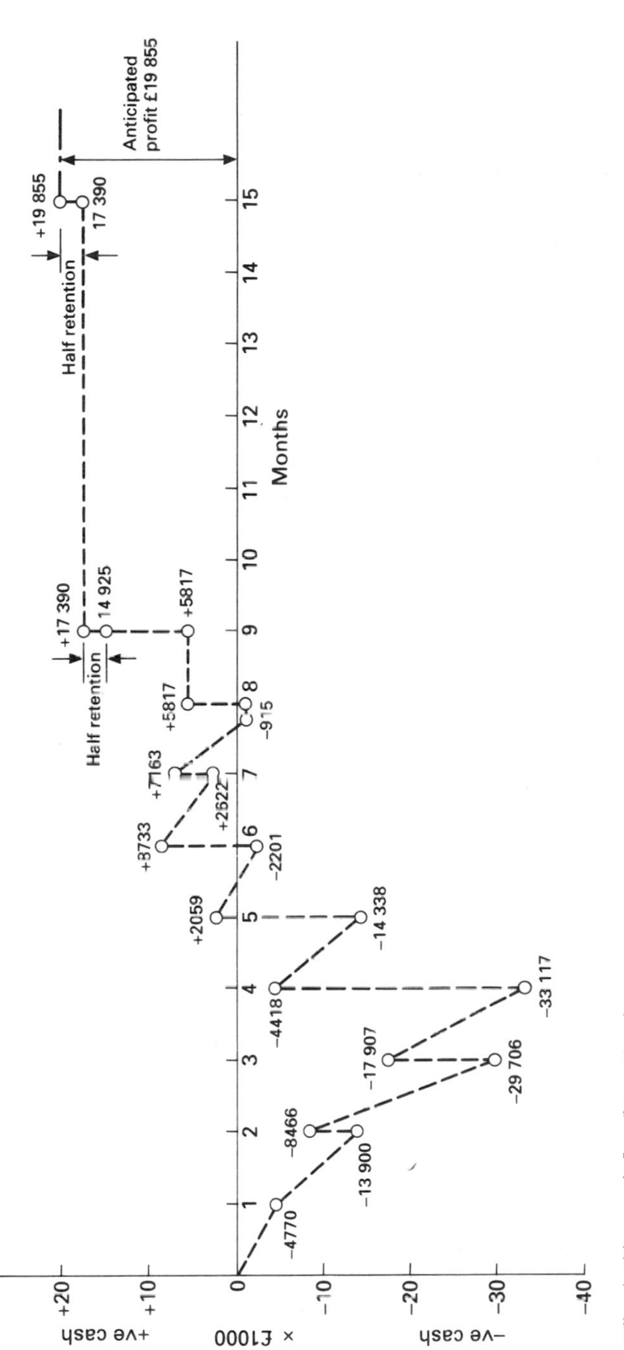


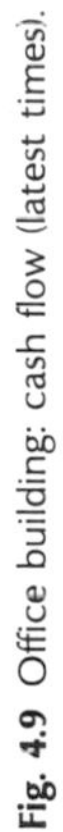
**Fig. 4.9** Office building: cash flow (latest times).

**Table 4.4** Monthly expenditure and income for all projects.

| Month | Project A | | Project B | | Project C | | Project D | |
|---|---|---|---|---|---|---|---|---|
| | Planned expenditure (£) | Anticipated income (£) | Planned expenditure (£) | Anticipated income (£) | Planned expenditure (£) | Anticipated income (£) | Planned expenditure (£) | Anticipated income (£) |
| 1 | 6 500 | | | | | | | |
| 2 | 17 500 | | | | | | | |
| 3 | 34 500 | 7 500 | 13 000 | | | | | |
| 4 | 36 500 | 20 100 | 35 000 | | | | | |
| 5 | 27 500 | 39 700 | 69 000 | 14 600 | | | | |
| 6 | 17 500 | 42 000 | 73 000 | 39 200 | | | 2 750 | |
| 7 | 5 500 | 31 600 | 55 000 | 77 300 | | | 8 000 | |
| 8 | 4 500 | 20 100 | 36 000 | 81 800 | 5 500 | | 17 000 | 3 300 |
| 9 | | 6 300 | 11 000 | 61 600 | 16 000 | | 18 750 | 9 500 |
| 10 | | 5 200 | 9 000 | 40 300 | 34 000 | 6 400 | 16 500 | 20 200 |
| 11 | | | | 12 300 | 37 500 | 18 700 | 13 500 | 22 300 |
| 12 | | | | 10 100 | 33 000 | 39 800 | 9 250 | 19 600 |
| 13 | | | | | 27 000 | 43 900 | 7 250 | 16 100 |
| 14 | | | | | 18 500 | 38 600 | 3 250 | 11 000 |
| 15 | | | | | 14 500 | 31 600 | 1 500 | 8 600 |
| 16 | | | | | 6 500 | 21 600 | 500 | 3 900 |
| 17 | | | | | 3 000 | 17 000 | | 1 800 |
| 18 | | | | | 1 000 | 7 600 | | 600 |
| 19 | | | | | | 3 500 | | |
| 20 | | | | | | 1 200 | | |

for each of these projects individually and the sum of all four are shown in Table 4.5, which can be shown graphically as in Fig. 4.10.

It can be seen that the maximum amount of working capital required to finance these projects is £169 650. It can also be seen that serious thought should be given to finding another project for commencement at about month 7, otherwise the available working capital is not going to be fully utilised.

In the preceding examples, the plotting of the actual expenditure and income as soon as figures become available highlights discrepancies, and enables control to be exercised.

### *4.1.6 Cost of financing projects*

Once the cash flow has been established for the project, it is a simple step to determine the cost of providing the working capital. Whether the cash resources are available within the company or have to be borrowed from external sources (such as a bank overdraft), the cost of interest payments should be taken into account. If the company's own funds are

**Table 4.5** Cumulative monthly expenditure, income and total cash flow for all projects.

| | Project A | | | | | Project B | | | | | Project C | | | | | Project D | | | | | |
|---|---|---|---|---|---|---|---|---|---|---|---|---|---|---|---|---|---|---|---|---|---|
| Month | Planned expenditure | Anticipated income | Cumulative planned expenditure | Cumulative anticipated income | Cash flow | Planned expenditure | Anticipated income | Cumulative planned expenditure | Cumulative anticipated income | Cash flow | Planned expenditure | Anticipated income | Cumulative planned expenditure | Cumulative anticipated income | Cash flow | Planned expenditure | Anticipated income | Cumulative planned expenditure | Cumulative anticipated income | Cash flow | Total cash flow for all projects |
| 1 | 6 500 | | 6 500 | | −6 500 | | | | | | | | | | | | | | | | −6 500 |
| 2 | 17 500 | | 24 000 | | −24 000 | | | | | | | | | | | | | | | | −24 000 |
| 3 | 34 500 | 7 500 | 58 500 | 7 500 | −51 000 | 13 000 | | 13 000 | | −13 000 | | | | | | | | | | | −64 000 |
| 4 | 36 500 | 20 100 | 95 000 | 27 600 | −67 400 | 35 000 | | 48 000 | | −48 000 | | | | | | | | | | | −115 400 |
| 5 | 27 500 | 39 700 | 122 500 | 67 300 | −55 200 | 69 000 | 14 600 | 117 000 | 14 600 | −102 400 | | | | | | | | | | | −157 600 |
| 6 | 17 500 | 42 000 | 140 000 | 109 300 | −30 700 | 73 000 | 39 200 | 190 000 | 53 800 | −136 200 | | | | | | 2 750 | | 2 750 | | −2 750 | −169 650 |
| 7 | 5 500 | 31 600 | 145 500 | 140 900 | −4 600 | 55 000 | 77 300 | 245 000 | 131 100 | −113 900 | | | | | | 8 000 | | 10 750 | | −10 750 | −129 250 |
| 8 | 4 500 | 20 100 | 150 000 | 161 000 | +11 000 | 36 000 | 81 800 | 281 000 | 212 900 | −68 100 | 5 500 | | 5 500 | | −5 500 | 17 000 | 3 300 | 27 750 | 3 300 | −24 450 | −87 050 |
| 9 | | 6 300 | | 167 300 | +17 300 | 11 000 | 61 600 | 292 000 | 274 500 | −17 500 | 16 000 | | 21 500 | | −21 500 | 18 750 | 9 500 | 46 500 | 12 800 | −33 700 | −55 400 |
| 10 | | 5 200 | | 172 500 | +22 500 | 9 000 | 40 300 | 301 000 | 314 800 | +13 800 | 34 000 | 6 400 | 55 500 | 6 400 | −49 100 | 16 500 | 20 200 | 63 000 | 33 000 | −30 000 | −42 800 |
| 11 | | | | | | | 12 300 | | 327 100 | +26 100 | 37 500 | 18 700 | 93 000 | 25 100 | −67 900 | 13 500 | 22 300 | 76 500 | 53 300 | −23 200 | −42 500 |
| 12 | | | | | | | 10 100 | | 337 200 | +36 200 | 33 000 | 39 800 | 125 000 | 64 900 | −61 100 | 9 250 | 19 600 | 85 750 | 74 900 | −10 850 | −13 250 |
| 13 | | | | | | | | | | | 27 000 | 43 900 | 153 000 | 108 800 | −44 200 | 7 250 | 16 100 | 93 000 | 91 000 | −2 000 | +12 500 |
| 14 | | | | | | | | | | | 18 500 | 38 600 | 171 500 | 147 400 | −24 100 | 3 250 | 11 000 | 96 250 | 102 000 | +5 750 | +40 350 |
| 15 | | | | | | | | | | | 14 500 | 31 600 | 186 000 | 179 000 | −7 000 | 1 500 | 8 600 | 97 750 | 110 600 | +12 850 | +64 550 |
| 16 | | | | | | | | | | | 6 500 | 21 600 | 192 500 | 200 600 | +8 100 | 500 | 3 900 | 98 250 | 114 500 | +16 250 | +83 050 |
| 17 | | | | | | | | | | | 3 000 | 17 000 | 195 500 | 217 600 | +22 100 | | 1 800 | | 116 300 | +18 050 | +98 850 |
| 18 | | | | | | | | | | | 1 000 | 7 600 | 195 500 | 225 200 | +28 700 | | 600 | | 116 900 | +18 650 | +106 050 |
| 19 | | | | | | | | | | | | 3 500 | | 228 700 | +32 200 | | | | | | +109 550 |
| 20 | | | | | | | | | | | | 1 200 | | 229 900 | +33 400 | | | | | | +110 750 |

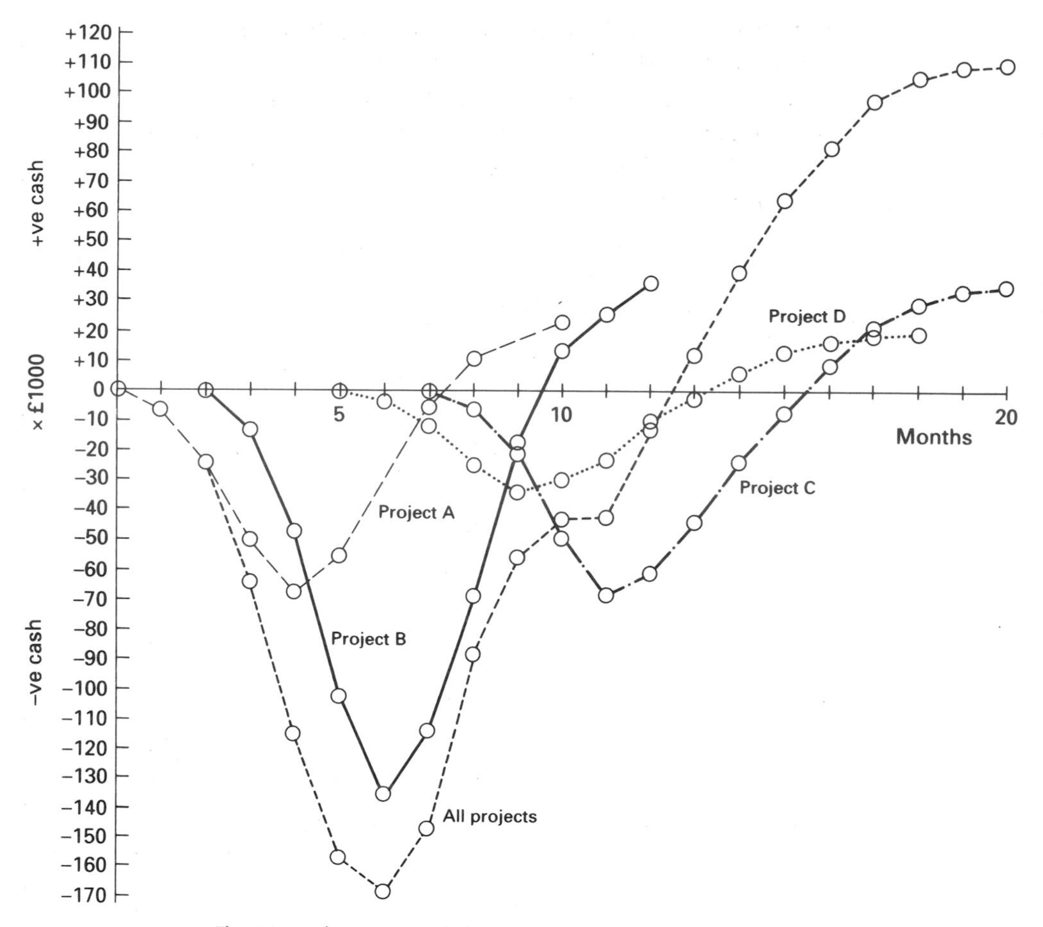


**Fig. 4.10** Multi-project cash flow.

used, it is equally important to allow for the cost of interest payments, as the money could have been available for investment elsewhere.

**Example 4.4: Cost of financing projects**

Taking the project shown in Fig. 4.3 as an example, it is possible to ascertain which of the solutions, earliest or latest start times, is the most effective when considering the cost of interest payments.

Take first the graph showing the cash flow for the earliest start times in Fig. 4.6. The area between the horizontal time axis and the graph below this axis represents the working capital requirement over the duration of the contract in £-months. The area above the line represents the surplus on the contract in £-months. We are at present interested only in the area below the line, as we are attempting to ascertain the cost of borrowing this working capital.

Table 4.6 shows the calculation of working capital for the programme with all activities taking place at their earliest times, and Table 4.7 shows a similar calculation with all activities taking place at their latest times.

**Table 4.6** Cost of working capital: earliest start times (based on Fig. 4.6).

| Month | Calculation for each month (£-months) | Area (£-months) |
|---|---|---|
| 1 | $\frac{(0 + 5070)}{2}$ | 2 535 |
| 2 | $\frac{(5070 + 14\,700)}{2}$ | 9 885 |
| 3 | $\frac{(8924 + 29\,844)}{2}$ | 19 384 |
| 4 | $\frac{(17\,475 + 37\,781)}{2}$ | 27 628 |
| 5 | $\frac{(9481 + 22\,029)}{2}$ | 15 755 |
| 6 | $\frac{(0 + 3678)}{2} \times \frac{3678}{3860}$ | 1 752 |
| | | 76 939 |

**Table 4.7** Cost of working capital: latest start times (based on Fig. 4.9).

| Month | Calculation for each month (£-months) | Area (£-months) |
|---|---|---|
| 1 | $\frac{(0 + 4710)}{2}$ | 2 355 |
| 2 | $\frac{(4710 + 13\,900)}{2}$ | 9 305 |
| 3 | $\frac{(8466 + 29\,706)}{2}$ | 19 086 |
| 4 | $\frac{(17\,907 + 33\,117)}{2}$ | 25 512 |
| 5 | $\frac{(4418 + 14\,338)}{2}$ | 9 378 |
| 6 | $\frac{(0 + 2201)}{2} \times \frac{2201}{4260}$ | 568.5 |
| 7 | nil | nil |
| 8 | $\frac{(915 + 915)}{2} \times \frac{1}{4}$ | 228.75 |
| | | 66 433.25 |

If the interest rate is 12% per annum then the cost of providing the working capital for this project under the earliest start time conditions is as follows:

$$76\,939 \times \frac{12\%}{12} = £769.39$$

If the interest rate is 12% per annum, then the cost of providing the working capital for this project under the latest start time conditions is as follows:

$$66\,433.25 \times \frac{12\%}{12} = £664.33$$

It can be seen that the latest start times solution, of the two extremes considered, gives the cheapest solution. Once the project becomes self-financing, the surpluses that result can either be invested on the money market on, say, 7 days recall, or used to finance other projects.

### *4.1.7 Tabular presentation of contract budgets*

Contract budgets can also be presented and controlled in tabular form as opposed to the graphical methods shown previously.

The elements of work selected for inclusion are those that need very careful control: that is, labour, material and plant. Cost data must be collected in such a way that they are readily identifiable with each of the elements. Overheads and profit can also be controlled using this method but usually only site overheads are included if the control is carried out on site. If head office overheads and profit are included, then this work will invariably be carried out by very senior levels of management. The chart is divided into two main parts. The right-hand side gives the current month's figures; the left-hand side gives the cumulative figures to date, including the current month. These charts can be a little more elaborate and show the current month and the cumulative figures before and after the current month. The technique enables a check to be kept on the current performance, total performance and trends. Any significant variances should be investigated and appropriate action taken to bring the work back into line with the budget.

Table 4.8 shows the early stages of a project presented in this manner. As this project is being controlled at site level, all the figures are costs, target costs being set by the production control department and the actual cost figures being collected on site as work proceeds.

The total cost column records the total cost in the whole project for the particular element. The target cost of work done records what the actual amount of work done should have cost, which can be compared with the actual cost shown in the next column. The variance column

**Table 4.8** Tabular presentation of a contract budget.

| Element | Cumulative to end of month 4 | | | | | Month 4 | | | |
|---|---|---|---|---|---|---|---|---|---|
| | Total cost of work (£) | Target cost of work done (£) | Actual cost of work done (£) | Variance (£) | Variance (%) | Target cost of work done (£) | Actual cost of work done (£) | Variance (£) | Variance (%) |
| Direct labour | | | | | | | | | |
| Excavation of basement and foundations | 1 800 | 1 800 | 1 973 | (173) | (9.61) | – | – | – | – |
| Concrete blinding | 1 600 | 1 600 | 1 437 | 163 | 10.19 | – | – | – | – |
| CB and fix reinf. in founds | 2 500 | 2 050 | 1 790 | 260 | 12.68 | 450 | 395 | 55 | 12.22 |
| Concrete foundations | 3 500 | 2 800 | 2 560 | 240 | 8.57 | 940 | 900 | 40 | 4.25 |
| CB and fix reinf. in slab | 3 100 | 1 320 | 1 560 | (240) | (18.18) | 540 | 530 | 10 | 1.85 |
| Concrete slab | 5 000 | 1 820 | 1 780 | 40 | 2.19 | 1 630 | 1 800 | (170) | (10.42) |
| Stop end and edge of formwork | 900 | 310 | 275 | 35 | 11.29 | 405 | 95 | 310 | 76.5 |
| Direct materials | | | | | | | | | |
| Concrete materials | | | | | | | | | |
| (blinding) | 4 800 | 4 800 | 5 150 | (350) | (7.29) | – | – | – | – |
| (foundations) | 10 500 | 8 400 | 7 680 | 720 | 8.57 | 2 800 | 2 700 | 100 | 3.57 |
| (slab) | 15 000 | 5 460 | 5 300 | 160 | 2.93 | 4 900 | 5 400 | (500) | (10.20) |
| Reinf. in foundations | 11 250 | 10 250 | 8 950 | 1 300 | 12.68 | 2 250 | 1 975 | 275 | 12.22 |
| Reinf. in slab | 13 950 | 5 940 | 7 020 | (1 080) | (18.18) | 2 400 | 2 390 | 10 | 0.41 |
| Timber in formwork | 700 | 245 | 300 | (55) | (22.44) | 450 | 600 | (150) | (33.33) |
| Plant | | | | | | | | | |
| Excavation of basement and foundations | 12 000 | 13 500 | 12 400 | 1 100 | 8.14 | – | – | – | – |
| Pumping water | Nil | Nil | 340 | (340) | – | – | 70 | (70) | – |
| Site on-costs | 8 660 | 6 030 | 6 450 | (420) | (6.96) | 1 685 | 1 720 | (35) | 2.07 |
| Totals | 95 260 | 66 325 | 64 965 | 1 360 | 2.05 | 18 450 | 18 575 | (125) | (0.68) |

records the difference between the figures shown in the previous two columns, and the variance % column is found by dividing the variance by the target figures.

The totals are arrived at by totalling the columns vertically. In the case of the variance column, the figures in brackets are unfavourable variances (negative values) and the ones without brackets are favourable variances (positive values). The variance % total is not arrived at by totalling this column but by dividing the total variance by the total target cost.

A more comprehensive version of the use of this technique is shown in the following example.

**Example 4.5: Tabular presentation**

The budget for the shell of a two-storey in situ concrete-framed building is shown in Table 4.9. The progress achieved on the project and the cumulative costs after the first four months are as shown in Table 4.10.

At the end of month 5 the progress achieved and the associated costs are as shown in Table 4.11. With this information to hand, the cost being incurred on the contract can be analysed, highlighting what variances are occurring and indicating where action is needed. This analysis is shown in Table 4.12.

### *4.1.8 Cash flow forecasting*

The cash flow calculations presented previously have all been based upon a detailed programme for the project, and detailed figures for the

**Table 4.9** Budget.

| Element | Labour (£) | Plant (£) | Material (£) | Site overheads (£) | Contribution (£) | Total (£) |
|---|---|---|---|---|---|---|
| Strip site | 1 650 | 8 250 | – | 1 200 | 2 160 | 13 260 |
| Excavate foundations and slab | 3 300 | 16 500 | – | 2 500 | 4 500 | 26 800 |
| Foundations | 14 400 | 3 200 | 19 800 | 4 700 | 8 460 | 50 560 |
| Ground floor slab | 7 400 | 3 300 | 9 900 | 2 600 | 4 680 | 27 880 |
| Ground floor columns, stairs and first floor slab | 15 400 | 5 700 | 14 900 | 4 500 | 8 100 | 48 600 |
| First floor columns and roof slabs | 16 900 | 5 900 | 15 800 | 4 800 | 8 640 | 52 040 |
| Roof cladding | 2 400 | 510 | 5 100 | 1 000 | 1 800 | 10 810 |
| Brick exterior walls | 19 000 | 9 400 | 25 000 | 6 700 | 1 200 | 61 300 |
| | 80 450 | 52 760 | 90 500 | 28 000 | 39 540 | 291 250 |

**Table 4.10** Progress at month 4.

| Element | Percentage completed (%) | Labour (£) | Plant (£) | Material (£) | Site overheads (£) | Contribution (£) | Total (£) |
|---|---|---|---|---|---|---|---|
| Strip site | 100 | 1 800 | 8 300 | – | 1 200 | 1 960 | 13 260 |
| Excavate for foundations and slab | 100 | 2 700 | 17 000 | – | 2 500 | 4 600 | 26 800 |
| Foundations | 100 | 14 500 | 2 900 | 19 600 | 4 700 | 8 860 | 50 560 |
| Ground floor slab | 80 | 6 620 | 2 440 | 7 800 | 2 080 | 3 364 | 22 304 |
| Ground floor columns, stairs and first floor slab | 60 | 9 940 | 3 120 | 9 000 | 2 700 | 4 400 | 29 160 |
| First floor columns and roof slab | 40 | 6 560 | 2 260 | 6 420 | 1 920 | 3 656 | 20 816 |
| Roof cladding | 25 | 550 | 148 | 1 275 | 250 | 479.5 | 2 702.5 |
| Brick exterior walls | 10 | 2 000 | 840 | 2 600 | 670 | 20 | 6 130 |
| | | 44 670 | 37 008 | 46 695 | 16 020 | 27 339.5 | 171 732.5 |

income and expenditure on the project. It is obvious therefore that this type of calculation would be carried out after the contract had been awarded to the contractor, this being the earliest time that such detailed information was available. It would, however, be very helpful if the contractor could produce an approximate forecast of the cash flow requirements for a potential contract without having to go to such elaborate lengths. If a forecast of the cash flow can be produced for a

**Table 4.11** Progress during month 5.

| Element | Percentage completed during the month (%) | Labour (£) | Plant (£) | Material (£) | Site overheads (£) | Contribution (£) | Total (£) |
|---|---|---|---|---|---|---|---|
| Strip site | – | – | – | – | – | – | – |
| Excavate for foundations and slab | – | – | – | – | – | – | – |
| Foundations | – | – | – | – | – | – | – |
| Ground floor slab | 20 | 1680 | 760 | 1880 | 520 | 736 | 5 576 |
| Ground floor columns, stairs and first floor slab | 15 | 2110 | 1000 | 2135 | 675 | 1370 | 7 290 |
| First floor columns and roof slab | 10 | 1890 | 540 | 1600 | 480 | 694 | 5 204 |
| Roof cladding | 10 | 270 | 81 | 500 | 100 | 130 | 1 081 |
| Brick exterior walls | 15 | 2650 | 1210 | 3750 | 1005 | 580 | 9 195 |
| | | 8600 | 3591 | 9865 | 2780 | 3510 | 28 346 |

**Table 4.12** Analysis of contract at end of month 5.

| Element | Total cost of work | Cumulative to end of month 4 | | | | Month 5 | | | | Cumulative to end of month 5 | | | |
|---|---|---|---|---|---|---|---|---|---|---|---|---|---|
| | | Target cost of work done | Actual cost of work done | Variance | | Target cost of work done | Actual cost of work done | Variance | | Target cost of work done | Actual cost of work done | Variance | |
| | (£) | (£) | (£) | (£) | (%) | (£) | (£) | (£) | (%) | (£) | (£) | (£) | (%) |
| *Costs* | | | | | | | | | | | | | |
| Labour | | | | | | | | | | | | | |
| Strip site | 1 650 | 1 650 | 1 800 | (150) | (9.09) | – | – | – | – | 1 650 | 1 800 | (150) | (9.09) |
| Excavate for foundations and slab | 3 300 | 3 300 | 2 700 | 600 | 18.18 | – | – | – | – | 3 300 | 2 700 | 600 | 18.18 |
| Foundations | 14 400 | 14 400 | 14 500 | (100) | (0.69) | – | – | – | – | 14 400 | 14 500 | (100) | (0.69) |
| Ground floor slab | 7 400 | 5 920 | 6 620 | (700) | (11.82) | 1 480 | 1 680 | (200) | (13.51) | 7 400 | 8 300 | (900) | (12.16) |
| Ground floor columns, stairs and first floor slab | 15 400 | 9 240 | 9 940 | (700) | (7.57) | 2 310 | 2 110 | 200 | 8.66 | 11 550 | 12 050 | (500) | (4.33) |
| First floor columns and roof slab | 16 900 | 6 760 | 6 560 | 200 | 2.96 | 1 690 | 1 890 | (200) | (11.83) | 8 450 | 8 450 | 0 | 0 |
| Roof cladding | 2 400 | 600 | 550 | 50 | 8.33 | 240 | 270 | (30) | (12.50) | 840 | 820 | 20 | 2.38 |
| Brick exterior walls | 19 000 | 1 900 | 2 000 | (100) | (5.26) | 2 850 | 2 650 | 200 | 7.02 | 4 750 | 4 650 | 100 | 2.11 |
| Plant | | | | | | | | | | | | | |
| Strip site | 8 250 | 8 250 | 8 300 | (50) | (0.61) | – | – | – | – | 8 250 | 8 300 | (50) | (0.61) |
| Excavate for foundations and slab | 16 500 | 16 500 | 17 000 | (500) | (3.03) | – | – | – | – | 16 500 | 17 000 | (500) | (3.03) |
| Foundations | 3 200 | 3 200 | 2 900 | 300 | 9.37 | – | – | – | – | 3 200 | 2 900 | 300 | 9.37 |
| Ground floor slab | 3 300 | 2 640 | 2 440 | 200 | 7.57 | 660 | 760 | (100) | (15.15) | 3 300 | 3 200 | 100 | 3.03 |
| Ground floor columns, stairs and first floor slab | 5 700 | 3 420 | 3 120 | 300 | 8.77 | 855 | 1 000 | (145) | (16.96) | 4 275 | 4 120 | 155 | 3.63 |
| First floor columns and roof slab | 5 900 | 2 360 | 2 260 | 100 | 4.23 | 590 | 540 | 50 | 8.47 | 2 950 | 2 800 | 150 | 5.08 |
| Roof cladding | 510 | 127.50 | 148 | (20.5) | (16.08) | 51 | 81 | (30) | (58.82) | 178.5 | 229 | (50.5) | (28.29) |
| Brick exterior walls | 9 400 | 940 | 840 | 100 | 10.63 | 1 410 | 1 210 | 200 | 14.18 | 2 350 | 2 050 | 300 | 12.76 |

| | | | | | | | | | | | | | |
|---|---|---|---|---|---|---|---|---|---|---|---|---|---|
| Materials | | | | | | | | | | | | | |
| Foundations | 19 800 | 19 800 | 19 600 | 200 | 1.01 | – | – | – | – | 19 800 | 19 600 | 200 | 1.01 |
| Ground floor slab | 9 900 | 7 920 | 7 800 | 120 | 1.52 | 1 980 | 1 880 | 100 | 5.05 | 9 900 | 9 680 | 220 | 2.22 |
| Ground floor columns, stairs and first floor slab | 14 900 | 8 940 | 9 000 | (60) | (0.67) | 2 235 | 2 135 | 100 | 4.47 | 11 175 | 11 135 | 40 | 0.36 |
| First floor columns and roof slab | 15 800 | 6 320 | 6 420 | (100) | (1.58) | 1 580 | 1 600 | (20) | (1.26) | 7 900 | 8 020 | (120) | (1.52) |
| Roof cladding | 5 100 | 1 275 | 1 275 | 0 | 0 | 510 | 500 | 10 | 1.96 | 1 785 | 1 775 | 10 | 0.56 |
| Brick exterior walls | 25 000 | 2 500 | 2 600 | (100) | (4.00) | 3 750 | 3 750 | 0 | 0 | 6 250 | 6 350 | (100) | (1.60) |
| Site overheads | | | | | | | | | | | | | |
| Strip site | 1 200 | 1 200 | 1 200 | 0 | 0 | – | – | – | – | 1 200 | 1 200 | 0 | 0 |
| Excavate for foundations and slab | 2 500 | 2 500 | 2 500 | 0 | 0 | – | – | – | – | 2 500 | 2 500 | 0 | 0 |
| Foundations | 4 700 | 4 700 | 4 700 | 0 | 0 | – | – | – | – | 4 700 | 4 700 | 0 | 0 |
| Ground floor slab | 2 600 | 2 080 | 2 080 | 0 | 0 | 520 | 520 | 0 | 0 | 2 600 | 2 600 | 0 | 0 |
| Ground floor columns, stairs and first floor slab | 4 500 | 2 700 | 2 700 | 0 | 0 | 675 | 675 | 0 | 0 | 3 375 | 3 375 | 0 | 0 |
| First floor columns and roof slab | 4 800 | 1 920 | 1 920 | 0 | 0 | 480 | 480 | 0 | 0 | 2 400 | 2 400 | 0 | 0 |
| Roof cladding | 1 000 | 250 | 250 | 0 | 0 | 100 | 100 | 0 | 0 | 350 | 350 | 0 | 0 |
| Brick exterior walls | 6 700 | 670 | 670 | 0 | 0 | 1 005 | 1 005 | 0 | 0 | 1 675 | 1 675 | 0 | 0 |
| Totals | | 143 982.5 | 144 393 | (410.5) | (0.29) | 24 971 | 24 836 | 13.5 | 0.54 | 168 953.5 | 169 229 | (275.5) | (0.16) |
| *Contribution* | | | | | | | | | | | | | |
| Strip site | 2 160 | 2 160 | 1 960 | (200) | (9.26) | – | – | – | – | 2 160 | 1 960 | (200) | (9.26) |
| Excavate for foundations and slab | 4 500 | 4 500 | 4 600 | 100 | 2.22 | – | – | – | – | 4 500 | 4 600 | 100 | 2.22 |
| Foundations | 8 460 | 8 460 | 8 860 | 400 | 4.73 | – | – | – | – | 8 460 | 8 860 | 400 | 4.73 |
| Ground floor slab | 4 680 | 3 744 | 3 364 | (380) | (10.15) | 936 | 736 | (200) | (21.36) | 4 680 | 4 100 | (580) | (12.39) |
| Ground floor columns, stairs and first floor slab | 8 100 | 4 860 | 4 400 | (460) | (9.47) | 1 215 | 1 370 | 155 | 12.75 | 6 075 | 5 770 | (305) | (5.02) |
| First floor columns and roof slab | 8 640 | 3 456 | 3 656 | 200 | 5.79 | 864 | 694 | (170) | (19.67) | 4 320 | 4 350 | 30 | 0.69 |
| Roof cladding | 1 800 | 450 | 479.5 | 29.5 | 6.55 | 180 | 130 | (50) | (27.77) | 630 | 609.5 | (20.5) | (3.25) |
| Brick exterior walls | 1 200 | 120 | 20 | (100) | (83.33) | 180 | 580 | 400 | 222.00 | 300 | 600 | 300 | 100.00 |
| Totals | 39 540 | 27 750 | 27 339.5 | (410.5) | (1.48) | 3 375 | 3 510 | 135 | 4.00 | 31 125 | 30 849.5 | (275.5) | (0.89) |

project, it will enable the contractor to decide whether he has sufficient financial resources to undertake the contract.

One method of forecasting cash flow for the project is based on establishing an ideal reference curve derived from cash flow curves for a reasonable-sized sample of similar projects carried out in the past, and using this reference curve to predict the likely cash flow for future projects. A typical ideal reference curve is shown in Fig. 4.11, in which it will be seen that the horizontal axis is divided into one hundred parts (percentile points) and represents any project duration. The vertical axis indicates the cash flow ratio: that is, a proportion of the contract sum. A simple example will illustrate how this reference curve can be used. Assume we have a contract, value £100 000, duration 24 months, and that we wish to know what the cash flow will be at the 12 month point: that is, after 50% of the project duration has elapsed. Referring to the reference curve, we find the 50 percentile point on the horizontal axis, drop a perpendicular down to the curve, and then read back onto the vertical cash flow ratio axis, arriving at a reading of $-0.24$. The contract sum is then multiplied by this cash flow ratio to give the absolute cash flow for this point in the contract duration (£100 000 × $-0.24 = -$£24 000).

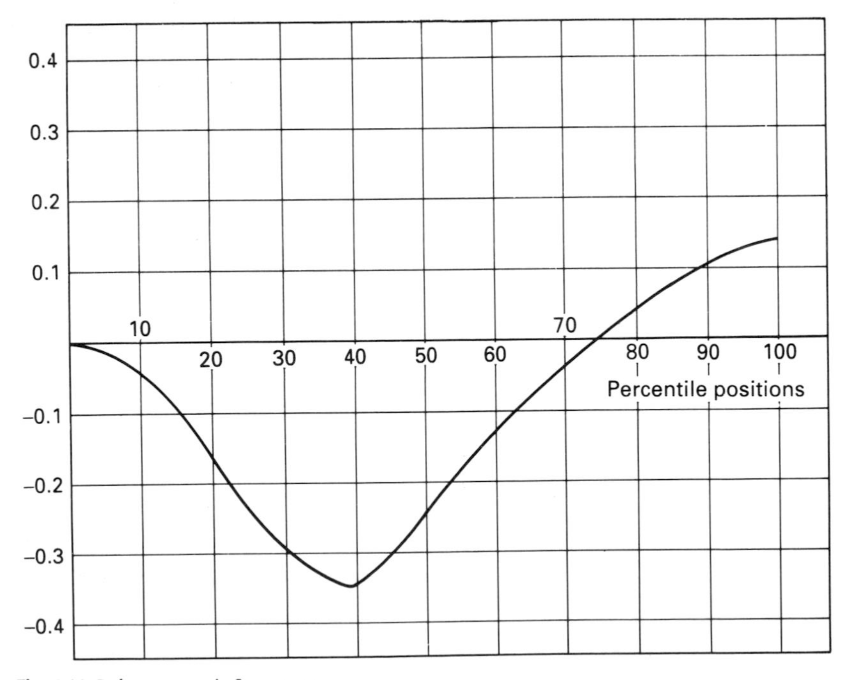


**Fig. 4.11** Reference cash flow curve.

**Example 4.6: Cash flow forecasting**

To take a more elaborate example, assume that the contractor has a predicted cash flow for all his current projects as shown in Table 4.5 (established from detailed analysis of the programmes and financial information for all his current projects). He now requires to know whether he can undertake two new projects, E and F. The maximum amount of working capital available is £170 000 and the details of projects E and F are as follows:

Project E is scheduled to start on month 7, the contract sum will be approximately £270 000 and the project duration will be 18 months. Project F is scheduled to start on month 9, the contract sum will be approximately £330 000 and the project duration will be 21 months.

The first step is to calculate the percentile position for each month of each project. This is shown in Table 4.13.

Having found the percentile points for each month on both projects, the next step is to find the cash flow for each project. This is shown in Tables 4.14 and 4.15.

**Table 4.13** Percentile points for each month.

| No. of months | Percentile points corresponding to end of each month for projects of duration: | |
|---|---|---|
| | 18 months | 21 months |
| 1 | 6 | 5 |
| 2 | 11 | 10 |
| 3 | 17 | 14 |
| 4 | 22 | 19 |
| 5 | 28 | 24 |
| 6 | 33 | 29 |
| 7 | 39 | 33 |
| 8 | 44 | 38 |
| 9 | 50 | 43 |
| 10 | 56 | 48 |
| 11 | 61 | 52 |
| 12 | 67 | 57 |
| 13 | 72 | 62 |
| 14 | 78 | 67 |
| 15 | 83 | 71 |
| 16 | 89 | 76 |
| 17 | 94 | 81 |
| 18 | 100 | 86 |
| 19 | | 90 |
| 20 | | 95 |
| 21 | | 100 |

**Table 4.14** Cash flow for Project E (contract sum £270 000; duration 18 months).

| Month relative to Table 4.5 | Number of months from start of project | Percentile position with respect to last column | Cash flow ratio from Fig. 4.11 | Absolute cash flow (£) |
|---|---|---|---|---|
| 7 | 1 | 6 | −0.018 | −0.018 × 270 000 = −4 860 |
| 8 | 2 | 11 | −0.05 | −0.05 × 270 000 = −13 500 |
| 9 | 3 | 17 | −0.115 | −0.115 × 270 000 = −31 050 |
| 10 | 4 | 22 | −0.195 | −0.195 × 270 000 = −52 650 |
| 11 | 5 | 28 | −0.273 | −0.273 × 270 000 = −73 710 |
| 12 | 6 | 33 | −0.32 | −0.32 × 270 000 = −86 400 |
| 13 | 7 | 39 | −0.347 | −0.347 × 270 000 = −93 690 |
| 14 | 8 | 44 | −0.31 | −0.31 × 270 000 = −83 700 |
| 15 | 9 | 50 | −0.237 | −0.237 × 270 000 = −63 990 |
| 16 | 10 | 56 | −0.17 | −0.17 × 270 000 = −45 900 |
| 17 | 11 | 61 | −0.12 | −0.12 × 270 000 = −32 400 |
| 18 | 12 | 67 | −0.065 | −0.065 × 270 000 = −17 550 |
| 19 | 13 | 72 | −0.025 | −0.025 × 270 000 = −6 750 |
| 20 | 14 | 78 | +0.025 | +0.025 × 270 000 = +6 750 |
| 21 | 15 | 83 | +0.065 | +0.065 × 270 000 = +17 550 |
| 22 | 16 | 89 | +0.105 | +0.105 × 270 000 = +28 350 |
| 23 | 17 | 94 | +0.125 | +0.125 × 270 000 = +33 750 |
| 24 | 18 | 100 | +0.14 | +0.14 × 270 000 = +37 800 |

**Table 4.15** Cash flow for project F (contract sum £330 000; duration 21 months).

| Month relative to Table 4.5 | Number of months from start of project | Percentile position with respect to last column | Cash flow ratio from Fig. 4.11 | Absolute cash flow (£) |
|---|---|---|---|---|
| 9 | 1 | 5 | −0.015 | −0.015 × 330 000 = −4 950 |
| 10 | 2 | 10 | −0.043 | −0.043 × 330 000 = −14 190 |
| 11 | 3 | 14 | −0.08 | −0.08 × 330 000 = −26 400 |
| 12 | 4 | 19 | −0.15 | −0.15 × 330 000 = −49 500 |
| 13 | 5 | 24 | −0.225 | −0.225 × 330 000 = −74 250 |
| 14 | 6 | 29 | −0.285 | −0.285 × 330 000 = −94 050 |
| 15 | 7 | 33 | −0.32 | −0.32 × 330 000 = −105 600 |
| 16 | 8 | 38 | −0.347 | −0.347 × 330 000 = −114 510 |
| 17 | 9 | 43 | −0.32 | −0.32 × 330 000 = −105 600 |
| 18 | 10 | 48 | −0.265 | −0.265 × 330 000 = −87 450 |
| 19 | 11 | 52 | −0.215 | −0.215 × 330 000 = −70 950 |
| 20 | 12 | 57 | −0.16 | −0.16 × 330 000 = −52 800 |
| 21 | 13 | 62 | −0.11 | −0.11 × 330 000 = −36 300 |
| 22 | 14 | 67 | −0.065 | −0.065 × 330 000 = −21 450 |
| 23 | 15 | 71 | −0.032 | −0.032 × 330 000 = −10 560 |
| 24 | 16 | 76 | +0.01 | +0.01 × 330 000 = +3 300 |
| 25 | 17 | 81 | +0.05 | +0.05 × 330 000 = +16 500 |
| 26 | 18 | 86 | +0.082 | +0.082 × 330 000 = +27 060 |
| 27 | 19 | 90 | +0.108 | +0.108 × 330 000 = +35 640 |
| 28 | 20 | 95 | +0.128 | +0.128 × 330 000 = +42 240 |
| 29 | 21 | 100 | +0.14 | +0.14 × 330 000 = +46 200 |

By inspecting it Table 4.16 can be seen that the maximum negative cash flow for all six projects is £169 650 for month 6 and does not therefore exceed the maximum amount of working capital available. It will be noticed that this occurs before projects E and F start, and that once these projects are both running the maximum working capital requirement does not exceed £155 440, which is on month 13. It will therefore be possible to undertake projects E and F with the present financial resources.

The cash flow for all six projects is therefore as shown in Table 4.16.

**Table 4.16** Cash flow for all six projects.

| Month | Projects A, B, C and D | Project E | Project F | Projects A, B, C, D, E and F |
|---|---|---|---|---|
| 1 | −6 500 | | | −6 500 |
| 2 | −24 000 | | | −24 000 |
| 3 | −64 000 | | | −64 000 |
| 4 | −115 400 | | | −115 400 |
| 5 | −157 600 | | | −157 600 |
| 6 | −169 650 | | | 169 650 |
| 7 | −120 250 | −4 860 | | −125 110 |
| 8 | −87 050 | −13 500 | | −100 550 |
| 9 | −55 400 | −31 050 | −4 950 | −91 400 |
| 10 | −42 800 | −52 650 | −14 190 | −109 640 |
| 11 | −42 500 | −73 710 | −26 400 | −142 610 |
| 12 | −13 250 | −86 400 | −49 500 | −149 150 |
| 13 | +12 500 | −93 690 | −74 250 | −155 440 |
| 14 | +40 350 | −83 700 | −94 050 | −137 400 |
| 15 | +64 550 | −63 990 | −105 600 | −105 040 |
| 16 | +83 050 | −45 900 | −114 510 | −77 360 |
| 17 | +98 850 | −32 400 | −105 600 | −39 150 |
| 18 | +106 050 | −17 550 | −87 450 | +1 050 |
| 19 | +109 550 | −6 750 | −70 950 | +31 850 |
| 20 | +110 750 | +6 750 | −52 800 | +64 700 |
| 21 | +110 750 | +17 550 | −36 300 | +92 000 |
| 22 | +110 750 | +28 350 | −21 450 | +117 650 |
| 23 | +110 750 | +33 750 | −10 560 | +133 940 |
| 24 | +110 750 | +37 800 | +3 300 | +151 850 |
| 25 | +110 750 | +37 800 | +16 500 | +165 050 |
| 26 | +110 750 | +37 800 | +27 060 | +175 610 |
| 27 | +110 750 | +37 800 | +35 640 | +184 190 |
| 28 | +110 750 | +37 800 | +42 240 | +190 790 |
| 29 | +110 750 | +37 800 | +46 200 | +194 750 |
| 30 | +110 750 | +37 800 | +46 200 | +194 750 |

## 4.2 Cost control

### 4.2.1 Introduction

Costing has been carried out in the construction industry for a number of years, but the type of costing used by some firms gives the total cost of the project only when it has been completed. This is then compared with the amount received, but if the project has lost money there is little that can be done about it. It is obviously a distinct advantage if a contractor can ascertain which section of a project is in deficit, and can know when it starts losing money. Cost control techniques have been developed for this purpose.

### 4.2.2 Objectives

The main objectives of cost control are:

- To see that the company's policy with regard to production is carried out, which in turn will ensure that planned profit margins are maintained.
- To arrive at the cost of each stage, operation (in the case of repetitive construction), or unit, and to carry out a continuous comparison with the target to ascertain the gain or loss on each. This information must be available early enough for corrective action to be taken wherever possible.
- To provide information on cost for use in future estimating.

On small projects, many operations are of insufficient duration to allow assessment of cost in time for corrective action to be taken. However, control still needs to be exercised, as the same operations are likely to occur on future projects.

### 4.2.3 Cost control systems

Before deciding upon the degree of sophistication of a system, consideration must be given to the cost of operating it and the benefits it provides. Clearly, a balance should be reached between the two.

Systems can vary from those that control the work on a section or stage basis to those that control it on a unit basis, and in some systems only certain sections of the project are selected for control. It is usual to limit site cost control to the control of labour and plant, as these are the areas where there is likely to be the greatest amount of variability.

### 4.2.4 *Timing of presentation of cost information*

When cost control techniques are used, the system must be administered in such a manner that the site manager is made aware of any deviations from planned cost as soon as they are recognised. Ideally the costs should be checked daily, but on most building operations this would be extremely expensive, and the cost of the work done is in consequence usually checked weekly.

### 4.2.5 *Uses of cost information*

Cost control information is invaluable for future estimating. Feedback must be accompanied by a full description of the conditions under which the work was carried out, as conditions can vary a great deal from one contract to another.

Another very important use of cost control information is in the pricing of variations. A separate record is kept of costs and the contractor will then have factual information to assist him in settling a rate for the work done.

The weekly measure carried out for the cost control system can also be used for the monthly valuation, and thus eliminates the need for the work to be re-measured at the end of each month, although materials on site would have to be ascertained separately for the purpose of the monthly valuation.

### 4.2.6 *Unit cost control*

There are many systems of cost control. In this chapter only one system will be considered. This system is in use at present by a small company. It is simple to operate, but is suitable for smaller companies as it gives reasonably accurate data at reasonable cost. More sophisticated systems would be more costly to operate.

Any costs in excess or below the target by more than a predetermined margin require further investigation and should be brought to the attention of management, thus applying the principle of management by exception. The information is collected by means of time sheets and weekly summary sheets, together with actual measurements of work done where necessary.

Preparation work - target for operative - targ prep - operatives Page 1

| Ref. | Description | Location |
|---|---|---|
| 10 | Strip down ceiling, denail joists and remove to skip | |
| 105 | Hack off wall plaster and remove to skip | |
| 305 | Remove hearthstone and deposit in skip | Bedrooms |
| 650 | Take out door linings and remove to skip | Internal |
| 905 | Take down corner flue and remove to skip | Bathroom |
| 910 | Take down stack and remove to skip | Bathroom |
| 1405 | Take down handrail and make good | Stair |
| 1505 | Take out fireplace and hearthstone and remove to skip | Living |
| 1805 | Hack up existing concrete floor (FD) | Ki/di/ut |
| 1810 | CA concrete floor | Ki/di/ut |
| 1820 | DPM turned 150 mm at walls | Ki/di/ut |
| 1825 | Secure DPM to wall with treated batten | Ki/di/ut |
| 1830 | Concrete slab floor | Ki/du/ut |
| 1835 | Insulation under solid floors | Ki/di/ut |
| 1840 | Insulation around edges of new solid floors | Ki/di/ut |
| 1905 | Take down chimney breast and remove to skip | Kit/din |
| 2600 | Remove and CA all rubbish internally and externally (RUB) | |
| 2605 | Take up floor coverings and remove to skip | |
| | Target 180 | |

**Fig. 4.12** Typical example of a target.

### *4.2.7 Targets*

Wherever possible, targets are set for operations: for example, preparation work. A full description of the work in the target is provided for the operative. A typical example of a target is shown in Fig. 4.12. Further examples of targets can be seen in Chapter 7. These targets provide the basis against which control will be exercised.

### *4.2.8 Time sheets (Fig. 4.13)*

Time sheets form the basis of the cost control system, and must therefore be as accurate as possible.

Because most of the projects undertaken by the company are between £5000 and £500 000, many of which are in the lower range, operatives fill in their own time sheets, which give details of the work carried out during the week. These are checked by the contracts manager. On the larger projects, the sheets are checked by a foreman on site.

### *4.2.9 Weekly summary sheet (Fig. 4.14)*

This sheet is used to collect all the information on operational hours from the time sheets. It shows the total hours spent on each operation in that week. This information is required for calculating the amount of bonus payable, in addition to providing information for the cost sheet.

TIME SHEET

NAME: M. Charlton

WEEK ENDING: 14. 10. 95

| JOB | DESCRIPTION OF WORK | Start and finish time | TOTAL OPERATIONAL TIME | | | | | | |
|---|---|---|---|---|---|---|---|---|---|
| | | | M | T | W | Th | F | Sa | Su |
| HARTLEY | target 1 - Demolish garage | 8.00<br>4.30 | 8 | | | | | | |
| HARTLEY | Target 1 - Strip footings | 8.00<br>4.30 | | 8 | | | | | |
| HARTLEY | Target 1 Complete footings<br>target 1 complete | 8.00<br>4.30 | | | 5 | | | | |
| HOLDSWORTH PACKAGING | travel to site & assist in setting out | | | | 3 | | | | |
| RADFORD BAKERY | Target 3 Break out concrete | 8.00<br>5.00 | | | | $8\frac{1}{2}$ | | | |
| BRADFORD BAKERY | target 3 - Demolish brickwork | 8.00<br>5.00 | | | | | $8\frac{1}{2}$ | | |
| | Remarks:-<br>Weather fine all week | | | | | | | | |

**Fig. 4.13** A time sheet.

### *4.2.10 Measurement of work done*

When operation targets are set, a site check that all the items in the target have been completed is all that is necessary. For operations such as brickwork the target may be in terms of hours per square metre, and

Week ending 14.10.95

| Project | Name | Trade | Operation | M | T | W | Th | F | Sa | Su | Travel time | Total | Remarks |
|---|---|---|---|---|---|---|---|---|---|---|---|---|---|
| Hartley | P. Rowbotham | Gen. Op. | Target 1 | 8 | 8 | 5 | | | | | | 21 | |
| | M. Charlton | Gen. Op. | Target 1 | 8 | 8 | 5 | | | | | | 21 | |
| Holdsworth | P. Rowbotham | Gen. Op. | Assist setting out | | | 3 | | | | | | 3 | |
| Packaging | M. Charlton | Gen. Op. | Assist setting out | | | 3 | | | | | | 3 | |
| Bradford | P. Rowbotham | Gen. Op. | Target 3 | | | | 8.5 | 8.5 | | | | 17 | |
| Bakery | M. Charlton | Gen. Op. | Target 3 | | | | 8.5 | 8.5 | | | | 17 | |

**Fig. 4.14** Weekly summary sheet.

in this case a measure will be necessary. A record of the weekly measure should be kept in the form of record drawings, which should be marked up weekly showing the amount of work done. This will ensure that gangs are not paid twice for the same piece of work, and will constitute a detailed record of progress.

### *4.2.11 Cost control sheet (Fig. 4.15)*

In this system the sheet can be filled in weekly or as each operation is completed. Information is collected from the weekly summary sheets and the records of measure. The cost of each operation can be calculated, and this facilitates control by comparison with the set targets. A description of the sheet is as follows:

- *Column A*: a record of the week in which the work was done.
- *Columns B and C*: the number of the operation or target, and a brief description of the operation.
- *Column D*: a measure of the work done in the week (if applicable).
- *Column E*: target rate, e.g. 4 hours per square metre.
- *Column F*: target hours. This is obtained from the set target (Fig. 4.12), or from the measure and target rate.
- *Column G*: target value

$$(\text{target hours}) \times (\text{all-in rate})$$

- *Column H*: time taken – from weekly summary.

| a | b | c | d | e | f | g | h | i | j | k | l | m | n | o |
|---|---|---|---|---|---|---|---|---|---|---|---|---|---|---|
| Week ending | Target no. | Operation | Measure | Target rate | Target hours | Target value | Time taken | Cost | Time saved | Bonus | Adjustment | Total cost | Saving/ loss | Remarks |
| 14/10/95 | 1 | Prep. work, excavation, conc. | | | 60 | 360.00 | 42 | 252.00 | 18 | 47.25 | | 299.25 | 60.75 | Weather fine |

**Fig 4.15** Cost control sheet.

- *Column I*: cost of time taken

  (time taken) × (all-in rate)

- *Column J*: time saved

  (target) – (time taken)

- *Column K*: bonus earned

  time saved × 50% of hourly rate paid

- *Column L*: travelling time, unclaimable daywork, etc.
- *Column M*: total cost

  cost + bonus + adjustment

- *Column N*: saving/loss

  (target value) – (total cost)

- *Column O*: remarks. Any comments relevant to the operation.

On larger projects, where operations such as carpentry second fix are broken down into separate targets, the results of earlier targets will clearly show whether remedial action needs to be taken, thus assisting the control process. On smaller projects the feedback clearly assists future estimating, and indicates whether the current rates are reasonable.

Chapter 5

# Bidding Strategy

## 5.1 Introduction

The majority of contracts in the UK are obtained via competitive tendering. If a contractor is to be successful he must always be aware of what competitors are doing. If he knew the tender figures of his competitors he would be in a very advantageous position indeed. As he does not know this information he must find ways of assessing what their tenders are likely to be to enable him to obtain contracts as profitably as possible: that is, to enable him to maximise his profit. He must therefore use bidding strategy based on scientific methods. In order to apply the scientific method of statistics to tendering, the contractor needs to know the performance of his competitors in previous competitions. It is the practice of some architects and local authorities to inform tendering contractors of the tenders submitted by all competitors, either by opening tenders while contractors are present or by informing them after opening the tenders. If the contractor collects this information over a period of time he can learn quite a lot about his competitors.

## 5.2 Factors affecting the approach to the problem

When a contractor is invited to tender for work, he is not normally told who his competitors are. However, it is often possible to find out who is competing for a particular project. Even if the contractor does not known precisely who his competitors are, in certain circumstances he at least knows approximately how many competitors he has, particularly if the Code of Procedure for Selective Tendering is being used. In the worst case he does not know who his competitors are nor how many there are. Each of these situations necessitates a slightly different approach.

## 5.3 When names of competing firms can be ascertained

When the contractor can ascertain who the competitors are, he can assess the optimum tender figure to give the maximum expected profit. This is the average profit he would make if he used the same profit margin on a large number of contracts where the estimate of cost was identical and the probability of being awarded the contracts remained the same.

**Example 5.1**

For example, assume the estimate of cost for a contract is £100 000 and the contractor submits a tender figure of £110 000. If he is awarded the contract he makes a profit of £10 000; if he is not awarded the contract he get nothing. If the probability of being awarded the contract is 0.4, the expected profit will be £100 000 × 0.4 = £4000. Assume that information has been collected over a period of time with respect to a particular competitor, and that these figures have been compared with the contractor's estimated cost before profit is added, as shown in Table 5.1.

**Table 5.1** Ratio of competitor's tender (*T*) to contractor's estimate (*E*).

| Contractor's estimate | Competitor's tender | (*T*/*E*) |
|---|---|---|
| £120 000 | £126 000 | 1.05 |
| £150 000 | £159 000 | 1.06 |
| £160 000 | £166 400 | 1.04 |
| £140 000 | £148 400 | 1.06 |
| £130 000 | £139 100 | 1.07 |
| £110 000 | £111 100 | 1.01 |
| £160 000 | £177 600 | 1.11 |
| £140 000 | £151 200 | 1.08 |
| £100 000 | £107 000 | 1.07 |
| etc. | etc. | etc. |

***Frequency of ratios***

The information from Table 5.1 can be tabulated as shown in Table 5.2, which gives the frequency of occurrence of the various ratios.

**Table 5.2** Frequency of ratios.

| T/E | Frequency |
|---|---|
| 0.98 | 1 |
| 0.99 | 2 |
| 1.00 | 2 |
| 1.01 | 3 |
| 1.02 | 4 |
| 1.03 | 6 |
| 1.04 | 9 |
| 1.05 | 13 |
| 1.06 | 18 |
| 1.07 | 22 |
| 1.08 | 23 |
| 1.09 | 19 |
| 1.10 | 14 |
| 1.11 | 10 |
| 1.12 | 7 |
| 1.13 | 4 |
| 1.14 | 2 |
| 1.15 | 1 |
| Total | 160 |

### *Probability of ratios*

From this table the probability of any particular ratio occurring can be found, as shown in Table 5.3.

The contractor has now got a considerable amount of information about this particular competitor. He knows that 0.144 or 14.4% of his competitor's tenders were 8% higher than his own estimate of cost, 0.6% were 2% lower than his estimate of cost, and 0.6% were 15% higher than his estimate.

### *Cumulative probability distribution*

From Table 5.3 the cumulative probability distribution for this competitor can be found. The contractor will then know the probability of any tender being lower than this competitor. For example, a tender of 0.98 or 98% of the estimate has a probability of being lower than this competitor of $1 - 0.006 = 0.994$. As the next tender/estimate ratio is 0.99 it is assumed that the probability of 0.994 applies for tenders from 0.98 to 0.989, and a tender less than 0.98 of the estimate has a probability of 1.00 of being lower than this competitor.

**Table 5.3** Probability of ratios.

| *T/E* | Probability of occurrence | $\frac{\text{frequency}}{\text{total}}$ |
|---|---|---|
| 0.98 | 1/160 | 0.006 |
| 0.99 | 2/160 | 0.013 |
| 1.00 | 2/160 | 0.013 |
| 1.01 | 3/160 | 0.019 |
| 1.02 | 4/160 | 0.025 |
| 1.03 | 6/160 | 0.038 |
| 1.04 | 9/160 | 0.056 |
| 1.05 | 13/160 | 0.081 |
| 1.06 | 18/160 | 0.112 |
| 1.07 | 22/160 | 0.137 |
| 1.08 | 23/160 | 0.144 |
| 1.09 | 19/160 | 0.119 |
| 1.10 | 14/160 | 0.087 |
| 1.11 | 10/160 | 0.062 |
| 1.12 | 7/160 | 0.044 |
| 1.13 | 4/160 | 0.025 |
| 1.14 | 2/160 | 0.013 |
| 1.15 | 1/160 | 0.006 |
| | | 1.000 |

**Table 5.4** Cumulative probability distribution.

| *T/E* | Probability that tender will be less than competitor | |
|---|---|---|
| 0.979 | | 1.000 |
| 0.989 | 1 − 0.006 | 0.994 |
| 0.999 | 1 − 0.019 | 0.981 |
| 1.009 | 1 − 0.032 | 0.968 |
| 1.019 | 1 − 0.051 | 0.949 |
| 1.029 | 1 − 0.076 | 0.924 |
| 1.039 | 1 − 0.114 | 0.886 |
| 1.049 | 1 − 0.170 | 0.830 |
| 1.059 | 1 − 0.251 | 0.749 |
| 1.069 | 1 − 0.363 | 0.637 |
| 1.079 | 1 − 0.500 | 0.500 |
| 1.089 | 1 − 0.644 | 0.356 |
| 1.099 | 1 − 0.763 | 0.237 |
| 1.109 | 1 − 0.850 | 0.150 |
| 1.119 | 1 − 0.912 | 0.088 |
| 1.129 | 1 − 0.956 | 0.044 |
| 1.139 | 1 − 0.981 | 0.019 |
| 1.149 | 1 − 0.994 | 0.006 |
| 1.159 | 1 − 1.000 | 0.000 |

It is of course possible to carry out the analysis in a more sophisticated way than has been shown with much smaller intervals than the 0.01 or 1% intervals used in this analysis. Table 5.4 shows the cumulative probability distribution for this competitor.

*Expected profit*

Once the probability of a tender's being less than that of the competitors is known, it is simple to calculate the expected profit for each ratio. For example, if the contractor tenders using a 4.9% profit, i.e. using a ratio of 1.049, then the probability of success is 0.83. His expected profit is therefore 0.049 × estimate × 0.83 or $0.049E \times 0.83 = 0.041E$.

Table 5.5 shows the expected profit for each ratio. From this it can be seen that the maximum expected profit is $0.442E$ or 4.42% using a ratio of 1.059: that is, a profit margin of 5.9% of the estimated cost.

**Table 5.5** Expected profit.

| $T/E$ | Expected profit |
|---|---|
| 0.979 | $1.000(0.979E - E) = -0.021E$ |
| 0.989 | $0.994(0.989E - E) = -0.0109E$ |
| 0.999 | $0.981(0.999E - E) = -0.0010E$ |
| 1.009 | $0.968(1.009E - E) = 0.0087E$ |
| 1.019 | $0.949(1.019E - E) = 0.0180E$ |
| 1.029 | $0.924(1.029E - E) = 0.0268E$ |
| 1.039 | $0.886(1.039E - E) = 0.0346E$ |
| 1.049 | $0.830(1.049E - E) = 0.0407E$ |
| 1.059 | $0.749(1.059E - E) = 0.0442E$ |
| 1.069 | $0.637(1.069E - E) = 0.0440E$ |
| 1.079 | $0.500(1.079E - E) = 0.0395E$ |
| 1.089 | $0.356(1.089E - E) = 0.0317E$ |
| 1.099 | $0.237(1.099E - E) = 0.0235E$ |
| 1.109 | $0.150(1.109E - E) = 0.0164E$ |
| 1.119 | $0.088(1.119E - E) = 0.0105E$ |
| 1.129 | $0.044(1.129E - E) = 0.0057E$ |
| 1.139 | $0.019(1.139E - E) = 0.0026E$ |
| 1.149 | $0.006(1.149E - E) = 0.0009E$ |
| 1.159 | $0.000(1.159E - E) = 0.0000E$ |
| | (to 4 decimal places) |

## 5.4 Competing against more than one known competitor

If there is more than one known competitor, the method of calculating the expected profit is very similar to that used for one competitor.

The result would be a number of cumulative probability distributions. The probability of being less than all competitors would be the product of the probabilities of being less than each individually.

If two such competitors are considered, the result may be as shown in Table 5.6.

When the probability of tendering lower than all the known competitors has been calculated, the expected profit for each tender/estimate ratio can be worked out, and the estimate showing the greatest expected profit can be selected, as shown in Table 5.7.

**Table 5.6** Cumulative probability distributions for two competitors.

| *T/E* | Probability that tender will be less than: | | |
|---|---|---|---|
| | Competitor 1 | Competitor 2 | Competitors 1 and 2 |
| 0.979 | 1.000 | 1.000 | 1.000 |
| 0.989 | 0.994 | 0.990 | 0.984 |
| 0.999 | 0.901 | 0.975 | 0.956 |
| 1.009 | 0.968 | 0.961 | 0.930 |
| 1.019 | 0.949 | 0.941 | 0.893 |
| 1.029 | 0.924 | 0.918 | 0.848 |
| 1.039 | 0.886 | 0.878 | 0.778 |
| 1.049 | 0.830 | 0.824 | 0.684 |
| 1.059 | 0.749 | 0.746 | 0.559 |
| 1.069 | 0.637 | 0.636 | 0.405 |
| 1.079 | 0.500 | 0.502 | 0.251 |
| 1.089 | 0.356 | 0.359 | 0.128 |
| 1.099 | 0.237 | 0.238 | 0.056 |
| 1.109 | 0.150 | 0.155 | 0.023 |
| 1.119 | 0.088 | 0.100 | 0.009 |
| 1.129 | 0.044 | 0.046 | 0.002 |
| 1.139 | 0.019 | 0.025 | 0.000 (to 4 decimal places) |
| 1.149 | 0.006 | 0.010 | 0.000 (to 4 decimal places) |
| 1.159 | 0.000 | 0.005 | 0.000 (to 4 decimal places) |

## 5.5 When names of competing firms cannot be ascertained

When the competitors are not known but the number is known a slight adjustment to the method is necessary. When using the Code of Procedure for Selective Tendering the maximum number of competitors will be known, as this is laid down in the code. Usually this figure will be the nearest the contractor will get to the actual number of competitors.

In this situation the only difference in method is that the probability of submitting a lower tender is calculated against the average of all

**Table 5.7** Expected profit for each tender/estimate ratio.

| $T/E$ | Expected profit |
|---|---|
| 0.979 | $1.000(0.979E-E) = -0.021E$ |
| 0.989 | $0.984(0.989E-E) = -0.0108E$ |
| 0.999 | $0.956(0.999E-E) = -0.0010E$ |
| 1.009 | $0.930(1.009E-E) = 0.0084E$ |
| 1.019 | $0.893(1.019E-E) = 0.0170E$ |
| 1.029 | $0.848(1.029E-E) = 0.0246E$ |
| 1.039 | $0.778(1.039E-E) = 0.0303E$ |
| 1.049 | $0.684(1.049E-E) = 0.0335E$ |
| 1.059 | $0.559(1.059E-E) = 0.0330E$ |
| 1.069 | $0.305(1.069E-E) = 0.021E$ |
| 1.079 | $0.251(1.079E-E) = 0.0198E$ |
| 1.089 | $0.128(1.089E-E) = 0.0114E$ |
| 1.099 | $0.056(1.099E-E) = 0.0055E$ |
| 1.109 | $0.023(1.109E-E) = 0.0025E$ |
| 1.119 | $0.009(1.119E-E) = 0.0011E$ |
| 1.129 | $0.002(1.129E-E) = 0.0002E$ |
| 1.139 | $0.000(1.139E-E) = 0.000E$ |
| 1.149 | $0.000(1.149E-E) = 0.000E$ |
| 1.159 | $0.000(1.159E-E) = 0.000E$ |

previous competitors. When this is ascertained, the probability of a tender's being lower than any number of competitors can be found as before. For example, if a tender that includes a profit margin of 6.9%, i.e. a tender figure of $1.069E$, has a probability of success of 0.600 against one competitor, the probability against three such competitors would be $0.600 \times 0.600 \times 0.600$ or $0.6^3 = 0.216$.

## 5.6 Tendering against an unknown number of competitors

If statistics are to be used in this case an estimate of the number of competitors must be made. It is considered bad practice to ask unlimited numbers of contractors to tender for the work, but if this is done and the number cannot be estimated, e.g. in open tendering, then statistics can be of very limited use.

## 5.7 Conclusions

It can be seen that by using statistics a contractor can gain an advantage over competitors who do not use analytical techniques in helping to determine profit margins on contracts. It should be appreciated that the main concern in this section has been to indicate the value of statistical methods for use in tendering. In practice other factors have to be taken into account besides the ones considered here. Indeed there are many other influencing factors in determining profit margins when tendering, such as the state of the market and the economic climate at the time of tendering, but this does not render statistical analysis irrelevant. Any method that assists in giving a more realistic appraisal of the situation should be taken into account.

# Chapter 6
# Computer Applications

## 6.1 Introduction

A considerable number of companies are now using computers for a variety of purposes, including those companies with a turnover as low as £500 000.00. There are many reasons for this, including the fact that microcomputers have become cheaper and much more powerful, and the availability of suitable software has increased considerably. This software includes general-purpose software such as that used for word processing, spreadsheets and databases, and software written for specific purposes such as estimating, accounting and costing, and the various aspects of planning.

The saving in time and the availability of information generated by computers is having a considerable influence on some construction companies, and this influence is likely to carry on increasing. Those companies that do not take advantage of the advances in information technology may well cease to be competitive. This chapter introduces and illustrates a range of applications relevant to the construction manager.

## 6.2 Integrated systems and databases

### *6.2.1 Introduction*

The arrival of computers has made it possible to re-examine the way in which companies are organised. The optimum organisation would be one in which the data for the company was totally integrated. Database management systems are designed to assist in the development of such systems.

### 6.2.2 *Database packages*

Many packages that are called 'database packages' are really information retrieval packages, which simply organise and retrieve data. There are many examples of manual information retrieval systems, and these include telephone directories and dictionaries. They are collections of information organised for a specific purpose. However, addresses and telephone numbers have little value on their own. They are only useful when related to a name (i.e. a key) to make them easy to find. Information retrieval systems can organise and find data very quickly compared with manual methods of retrieval. For example, to find a person who has a particular telephone number could take hours or even days manually, whereas a computer would find that person in seconds.

In addition to the above, a true database package in computing terms should:

- merge information in such a way that it appears to the user that only one file is in use;
- eliminate redundant or unnecessary data;
- make the data independent of the application programmes.

Figure 6.1 shows a comparison of a database approach and a conventional file approach for accessing data.

### 6.2.3 *Integrated systems*

Fully integrated systems take a considerable time to develop, and as they involve a complete new look at the way processes are carried out, they often necessitate considerable changes in organisation and procedures.

A good starting point is to concentrate on a subsystem within an overall framework, and to build on this at a later stage.

Much has been written on the need to integrate production information with that required for estimating. This can be difficult when using traditional bills of quantities based on the Standard Method of Measurement, where bill items are not compatible with operational data, particularly when these data are required for short-term planning and for setting incentive targets.

A system has been developed by Ray Oxley, Steve Westgate and Mike Gallagher for use on projects using quantities based on operations. The system is called MASTER.

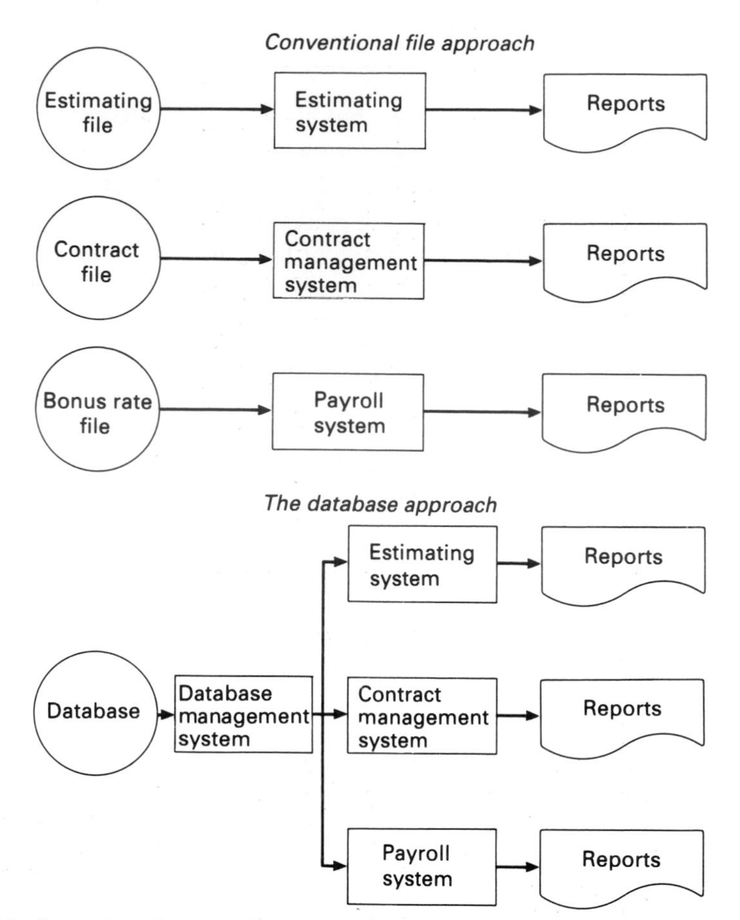


**Fig. 6.1** Comparison of systems for accessing data: (a) the conventional file approach; (b) the database approach.

## 6.3 MASTER

### *6.3.1 The total system*

The system consists of an integrated package covering estimating, planning and control. The system is menu driven, and the links between the menus are shown in Fig. 6.2. Screens are displayed whenever data entry is required.

### *6.3.2 The standard operations file*

The system includes a standard operations file or operation data bank, which is:

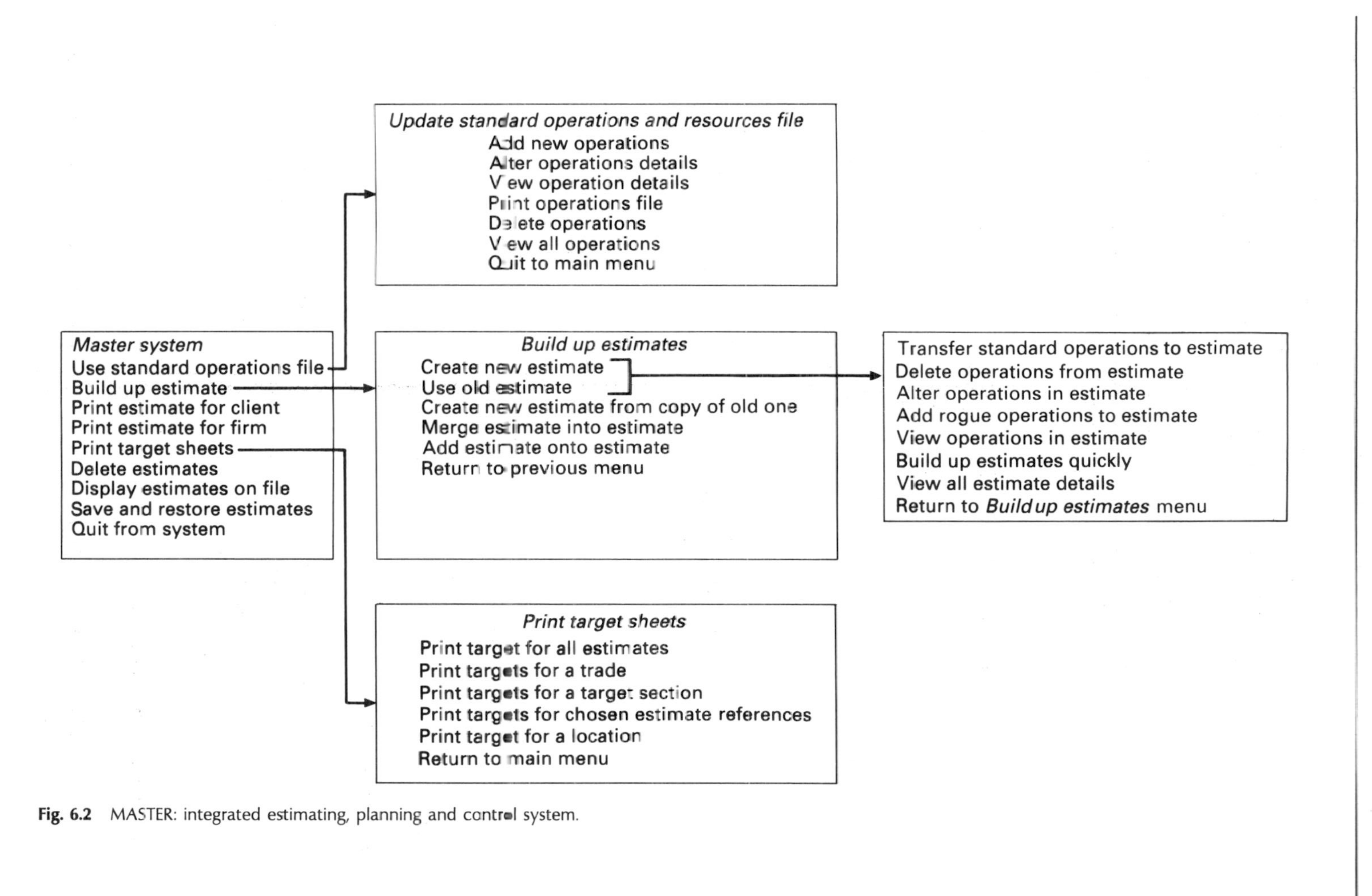


**Fig. 6.2** MASTER: integrated estimating, planning and control system.

- ❑ Flexible to enable extension or amendment of the data bank and to allow contractors to set up their own data bank, either operationally or in terms of SMM items;
- ❑ independent of financial considerations, i.e. based on outputs.

Part of the standard operations file is shown in Fig. 6.3.

Standard operations file Page 2

| Trad | Subt | No | Output | Unit | Description |
|---|---|---|---|---|---|
| Carp | 1fix | 127 | 0.25 | m | Bearers |
| Carp | 1fix | 60 | 1.00 | m | Bracketing around beams in preparation for plasterboard |
| Carp | 1fix | 45 | 0.20 | m | Door linings - assemble and fix |
| Carp | 1fix | 8 | 0.10 | m | Floor joists - take out |
| Carp | 1fix | 9 | 0.10 | m | Handrail - take off |
| Carp | 1fix | 42 | 0.35 | m | Mastic pointing |
| Carp | 1fix | 6 | 0.25 | m | Stud partition |
| Carp | 1fix | 46 | 1.00 | m | Window boards on bearers |
| Carp | 1fix | 59 | 0.80 | m | Windows constructed on site from framing |
| Carp | 1fix | 5 | 0.40 | $m^2$ | Flooring in chipboard |
| Carp | 1fix | 57 | 0.75 | $m^2$ | Floor boarding 25 mm thick - tongued and grooved |
| Carp | 1fix | 62 | 0.50 | $m^2$ | Floor boards - take up |
| Carp | 1fix | 108 | 1.50 | $m^2$ | Floor boards in isolated areas |
| Carp | 1fix | 50 | 1.00 | $m^2$ | Matchboarding in areas exceeding 2 $m^2$ |
| Carp | 1fix | 86 | 1.00 | $m^2$ | Panelling, fix grounds |
| Carp | 1fix | 49 | 0.40 | $m^2$ | Styroliner or similar boarding on battens |
| Carp | 1fix | 97 | 0.05 | $m^2$ | Vapour barrier |
| Carp | 1fix | 99 | 10.00 | no | Airing cupd of 1 side and front built up from std units and bbd |
| Carp | 1fix | 61 | 1.00 | no | Door frame - fit into existing opening |
| Carp | 1fix | 64 | 1.25 | no | Door, internal - take off and make good lining |
| Carp | 1fix | 90 | 0.50 | no | Door and lining - remove |
| Carp | 1fix | 100 | 1.00 | no | Door and frame - remove and prepare for new frame |
| Carp | 1fix | 101 | 0.75 | no | Door frame |
| Carp | 1fix | 102 | 2.50 | no | Door complete with mortice lock and bolt |
| Carp | 1fix | 111 | 1.50 | no | Door - take out and make good frame |
| Carp | 1fix | 129 | 1.50 | no | Door, external |
| Carp | 1fix | 93 | 0.33 | no | Holes in wall for joists |
| Carp | 1fix | 103 | 2.00 | no | Letter box and inner flap including cutting hole |
| Carp | 1fix | 113 | 2.20 | no | Roof hatch - hack off plast, form OPG and fix lining |
| Carp | 1fix | 119 | 15.00 | no | Staircase, single flight |
| Carp | 1fix | 112 | 0.75 | no | Stormguard |
| Carp | 1fix | 118 | 0.50 | no | Weather board |
| Carp | 1fix | 65 | 1.00 | no | Window - take out and prepare opening for new |
| Carp | 1fix | 85 | 0.25 | no | Window boards - fitted ends |
| Carp | 1fix | 128 | 1.00 | no | Windows |
| Carp | 2fix | 54 | 0.16 | m | Architraves up to 75 mm wide |
| Carp | 2fix | 55 | 0.20 | m | Architraves over 75 mm wide |
| Carp | 2fix | 56 | 0.20 | m | Curtain rail |
| Carp | 2fix | 107 | 0.50 | m | Handrail |
| Carp | 2fix | 66 | 1.00 | m | Pipe boxing to water pipes |
| Carp | 2fix | 67 | 1.50 | m | Pipe boxing to soil pipe |
| Carp | 2fix | 7 | 0.75 | m | Shelving including bearers and brackets |
| Carp | 2fix | 52 | 0.15 | m | Skirting up to 100 mm deep |
| Carp | 2fix | 53 | 0.20 | m | Skirtings over 100 mm deep |
| Carp | 2fix | 21 | 0.10 | $m^2$ | Insulation in roof |
| Carp | 2fix | 51 | 1.50 | no | Door - internal including mortice latch furniture |
| Carp | 2fix | 98 | 0.25 | no | Door stops |
| Carp | 2fix | 104 | 0.12 | no | Door numbers |
| Carp | 2fix | 130 | 1.00 | no | Door, internal |
| Carp | 2fix | 11 | 12.00 | no | Patio door and frame - two leaves |
| Carp | 2fix | 114 | 1.50 | no | Roof hatch - door and furniure |

**Fig. 6.3** Part of the standard operations file.

### 6.3.3 *The estimating module*

*Data preparation*

Quantities are taken off by the contractor, using his existing methods, but a code from the data bank is added for standard operations, or in the case of rogue items an 'R' is recorded. All-in materials rates and plant rates are calculated for each operation, and sums or rates to be included for subcontractors are obtained. If quantities are included, as is sometimes the case when schedules of works are provided, they can of course be used.

*Entering data*

The information (trade and operation number from the standard operations file) is first keyed into the computer for standard operations, followed by information for rogue operations. Materials, plant and subcontractor data are also entered for each operation. The target description, e.g. Car 1, is entered together with the location of the operation. These latter two items can be entered later if preferred. The operations can then be altered or deleted, and new operations can be added as required.

*The printout for management*

An estimate can then be printed out giving the cost of labour, materials, plant, subcontractors and total cost for each operation, together with the total cost of each resource for presentation to management for conversion into a tender.

Part of a printout for the use of management is shown in Fig. 6.4.

Extension to house Page 1

| Ref | Trad | No | Targ | Output | Quantity | Unit | Description | Lab cost | Mat cost | Plt cost | Scn cost | Totals |
|---|---|---|---|---|---|---|---|---|---|---|---|---|
| 6 | Carp | 9107 | Fdns | 0.50 | 1.00 | no | Carefully remove gate and store for refixing | 2.84 | 0.00 | 0.00 | 0.00 | 2.84 |
| 9 | Lab | 9110 | Fdns | 0.00 | 1.00 | item | Transfer blocks and other materials from yard to site | 0.00 | 0.00 | 0.00 | 0.00 | 0.00 |
| 10 | Lab | 3 | Fdns | 0.25 | 13.00 | m$^2$ | Take up paving flags and store for re-use | 15.54 | 0.00 | 0.00 | 0.00 | 15.54 |
| 15 | Lab | 5 | Fdns | 4.00 | 7.00 | m$^3$ | Excavate trench | 133.84 | 0.00 | 0.00 | 0.00 | 133.84 |
| 17 | Lab | 9113 | Fdns | 2.00 | 3.80 | m$^3$ | Wheel material up to 40 m and deposit in skip | 36.33 | 0.00 | 0.00 | 0.00 | 36.33 |
| 20 | Lab | 60 | Fdns | 2.00 | 1.10 | m$^3$ | Concrete in foundations | 10.52 | 55.00 | 0.00 | 0.00 | 65.52 |
| 30 | Blr | 9075 | Fdns | 3.00 | 10.00 | m$^2$ | Walling below DPC 2 leaves blockwork | 170.10 | 70.00 | 0.00 | 0.00 | 240.10 |
| 35 | Blr | 47 | Fdns | 0.45 | 20.00 | m$^2$ | Unload blocks 7N/mm wheel up to 20 m | 43.02 | 0.00 | 0.00 | 0.00 | 43.02 |
| 40 | Blr | 9099 | Fdns | 0.25 | 18.10 | m | Fix furfix to existing wall | 25.66 | 112.40 | 0.00 | 0.00 | 138.06 |
| 45 | Blr | 9100 | Fdns | 0.67 | 9.10 | m | Saw cut to form continuous cavity | 34.57 | 9.10 | 0.00 | 0.00 | 43.67 |
| 47 | Blr | 9115 | Fdns | 2.00 | 1.00 | item | Thicken concrete under and build sleeper wall (see JP) | 11.34 | 3.00 | 0.00 | 0.00 | 14.34 |
| 50 | Lab | 13 | Fdns | 1.00 | 3.10 | m$^3$ | Backfill excavation | 14.82 | 0.00 | 0.00 | 0.00 | 14.82 |
| 70 | Lab | 19 | Fdns | 8.00 | 0.30 | m$^3$ | Cavity fill | 11.47 | 15.00 | 0.00 | 0.00 | 26.47 |
| 71 | Blr | 9003 | Fdns | 0.25 | 4.00 | no | Lintel over drain | 5.40 | 18.00 | 0.00 | 0.00 | 23.40 |

**Fig. 6.4** Part of a typical printout for the use of management.

*Conversion into a tender*

At this stage, preliminaries are added, and the estimate is converted into a tender by management.

*The printout for the client*

After management have carried out the conversion and arrived at a mark-up, a printout can be obtained in the form of a priced bill for the client.

Part of a printout for issuing to the client is shown in Fig. 6.5. Alternatively the printout for the client can be produced excluding the prices but stating the total only.

*Standard estimates*

Many firms carry out projects that are very similar in terms of method of construction and breakdown of operations. Standard estimate files can be set up when this occurs, and new estimates can then be produced very quickly. This is considered to be a very important feature of the MASTER system.

For example, one of the areas of work undertaken by one company is refurbishment. A model was set up specifically for this type of work, based totally on rogue items, tailored particularly for the work.

*Variations*

Variations can be produced using the system together with separate targets for the variations if required, or they can be incorporated into the main estimate and therefore be included in targets, etc.

Extension to house Page 1

| Ref | Description | Quantity | Unit | Rate | Totals | Location |
|---|---|---|---|---|---|---|
| 6 | Carefully remove gate and store for refixing | 1.00 | no | 3.97 | 3.97 | |
| 9 | Transfer blocks and other materials from yard to site | 1.00 | item | 0.00 | 0.00 | |
| 10 | Take up paving flags and store for re-use | 13.00 | $m^2$ | 1.67 | 21.71 | |
| 15 | Excavate trench | 7.00 | $m^3$ | 26.77 | 187.39 | |
| 17 | Wheel material up to 40 m and deposit in skip | 3.80 | $m^3$ | 13.38 | 50.84 | |
| 20 | Concrete in foundations | 1.10 | $m^3$ | 70.88 | 77.97 | |
| 30 | Walling below DPC 2 leaves blockwork | 10.00 | $m^2$ | 31.86 | 318.60 | |
| 35 | Unload blocks 7 N/mm wheel up to 20 m | 20.00 | $m^2$ | 3.01 | 60.20 | |
| 40 | Fix furfix to existing wall | 18.10 | m | 9.13 | 165.25 | |
| 45 | Saw cut to form continuous cavity | 9.10 | m | 6.47 | 58.88 | |
| 47 | Thicken concrete under and build sleeper wall (see JP) | 1.00 | item | 19.33 | 19.33 | |
| 50 | Backfill excavation | 3.10 | $m^3$ | 6.69 | 20.74 | |
| 70 | Cavity fill | 0.30 | $m^3$ | 111.03 | 33.31 | |
| 71 | Lintel over drain | 4.00 | no | 7.07 | 28.28 | |
| 85 | Oversite concrete | 1.40 | $m^3$ | 77.58 | 108.61 | |
| 88 | Floor joists at ground level | 29.02 | m | 2.61 | 75.74 | |
| 89 | Erect scaffolding | 1.00 | item | 65.07 | 65.07 | |

**Fig. 6.5** Part of a typical printout for issuing to the client.

### 6.3.4 *The planning control and incentive module*

This module provides information for the supervisor of the project and for the operatives.

*Grouping the operations*

Operations can be generated in various ways for convenience of planning, setting targets or control as follows:

- Man-hours for a whole project (useful for small projects to be carried out by one team of operatives).
- Man-hours for a complete trade, e.g. brickwork (useful when a trade visits the site only once).
- Man-hours for a section of the project, e.g. carpentry first fix (useful for projects of a reasonable size when trades visit the site more than once). This is the most commonly used method.
- Man-hours for individual operations, e.g. hang doors, fit units, etc. (useful when the supervisor wishes to build up his own targets, taking account of the current position on site).
- Man-hours for work in a particular location.

*Information for management*

Man hours or costs included in the estimate can be provided in a number of ways as follows:

- Actual man-hours or costs included in the estimate for each operation.
- Man-hours or costs for each operation adjusted to take account of the type of incentive scheme being used, e.g. different rates of pay-back to operatives.
- Man-hours or costs for each operation adjusted to allow for certain 'preliminary' items, e.g. travelling time, clearing up at each stage, etc.

In each case the printout shows the estimate reference number, operation description, quantity, output per unit, adjusted and unadjusted target hours or costs for each item in the target and the adjusted and unadjusted total hours for the target set. It also shows the total material, labour, subcontractor and plant costs. This provides all the information necessary for programming and preparing schedules. A printout for foundation work for issuing to the site supervisor is shown in Fig. 6.6.

*Information for operatives*

A printout giving the estimate reference, operation description, quantity and total target hours or cost is provided for handing to the operatives.

Extension to house - foundations, unloading materials, etc. - targ fdns - managment Page 1

| Ref | Description | Hours | Targ | Sub cost | Plt cost | Mat cost |
|---|---|---|---|---|---|---|
| 6 | Carefully remove gate and store for refixing | 0.50 | | | | |
| | 1.00 no 0.50 h/no | | 0.45 | | | |
| 9 | Transfer blocks and other materials from yard to site | 4.00 | | | | |
| | 1.00 item 4.00 h/item | | 3.60 | | | |
| 10 | Take up paving flags and store for re-use | 3.25 | | | | |
| | 13.00 $m^2$ 0.25 h/$m^2$ | | 2.93 | | | |
| 15 | Excavate trench | 28.00 | | | | |
| | 7.00 $m^3$ 4.00 h/$m^2$ | | 25.20 | | | |
| 17 | Wheel material up to 40 m and deposit in skip | 7.60 | | | | |
| | 3.80 $m^3$ 2.00 h/$m^3$ | | 6.84 | | | |
| 20 | Concrete in foundations | 2.20 | | | | |
| | 1.10 $m^3$ 2.00 h/$m^3$ | | 1.98 | | | 55.00 |
| 35 | Unload blocks 7 N/mm wheel up to 20 m | 9.00 | | | | |
| | 2.00 $m^2$ 0.45 h/$m^2$ | | 8.10 | | | |
| 97 | Unload bricks and blocks for superstructure and wheel to rear | 9.40 | | | | |
| | 18.80 $m^2$ 0.50 h/$m^2$ | | 8.46 | | | |

| | | | | | |
|---|---|---|---|---|---|
| | | Labour hours | 63.95 | Tot mat cost | 55.00 |
| Factor | 0.90 | Factor total | 57.56 | Tot lab cost | 306.13 |
| | | | | Tot sub cost | 0.00 |
| | | | | Tot plt cost | 0.00 |

**Fig. 6.6** A printout for foundation work for issuing to the site supervisor.

A printout for foundation work for issuing to the operatives is shown in Fig. 6.7.

The operations target can be generated omitting the quantities.

### *6.3.5 Integrated estimating and production data*

By integrating the estimating and production processes, data are provided to control labour output and costs, plant output and costs, and to update the database as appropriate. The system provides all the data for an incentive scheme based on realistic outputs, which will help to increase productivity.

Production of data is available immediately a contract is secured, and site staff are equipped with all the production data they need.

Extension to house - foundations, unloading materials, etc. - targ fdns - operatives Page 1

| Ref | Description | Quantity | Unit | Location |
|---|---|---|---|---|
| 6 | Carefully remove gate and store for refixing | 1.00 | no | |
| 9 | Transfer blocks and other materials from yard to site | 1.00 | item | |
| 10 | Take up paving flags and store for re-use | 13.00 | $m^2$ | |
| 15 | Excavate trench | 7.00 | $m^3$ | |
| 17 | Wheel material up to 40 m and deposit in skip | 3.80 | $m^3$ | |
| 20 | Concrete in foundations | 1.10 | $m^3$ | |
| 35 | Unload blocks 7 N/mm wheel up to 20 m | 20.00 | $m^2$ | |
| 97 | Unload bricks and blocks for superstructure and wheel to rear | 18.80 | $m^2$ | |
| | | Target | 58 | |

**Fig. 6.7** A printout for foundation work for issuing to the operatives.

### 6.3.6 *Case study*

A case study using the MASTER system is shown in case study 1 in Chapter 7.

## 6.4 Spreadsheets

### 6.4.1 *Introduction*

There are several spreadsheet packages available, which are very similar in principle.

A spreadsheet is a computerised version of a large sheet of paper, of which a small portion is visible on the screen. It consists of rows and columns, as shown in Fig. 6.8. Cells are identified by reference to the row and the column. Information entered into a cell is retained in the computer's memory, and the cells can be related to each other. If the information in one cell changes, the information in all relative cells also changes. In Fig. 6.8 cell E1 contains a value of 750, F1 will contain a value of 750 × 1.05 = 787.5. If E1 is altered to 800, F1 will change to 800 × 1.05 = 840. The screen acts as a window viewing a small part of the spreadsheet, and can be moved from side to side and up and down to view all cells. Data and text can be changed very easily, e.g. deleted, inserted, altered, etc.

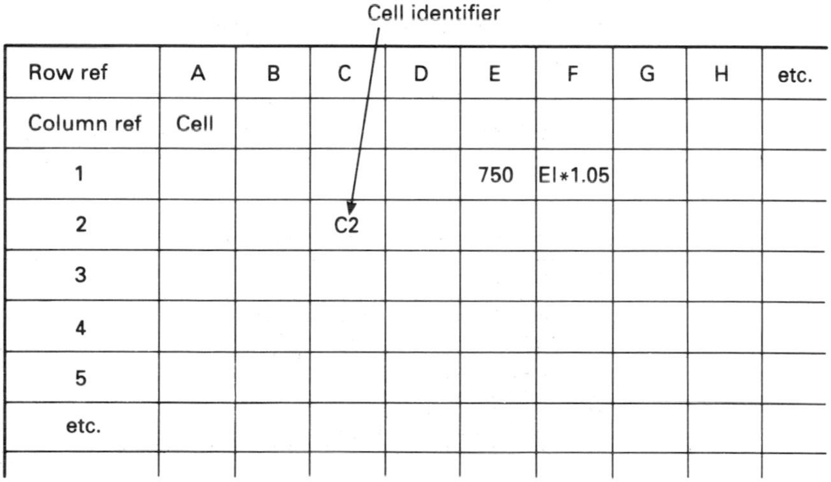

Cell identifier

| Row ref | A | B | C | D | E | F | G | H | etc. |
|---|---|---|---|---|---|---|---|---|---|
| Column ref | Cell | | | | | | | | |
| 1 | | | | | 750 | E1*1.05 | | | |
| 2 | | | C2 | | | | | | |
| 3 | | | | | | | | | |
| 4 | | | | | | | | | |
| 5 | | | | | | | | | |
| etc. | | | | | | | | | |

**Fig. 6.8** Small part of a spreadsheet.

Programme for small office building
Expenditure and income at earliest time 1 Retention 5%
Expenditure and income at latest time

| Operation | Duration (days) | Planned expenditure | Planned income | Weeks 1 | 2 | 3 | 4 | 5 | 6 | 7 | 8 | 9 | 10 | 11 | 12 | 13 | 14 | 15 | 16 |
|---|---|---|---|---|---|---|---|---|---|---|---|---|---|---|---|---|---|---|---|
| Site set up | 3 | 210 | 240 | 210 | | | | | | | | | | | | | | | |
| | | | | 240 | | | | | | | | | | | | | | | |
| Setting out | 4 | 60 | 80 | 60 | | | | | | | | | | | | | | | |
| | | | | 80 | | | | | | | | | | | | | | | |
| Excavate to reduce level | 10 | 1000 | 1200 | 100 | 500 | 400 | | | | | | | | | | | | | |
| | | | | 120 | 600 | 480 | | | | | | | | | | | | | |
| Piling | 12 | 3600 | 4800 | | | 300 | 1500 | 1500 | 300 | | | | | | | | | | |
| | | | | | | 400 | 2000 | 2000 | 400 | | | | | | | | | | |
| Drains | 6 | 300 | 360 | | | 50 | 250 | 0 | 0 | 0 | 0 | 0 | 0 | 0 | | | | | |
| | | | | | | 60 | 300 | 0 | 0 | 0 | 0 | 0 | 0 | 0 | | | | | |
| Pad foundation | 9 | 630 | 720 | | | | | | 280 | 350 | | | | | | | | | |
| | | | | | | | | | 320 | 400 | | | | | | | | | |
| Strip foundation | 5 | 500 | 600 | | | | | | 400 | 100 | 0 | 0 | 0 | 0 | | | | | |
| | | | | | | | | | 480 | 120 | 0 | 0 | 0 | 0 | | | | | |
| In situ concrete frame | 22 | 22000 | 33000 | | | | | | | | 5000 | 5000 | 5000 | 5000 | 2000 | | | | |
| | | | | | | | | | | | 7500 | 7500 | 7500 | 7500 | 3000 | | | | |
| Ground floor slab | 5 | 800 | 900 | | | | | | | | | | | | 480 | 320 | 0 | | |
| | | | | | | | | | | | | | | | 540 | 360 | 0 | | |
| Brick infill panels | 15 | 8700 | 9750 | | | | | | | | | | | | 1740 | 2900 | 2900 | 1160 | |
| | | | | | | | | | | | | | | | 1950 | 3250 | 3250 | 1300 | |
| Precast stairs and floors | 3 | 2400 | 2700 | | | | | | | | | | | | | 2400 | 0 | 0 | |
| | | | | | | | | | | | | | | | | 2700 | 0 | 0 | |
| Hardwood frames | 10 | 1000 | 1200 | | | | | | | | | | | | | | | 300 | 500 |
| | | | | | | | | | | | | | | | | | | 360 | 600 |
| Precast roof slab | 2 | 2000 | 2500 | | | | | | | | | | | | | | 2000 | 0 | 0 |
| | | | | | | | | | | | | | | | | | 2500 | 0 | 0 |
| Block partitions ground floor | 7 | 1750 | 2100 | | | | | | | | | | | | | | 1250 | 500 | 0 |
| | | | | | | | | | | | | | | | | | 1500 | 600 | 0 |
| Glazing | 6 | 240 | 300 | | | | | | | | | | | | | | | | |
| External doors | 4 | 1600 | 1800 | | | | | | | | | | | | | | | | |
| Roof covering | 5 | 1000 | 1200 | | | | | | | | | | | | | | 600 | 400 | 0 |
| | | | | | | | | | | | | | | | | | 720 | 480 | 0 |
| External plumbing | 6 | 180 | 210 | | | | | | | | | | | | | | 90 | 90 | 0 |
| | | | | | | | | | | | | | | | | | 105 | 105 | 0 |
| Block partitions first floor | 7 | 2100 | 2450 | | | | | | | | | | | | | | | 900 | 1200 |
| | | | | | | | | | | | | | | | | | | 1050 | 1400 |
| Plumbing first fix | 5 | 900 | 1050 | | | | | | | | | | | | | | | | 300 |
| | | | | | | | | | | | | | | | | | | | 350 |
| Electrical first fix | 8 | 400 | 480 | | | | | | | | | | | | | | | | |
| External painting | 12 | 240 | 300 | | | | | | | | | | | | | | | | 40 |
| | | | | | | | | | | | | | | | | | | | 50 |
| External works | 20 | 7564 | 9000 | | | | | | | | | | | | | | | | 756 |
| | | | | | | | | | | | | | | | | | | | 900 |
| Joinery first fix | 8 | 800 | 880 | | | | | | | | | | | | | | | | |
| Plastering | 18 | 1800 | 2160 | | | | | | | | | | | | | | | | |
| Floor screeding | 10 | 600 | 700 | | | | | | | | | | | | | | | | |
| Quarry tiling | 1 | 100 | 140 | | | | | | | | | | | | | | | | |
| Plumbing second fix | 10 | 1000 | 1200 | | | | | | | | | | | | | | | | |
| Joinery second fix | 10 | 1000 | 1200 | | | | | | | | | | | | | | | | |
| Internal painting | 14 | 420 | 504 | | | | | | | | | | | | | | | | |
| Vinyl tiling | 6 | 240 | 300 | | | | | | | | | | | | | | | | |
| Electrical second fix | 6 | 360 | 540 | | | | | | | | | | | | | | | | |
| Cleaning and handover | 1 | 50 | 60 | | | | | | | | | | | | | | | | |
| Preliminaries | | 13175 | 13950 | 425 | 425 | 425 | 425 | 425 | 425 | 425 | 425 | 425 | 425 | 425 | 425 | 425 | 425 | 425 | 425 |
| | | | | 450 | 450 | 450 | 450 | 450 | 450 | 450 | 450 | 450 | 450 | 450 | 450 | 450 | 450 | 450 | 450 |
| Planned expenditure | | 78719 | | 795 | 925 | 1175 | 2175 | 1925 | 1405 | 875 | 5425 | 5425 | 5425 | 5425 | 4645 | 6045 | 7265 | 3775 | 3221 |
| Planned income | | | 98574 | 890 | 1050 | 1390 | 2750 | 2450 | 1650 | 970 | 7950 | 7950 | 7950 | 7950 | 5940 | 6760 | 8525 | 4345 | 3750 |
| Planned expenditure monthly | | | | | | | 5070 | | | | 9630 | | | | 20920 | | | | 20306 |
| Planned income monthly | | | | | | | 6080 | | | | 13020 | | | | 29790 | | | | 23380 |
| Retention on planned income | | | | | | | 304 | | | | 651 | | | | 1490 | | | | 1169 |
| Income less retention time adjusted | | | | | | | 0 | | | | 5776 | | | | 12369 | | | | 28301 |
| Accumulative expenditure | | | | | | | 5070 | | | | 14700 | | | | 35620 | | | | 55926 |
| Accumulative income time adjusted | | | | | | | 0 | | | | 5776 | | | | 18145 | | | | 46446 |
| Valuation number | | | | | | | 1 | 1 | | | 2 | 2 | | | 3 | 3 | | | 4 |
| Difference between expenditure and income | | | | | | | -5070 | -5070 | | | -14700 | -8924 | | | -29844 | -17475 | | | -37781 |

Graph plottings

| | | | | | | | | | | | | | | | | | | | |
|---|---|---|---|---|---|---|---|---|---|---|---|---|---|---|---|---|---|---|---|
| 0 | 1 | 1 | 2 | 2 | 3 | 3 | 4 | 4 | 5 | 5 | 6 | 6 | 7 | 7 | 7.75 | 8 | 8 | 9 | 9 |
| 0 | -5070 | -5070 | -14700 | -8924 | -29844 | -17475 | -37781 | -9481 | -22029 | 183 | -3678 | 10183 | 5963 | 10086 | 7921 | 7921 | 12515 | 12515 | 17391 |

**Fig. 6.9** Printout showing all activities at their earliest times; 5% retention.

| 17 | 18 | 19 | 20 | 21 | 22 | 23 | 24 | 25 | 26 | 27 | 28 | 29 | 30 | 31 | 32 | 33 | 34 | 35 | 36 | 37 | 38 | 39 | 59 | 60 |
|---|---|---|---|---|---|---|---|---|---|---|---|---|---|---|---|---|---|---|---|---|---|---|---|---|
| 200 | | | | | | | | | | | | | | | | | | | | | | | | |
| 240 | | | | | | | | | | | | | | | | | | | | | | | | |
| 0 | | | | | | | | | | | | | | | | | | | | | | | | |
| 0 | | | | | | | | | | | | | | | | | | | | | | | | |
| 0 | | | | | | | | | | | | | | | | | | | | | | | | |
| 0 | | | | | | | | | | | | | | | | | | | | | | | | |
| 120 | 120 | | | | | | | | | | | | | | | | | | | | | | | |
| 150 | 150 | | | | | | | | | | | | | | | | | | | | | | | |
| 1200 | 400 | | | | | | | | | | | | | | | | | | | | | | | |
| 1350 | 450 | | | | | | | | | | | | | | | | | | | | | | | |
| 0 | 0 | | | | | | | | | | | | | | | | | | | | | | | |
| 0 | 0 | | | | | | | | | | | | | | | | | | | | | | | |
| 0 | 0 | | | | | | | | | | | | | | | | | | | | | | | |
| 0 | 0 | | | | | | | | | | | | | | | | | | | | | | | |
| 0 | 0 | | | | | | | | | | | | | | | | | | | | | | | |
| 0 | 0 | | | | | | | | | | | | | | | | | | | | | | | |
| 600 | | | | | | | | | | | | | | | | | | | | | | | | |
| 700 | | | | | | | | | | | | | | | | | | | | | | | | |
| | 100 | 250 | 50 | | | | | | | | | | | | | | | | | | | | | |
| | 120 | 300 | 60 | | | | | | | | | | | | | | | | | | | | | |
| 100 | 100 | 0 | 0 | 0 | 0 | 0 | 0 | 0 | 0 | 0 | 0 | 0 | 0 | 0 | | | | | | | | | | |
| 125 | 125 | 0 | 0 | 0 | 0 | 0 | 0 | 0 | 0 | 0 | 0 | 0 | 0 | 0 | | | | | | | | | | |
| 1891 | 1891 | 1891 | 1135 | 0 | 0 | 0 | 0 | 0 | 0 | 0 | 0 | 0 | 0 | 0 | | | | | | | | | | |
| 2250 | 2250 | 2250 | 1350 | 0 | 0 | 0 | 0 | 0 | 0 | 0 | 0 | 0 | 0 | 0 | | | | | | | | | | |
| | | | 400 | 400 | 0 | 0 | | | | | | | | | | | | | | | | | | |
| | | | 440 | 440 | 0 | 0 | | | | | | | | | | | | | | | | | | |
| | | | 400 | 500 | 500 | 400 | | | | | | | | | | | | | | | | | | |
| | | | 480 | 600 | 600 | 480 | | | | | | | | | | | | | | | | | | |
| | | | | | | 60 | 300 | 240 | | | | | | | | | | | | | | | | |
| | | | | | | 70 | 350 | 280 | | | | | | | | | | | | | | | | |
| | | | | | | | | 100 | 0 | 0 | | | | | | | | | | | | | | |
| | | | | | | | | 140 | 0 | 0 | | | | | | | | | | | | | | |
| | | | | | | | | 100 | 500 | 400 | | | | | | | | | | | | | | |
| | | | | | | | | 120 | 600 | 480 | | | | | | | | | | | | | | |
| | | | | | | | | 100 | 500 | 400 | | | | | | | | | | | | | | |
| | | | | | | | | 120 | 600 | 480 | | | | | | | | | | | | | | |
| | | | | | | | | | | 30 | 150 | 150 | 90 | | | | | | | | | | | |
| | | | | | | | | | | 36 | 180 | 180 | 108 | | | | | | | | | | | |
| | | | | | | | | | | | | | 80 | 160 | | | | | | | | | | |
| | | | | | | | | | | | | | 100 | 200 | | | | | | | | | | |
| | | | | | | | | | | | | | 120 | 240 | | | | | | | | | | |
| | | | | | | | | | | | | | 180 | 360 | | | | | | | | | | |
| | | | | | | | | | | | | | | 50 | | | | | | | | | | |
| | | | | | | | | | | | | | | 60 | | | | | | | | | | |
| 425 | 425 | 425 | 425 | 425 | 425 | 425 | 425 | 425 | 425 | 425 | 425 | 425 | 425 | 425 | | | | | | | | | | |
| 450 | 450 | 450 | 450 | 450 | 450 | 450 | 450 | 450 | 450 | 450 | 450 | 450 | 450 | 450 | | | | | | | | | | |
| 4536 | 3036 | 2566 | 2410 | 1325 | 925 | 885 | 725 | 965 | 1425 | 1255 | 575 | 575 | 715 | 875 | | | | | | | | | | |
| 5265 | 3545 | 3000 | 2780 | 1490 | 1050 | 1000 | 800 | 1110 | 1650 | 1446 | 630 | 630 | 838 | 1070 | | | | | | | | | | |
| | | | 12548 | | | | 3860 | | | | 4220 | | | 2165 | | | | | | | | | | |
| | | | 14590 | | | | 4340 | | | | 4836 | | | 2538 | | | | | | | | | | |
| | | | 730 | | | | 217 | | | | 242 | | | 127 | | | | | | | | | | |
| | | | 22211 | | | | 13861 | | | | 4123 | | | | 4594 | | | | 2411 | | | | | |
| | | | 68474 | | | | 72334 | | | | 76554 | | | 78719 | | | | | | | | | | |
| | | | 68657 | | | | 82517 | | | | 86640 | | | | 91234 | | | | 96110 | | | | | |
| 4 | | | 5 | 5 | | | 6 | 6 | | | 7 | 7 | | 7.75 | 8 | 8 | | | 9 | 9 | | | 10 | 10 |
| -9481 | | | -22029 | 183 | | | -3678 | 10183 | | | 5963 | 10086 | | 7921 | 7921 | 12515 | | | 12515 | 17391 | | | 17391 | 19855 |
| 10 | 10 | | | | | | | | | | | | | | | | | | | | | | | |
| 17391 | 19855 | | | | | | | | | | | | | | | | | | | | | | | |

### 6.4.2 *Uses in the construction industry*

There are many uses for spreadsheets, which are ideal for solving problems and carrying out calculations on data that can be presented in tabular form. Spreadsheets are particularly useful for financial modelling and presenting contract and company cash flows. Spreadsheets were used for the following calculations in this book.

*Line-of-balance calculations*

This was set up as a standard model requiring general information such as handover rates, number of hours worked per week, number of units and activity data in the form of the operations, productive hours, minimum gang size and number of gangs in each unit to be filled in. All calculations were then produced automatically using the spreadsheet.

*Calculations on the effects of incentives on costs and earnings*

A standard model was set up to produce all the tables starting at Table 3.10, requiring only hours at 100 rating, percentage added (to set target), percentage payback, basic hourly rate, all-in rate and output rating assumed for the estimate. The graphs could also be produced by computer.

**Example 6.1: Project budget and cash flow**

Example 4.2 will be used as an example of the detailed application of spreadsheets.

Lotus 1-2-3 will be used. This is a very powerful spreadsheet/graphics/database package, which is currently used throughout industry. Graphs can be produced in colour when a colour printer is used, and a number of different typefaces (fonts) are available.

A model was first set up, and headings, retention, operations, duration, planned expenditure and income, and the program were entered using the monetary values in each week, the floats being represented by a zero. The planned expenditure and income and monthly planned expenditure and income was then calculated using simple '@ sum' functions.

The remainder of the calculations were then carried out using formulae to relate the cells. A printout showing all activities at their earliest time and 5% retention is shown in Fig. 6.9. A printout of some of the formulae that determined this chart is shown in Fig. 6.10. The

| 1 | A | B | C | D | E | F | G | H | I | J | K | L | M | N | O | P | Q | R | S | T |
|---|---|---|---|---|---|---|---|---|---|---|---|---|---|---|---|---|---|---|---|---|
| 2 | | Programme for small office building | | | | | | | | | | | | | | | | | | |
| 3 | | Expenditure and income at earliest time | 1 | Retention 5% | | | | | | | | | | | | | | | | |
| 4 | | Expenditure and income at latest time | | | | | | | | | | | | | | | | | | |
| 5 | | | | | | | | | | | | | | | | | | | | |
| 6 | | Operation | Duration | Planned | Planned | Weeks | | | | | | | | | | | | | | |
| 7 | | | (days) | expenditure | income | 1 | 2 | 3 | 4 | 5 | 6 | 7 | 8 | 9 | 10 | 11 | 12 | 13 | 14 | 15 |
| 8 | | | | | | | | | | | | | | | | | | | | |
| 9 | | Site set up | 3 | 210 | 240 | 210 | | | | | | | | | | | | | | |
| 10 | | | | | | 240 | | | | | | | | | | | | | | |
| 11 | | Setting out | 4 | 60 | 80 | 60 | | | | | | | | | | | | | | |
| 12 | | | | | | 80 | | | | | | | | | | | | | | |
| 13 | | Excavate to reduce level | 10 | 1000 | 1200 | 100 | 500 | 400 | | | | | | | | | | | | |
| 14 | | | | | | 120 | 600 | 480 | | | | | | | | | | | | |
| 15 | | Piling | 12 | 3600 | 4800 | | | 300 | 1500 | 1500 | 300 | | | | | | | | | |
| 16 | | | | | | | | 400 | 2000 | 2000 | 400 | | | | | | | | | |
| 17 | | Drains | 6 | 300 | 360 | | | 50xD3 | 250xD3 | 0 | 0 | 0 | 0 | 0 | 300xD4 | 250xD4 | | | | |
| 18 | | | | | | | | 60xD3 | 300xD3 | 0 | 0 | 0 | 0 | 0 | 360xD4 | 300xD4 | | | | |
| 19 | | Pad foundation | 9 | 630 | 720 | | | | | | 280 | 350 | | | | | | | | |
| 20 | | | | | | | | | | | 320 | 400 | | | | | | | | |
| 21 | | Strip foundation | 5 | 500 | 600 | | | | | | 400xD3 | 100xD3 | 0 | 0 | 400xD4 | 100xD4 | | | | |
| 22 | | | | | | | | | | | 480xD3 | 120xD3 | 0 | 0 | 480xD4 | 120xD4 | | | | |
| 23 | | In situ concrete frame | 22 | 22000 | 23000 | | | | | | | | 5000 | 5000 | 5000 | 5000 | 2000 | | | |
| 24 | | | | | | | | | | | | | 7500 | 7500 | 7500 | 7500 | 3000 | | | |
| 25 | | Ground floor slab | 5 | 800 | 900 | | | | | | | | | | | | 480xD3 | 320xD3 | 800xD4 | |
| 26 | | | | | | | | | | | | | | | | | 540xD3 | 360xD3 | 900xD4 | |
| 27 | | Brick infill panels | 15 | 8700 | 9750 | | | | | | | | | | | | 1740 | 2900 | 2900 | 1160 |
| 28 | | | | | | | | | | | | | | | | | 1950 | 3250 | 3250 | 1300 |
| 29 | | Precast stairs and floors | 3 | 2400 | 2700 | | | | | | | | | | | | | 2400xD3 | 0 | 2400xD4 |
| 30 | | | | | | | | | | | | | | | | | | 2700xD3 | 0 | 2700xD4 |

**Figure 6.10** Formulae that determine whether activity is at earliest or latest time; used to produce the printout in Fig. 6.9.

Programme for small office building
Expenditure and income at earliest time Retention 5%
Expenditure and income at latest time

| Operation | Duration (days) | Planned expenditure | Planned income | Weeks 1 | 2 | 3 | 4 | 5 | 6 | 7 | 8 | 9 | 10 | 11 | 12 | 13 | 14 | 15 | 16 |
|---|---|---|---|---|---|---|---|---|---|---|---|---|---|---|---|---|---|---|---|
| Site set up | 3 | 210 | 240 | 210 | | | | | | | | | | | | | | | |
| | | | | 240 | | | | | | | | | | | | | | | |
| Setting out | 4 | 60 | 80 | 60 | | | | | | | | | | | | | | | |
| | | | | 80 | | | | | | | | | | | | | | | |
| Excavate to reduce level | 10 | 1000 | 1200 | 100 | 500 | 400 | | | | | | | | | | | | | |
| | | | | 120 | 600 | 480 | | | | | | | | | | | | | |
| Piling | 12 | 3600 | 4800 | | | 300 | 1500 | 1500 | 300 | | | | | | | | | | |
| | | | | | | 400 | 2000 | 2000 | 400 | | | | | | | | | | |
| Drains | 6 | 300 | 360 | | | 0 | 0 | 0 | 0 | 0 | 0 | 0 | 250 | 50 | | | | | |
| | | | | | | 0 | 0 | 0 | 0 | 0 | 0 | 0 | 300 | 60 | | | | | |
| Pad foundation | 9 | 630 | 720 | | | | | | 280 | 350 | | | | | | | | | |
| | | | | | | | | | 320 | 400 | | | | | | | | | |
| Strip foundation | 5 | 500 | 600 | | | | | | 0 | 0 | 0 | 0 | 400 | 100 | | | | | |
| | | | | | | | | | 0 | 0 | 0 | 0 | 480 | 120 | | | | | |
| In situ concrete frame | 22 | 22000 | 33000 | | | | | | | | 5000 | 5000 | 5000 | 5000 | 2000 | | | | |
| | | | | | | | | | | | 7500 | 7500 | 7500 | 7500 | 3000 | | | | |
| Ground floor slab | 5 | 800 | 900 | | | | | | | | | | | | 0 | 0 | 800 | | |
| | | | | | | | | | | | | | | | 0 | 0 | 900 | | |
| Brick infill panels | 15 | 8700 | 9750 | | | | | | | | | | | | 1740 | 2900 | 2900 | 1160 | |
| | | | | | | | | | | | | | | | 1950 | 3250 | 3250 | 1300 | |
| Precast stairs and floors | 3 | 2400 | 2700 | | | | | | | | | | | | | 0 | 0 | 2400 | |
| | | | | | | | | | | | | | | | | 0 | 0 | 2700 | |
| Hardwood frames | 10 | 1000 | 1200 | | | | | | | | | | | | | | | 300 | 500 |
| | | | | | | | | | | | | | | | | | | 360 | 600 |
| Precast roof slab | 2 | 2000 | 2500 | | | | | | | | | | | | | | 0 | 0 | 1000 |
| | | | | | | | | | | | | | | | | | 0 | 0 | 1250 |
| Block partitions ground floor | 7 | 1750 | 2100 | | | | | | | | | | | | | | 0 | 0 | 1250 |
| | | | | | | | | | | | | | | | | | 0 | 0 | 1500 |
| Glazing | 6 | 240 | 300 | | | | | | | | | | | | | | | | |
| External doors | 4 | 1600 | 1800 | | | | | | | | | | | | | | | | |
| Roof covering | 5 | 1000 | 1200 | | | | | | | | | | | | | | 0 | 0 | 0 |
| | | | | | | | | | | | | | | | | | 0 | 0 | 0 |
| External plumbing | 6 | 180 | 210 | | | | | | | | | | | | | | 0 | 0 | 0 |
| | | | | | | | | | | | | | | | | | 0 | 0 | 0 |
| Block partitions first floor | 7 | 2100 | 2450 | | | | | | | | | | | | | | | 0 | 0 |
| | | | | | | | | | | | | | | | | | | 0 | 0 |
| Plumbing first fix | 5 | 900 | 1050 | | | | | | | | | | | | | | | | 300 |
| | | | | | | | | | | | | | | | | | | | 350 |
| Electrical first fix | 8 | 400 | 480 | | | | | | | | | | | | | | | | |
| External painting | 12 | 240 | 300 | | | | | | | | | | | | | | | | 0 |
| | | | | | | | | | | | | | | | | | | | 0 |
| External works | 20 | 7564 | 9000 | | | | | | | | | | | | | | | | 0 |
| | | | | | | | | | | | | | | | | | | | 0 |
| Joinery first fix | 8 | 800 | 880 | | | | | | | | | | | | | | | | |
| Plastering | 18 | 1800 | 2160 | | | | | | | | | | | | | | | | |
| Floor screeding | 10 | 600 | 700 | | | | | | | | | | | | | | | | |
| Quarry tiling | 1 | 100 | 140 | | | | | | | | | | | | | | | | |
| Plumbing second fix | 10 | 1000 | 1200 | | | | | | | | | | | | | | | | |
| Joinery second fix | 10 | 1000 | 1200 | | | | | | | | | | | | | | | | |
| Internal painting | 14 | 420 | 504 | | | | | | | | | | | | | | | | |
| Vinyl tiling | 6 | 240 | 300 | | | | | | | | | | | | | | | | |
| Electrical second fix | 6 | 360 | 540 | | | | | | | | | | | | | | | | |
| Cleaning and handover | 1 | 50 | 60 | | | | | | | | | | | | | | | | |
| Preliminaries | | 13175 | 13950 | 425 | 425 | 425 | 425 | 425 | 425 | 425 | 425 | 425 | 425 | 425 | 425 | 425 | 425 | 425 | 425 |
| | | | | 450 | 450 | 450 | 450 | 450 | 450 | 450 | 450 | 450 | 450 | 450 | 450 | 450 | 450 | 450 | 450 |
| Planned expenditure | | 78719 | | 795 | 925 | 1125 | 1925 | 1925 | 1005 | 775 | 5425 | 5425 | 6075 | 5575 | 4165 | 3325 | 4125 | 4285 | 3475 |
| Planned income | | | 98574 | 890 | 1050 | 1330 | 2450 | 2450 | 1170 | 850 | 7950 | 7950 | 8730 | 8130 | 5400 | 3700 | 4600 | 4810 | 4150 |
| Planned expenditure monthly | | | | | | | 4770 | | | | 9130 | | | | 21240 | | | | 15210 |
| Planned income monthly | | | | | | | 5720 | | | | 12420 | | | | 30210 | | | | 17260 |
| Retention on planned income | | | | | | | 286 | | | | 621 | | | | 1511 | | | | 863 |
| Income less retention time adjusted | | | | | | | 0 | | | | 5434 | | | | 11799 | | | | 28700 |
| Accumulative expenditure | | | | | | | 4770 | | | | 13900 | | | | 35140 | | | | 50350 |
| Accumulative income time adjusted | | | | | | | 0 | | | | 5434 | | | | 17233 | | | | 45933 |
| Valuation number | | | | | | | 1 | 1 | | | 2 | 2 | | | 3 | 3 | | | 4 |
| Difference between expenditure and income | | | | | | | -4770 | -4770 | | | -13900 | -8466 | | | -29706 | -17907 | | | -33117 |

Graph plottings

| | | | | | | | | | | | | | | | | | | | |
|---|---|---|---|---|---|---|---|---|---|---|---|---|---|---|---|---|---|---|---|
| 0 | 1 | 1 | 2 | 2 | 3 | 3 | 4 | 4 | 5 | 5 | 6 | 6 | 7 | 7 | 7.75 | 8 | 8 | 9 | 9 |
| 0 | -4770 | -4770 | -13960 | -8466 | -29706 | -17907 | -33117 | -4418 | -14338 | 2060 | -2201 | 8734 | 2623 | 7164 | -914 | -914 | 5818 | 5818 | 17391 |

**Fig. 6.11** Activities at latest times; 5% retention.

| 17 | 18 | 19 | 20 | 21 | 22 | 23 | 24 | 25 | 26 | 27 | 28 | 29 | 30 | 31 | 32 | 33 | 34 | 35 | 36 | 37 | 38 | 39 | 59 | 60 |
|---|---|---|---|---|---|---|---|---|---|---|---|---|---|---|---|---|---|---|---|---|---|---|---|---|
| 200 | | | | | | | | | | | | | | | | | | | | | | | | |
| 240 | | | | | | | | | | | | | | | | | | | | | | | | |
| 1000 | | | | | | | | | | | | | | | | | | | | | | | | |
| 1250 | | | | | | | | | | | | | | | | | | | | | | | | |
| 500 | | | | | | | | | | | | | | | | | | | | | | | | |
| 600 | | | | | | | | | | | | | | | | | | | | | | | | |
| 120 | 20 | | | | | | | | | | | | | | | | | | | | | | | |
| 150 | 50 | | | | | | | | | | | | | | | | | | | | | | | |
| 1200 | 20 | | | | | | | | | | | | | | | | | | | | | | | |
| 1350 | 50 | | | | | | | | | | | | | | | | | | | | | | | |
| 600 | 100 | | | | | | | | | | | | | | | | | | | | | | | |
| 720 | 80 | | | | | | | | | | | | | | | | | | | | | | | |
| 90 | 90 | | | | | | | | | | | | | | | | | | | | | | | |
| 105 | 105 | | | | | | | | | | | | | | | | | | | | | | | |
| 900 | 200 | | | | | | | | | | | | | | | | | | | | | | | |
| 1050 | 400 | | | | | | | | | | | | | | | | | | | | | | | |
| 600 | | | | | | | | | | | | | | | | | | | | | | | | |
| 700 | | | | | | | | | | | | | | | | | | | | | | | | |
| | 100 | 250 | 50 | | | | | | | | | | | | | | | | | | | | | |
| | 120 | 300 | 60 | | | | | | | | | | | | | | | | | | | | | |
| 0 | 0 | 0 | 0 | 0 | 0 | 0 | 0 | 0 | 0 | 0 | 0 | 40 | 100 | 100 | | | | | | | | | | |
| 0 | 0 | 0 | 0 | 0 | 0 | 0 | 0 | 0 | 0 | 0 | 0 | 50 | 125 | 125 | | | | | | | | | | |
| 0 | 0 | 0 | 0 | 0 | 0 | 0 | 0 | 0 | 0 | 0 | 1891 | 1891 | 1891 | 1891 | | | | | | | | | | |
| 0 | 0 | 0 | 0 | 0 | 0 | 0 | 0 | 0 | 0 | 0 | 2250 | 2250 | 2250 | 2250 | | | | | | | | | | |
| | | | 0 | 0 | 300 | 500 | | | | | | | | | | | | | | | | | | |
| | | | 0 | 0 | 330 | 550 | | | | | | | | | | | | | | | | | | |
| | | | 400 | 500 | 500 | 400 | | | | | | | | | | | | | | | | | | |
| | | | 480 | 600 | 600 | 480 | | | | | | | | | | | | | | | | | | |
| | | | | | | 60 | 300 | 240 | | | | | | | | | | | | | | | | |
| | | | | | | 70 | 350 | 280 | | | | | | | | | | | | | | | | |
| | | | | | | | | 0 | 0 | 100 | | | | | | | | | | | | | | |
| | | | | | | | | 0 | 0 | 140 | | | | | | | | | | | | | | |
| | | | | | | | | 100 | 500 | 400 | | | | | | | | | | | | | | |
| | | | | | | | | 120 | 600 | 480 | | | | | | | | | | | | | | |
| | | | | | | | | 100 | 500 | 400 | | | | | | | | | | | | | | |
| | | | | | | | | 120 | 600 | 480 | | | | | | | | | | | | | | |
| | | | | | | | | | | 30 | 150 | 150 | 90 | | | | | | | | | | | |
| | | | | | | | | | | 36 | 180 | 180 | 108 | | | | | | | | | | | |
| | | | | | | | | | | | | | 80 | 160 | | | | | | | | | | |
| | | | | | | | | | | | | | 100 | 200 | | | | | | | | | | |
| | | | | | | | | | | | | | 120 | 240 | | | | | | | | | | |
| | | | | | | | | | | | | | 180 | 360 | | | | | | | | | | |
| | | | | | | | | | | | | | | 50 | | | | | | | | | | |
| | | | | | | | | | | | | | | 60 | | | | | | | | | | |
| 425 | 425 | 425 | 425 | 425 | 425 | 425 | 425 | 425 | 425 | 425 | 425 | 425 | 425 | 425 | | | | | | | | | | |
| 450 | 450 | 450 | 450 | 450 | 450 | 450 | 450 | 450 | 450 | 450 | 450 | 450 | 450 | 450 | | | | | | | | | | |
| 5635 | 2735 | 675 | 875 | 925 | 1225 | 1385 | 725 | 865 | 1425 | 1355 | 2466 | 2506 | 2706 | 2866 | | | | | | | | | | |
| 6615 | 3155 | 750 | 990 | 1050 | 1380 | 1550 | 800 | 970 | 1650 | 1586 | 2880 | 2930 | 3213 | 3445 | | | | | | | | | | |
| | | | 9920 | | | | 4260 | | | | 6111 | | | 8078 | | | | | | | | | | |
| | | | 11510 | | | | 4780 | | | | 7086 | | | 9588 | | | | | | | | | | |
| | | | 576 | | | | 239 | | | | 354 | | | 479 | | | | | | | | | | |
| | | | 16397 | | | | 10935 | | | | 4541 | | | | 6732 | | | | 9109 | | | | | |
| | | | 60270 | | | | 64530 | | | | 70641 | | | 78719 | | | | | | | | | | |
| | | | 62330 | | | | 73264 | | | | 77805 | | | | 84537 | | | | 96110 | | | | | |
| 4 | | | 5 | 5 | | | 6 | 6 | | | 7 | 7 | | 7.75 | 8 | 8 | | | 9 | 9 | | | 10 | 10 |
| -4418 | | | -14338 | 2060 | | | -2201 | 8734 | | | 2623 | 7164 | | -914 | -914 | 5818 | | | 5818 | 17391 | | | 17391 | 19855 |
| 10 | 10 | | | | | | | | | | | | | | | | | | | | | | | |
| 17391 | 19855 | | | | | | | | | | | | | | | | | | | | | | | |

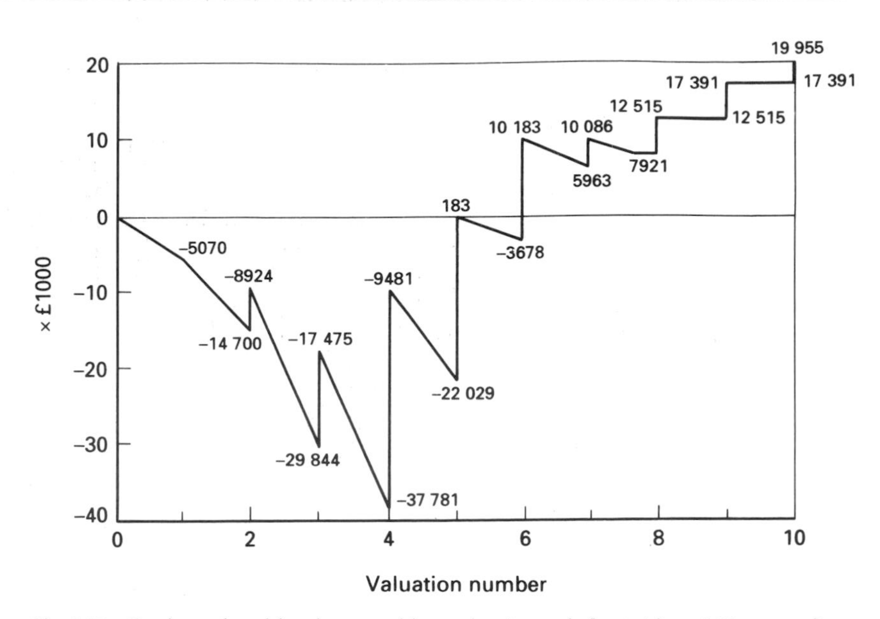


**Fig. 6.12** Graph produced by the spreadsheet, showing cash flow with activities at earliest times; 5% retention.

graph showing cash flow at earliest times is shown in Fig. 6.12. The power of the package is the manipulation, which can be carried out at tremendous speed. By simply omitting the '1' from 'Expenditure and income at earliest time' and inserting a '1' in 'Expenditure and income at latest time' the whole spreadsheet can be changed. Any slight adjustments resulting from operations starting part way through a week can then be made. The result is shown in Fig. 6.11, and the graph produced is shown in Fig. 6.13. These changed calculations and the production of the graph would have taken at least one and a half hours manually, and there is a high probability of mathematical errors being made.

Again, simply by altering the retention figure to 3% a complete new spreadsheet and graph can be produced, as shown in Figs 6.15 and 6.14 showing the cash flow with all the activities at their latest times. By altering the '1' back to its original position a further spreadsheet and graph can be produced showing the cash flow with activities at their earliest time, as shown in Figs 6.16 and 6.17.

Many other alternatives could be tried, requiring very little effort but with very fast results.

It can be seen that spreadsheets have much to offer the construction manager.

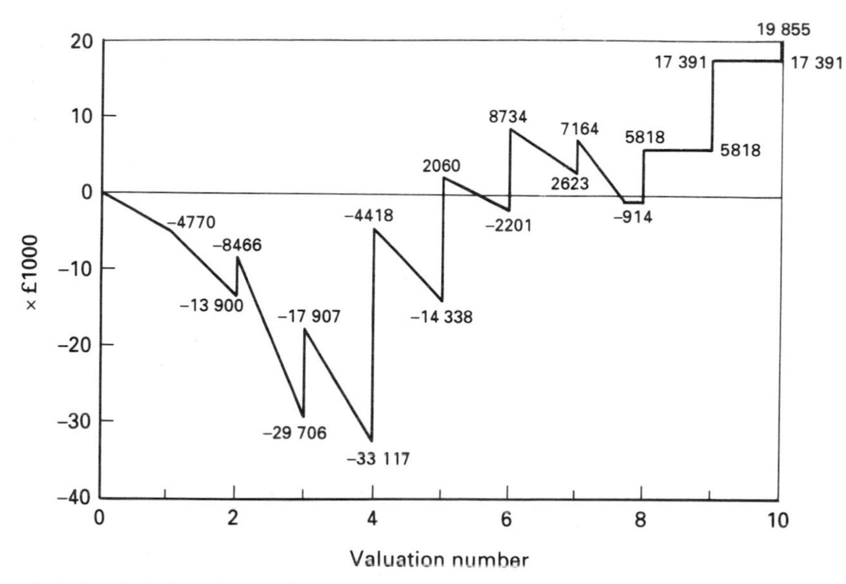

Fig. 6.13 Cash flow with activities at latest times; 5% retention.

## 6.5 Word processing

### *6.5.1 Introduction*

Word processing can make a valuable contribution to the efficiency of a construction company. Using word processing it is possible to do

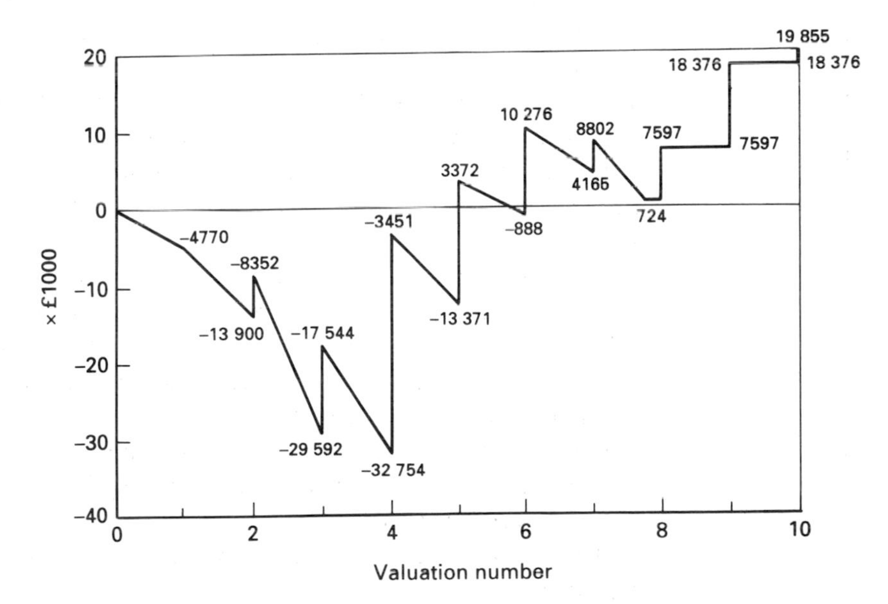

Fig. 6.14 Cash flow with activities at latest times; 3% retention.

Programme for small office building
Expenditure and income at earliest time Retention 3%
Expenditure and income at latest time

| Operation | Duration (days) | Planned expenditure | Planned income | Weeks 1 | 2 | 3 | 4 | 5 | 6 | 7 | 8 | 9 | 10 | 11 | 12 | 13 | 14 | 15 | 16 |
|---|---|---|---|---|---|---|---|---|---|---|---|---|---|---|---|---|---|---|---|
| Site set up | 3 | 210 | 240 | 210 | | | | | | | | | | | | | | | |
| | | | | 240 | | | | | | | | | | | | | | | |
| Setting out | 4 | 60 | 80 | 60 | | | | | | | | | | | | | | | |
| | | | | 80 | | | | | | | | | | | | | | | |
| Excavate to reduce level | 10 | 1000 | 1200 | 100 | 500 | 400 | | | | | | | | | | | | | |
| | | | | 120 | 600 | 480 | | | | | | | | | | | | | |
| Piling | 12 | 3600 | 4800 | | | 300 | 1500 | 1500 | 300 | | | | | | | | | | |
| | | | | | | 400 | 2000 | 2000 | 400 | | | | | | | | | | |
| Drains | 6 | 300 | 360 | | | 0 | 0 | 0 | 0 | 0 | 0 | 0 | 250 | 50 | | | | | |
| | | | | | | 0 | 0 | 0 | 0 | 0 | 0 | 0 | 300 | 60 | | | | | |
| Pad foundation | 9 | 630 | 720 | | | | | | 280 | 350 | | | | | | | | | |
| | | | | | | | | | 320 | 400 | | | | | | | | | |
| Strip foundation | 5 | 500 | 600 | | | | | | 0 | 0 | 0 | 0 | 400 | 100 | | | | | |
| | | | | | | | | | 0 | 0 | 0 | 0 | 480 | 120 | | | | | |
| In situ concrete frame | 22 | 22000 | 33000 | | | | | | | | 5000 | 5000 | 5000 | 5000 | 2000 | | | | |
| | | | | | | | | | | | 7500 | 7500 | 7500 | 7500 | 3000 | | | | |
| Ground floor slab | 5 | 800 | 900 | | | | | | | | | | | | 0 | 0 | 800 | | |
| | | | | | | | | | | | | | | | 0 | 0 | 900 | | |
| Brick infill panels | 15 | 8700 | 9750 | | | | | | | | | | | | 1740 | 2900 | 2900 | 1160 | |
| | | | | | | | | | | | | | | | 1950 | 3250 | 3250 | 1300 | |
| Precast stairs and floors | 3 | 2400 | 2700 | | | | | | | | | | | | | 0 | 0 | 2400 | |
| | | | | | | | | | | | | | | | | 0 | 0 | 2700 | |
| Hardwood frames | 10 | 1000 | 1200 | | | | | | | | | | | | | | | 300 | 500 |
| | | | | | | | | | | | | | | | | | | 360 | 600 |
| Precast roof slab | 2 | 2000 | 2500 | | | | | | | | | | | | | | 0 | 0 | 1000 |
| | | | | | | | | | | | | | | | | | 0 | 0 | 1250 |
| Block partitions ground floor | 7 | 1750 | 2100 | | | | | | | | | | | | | | 0 | 0 | 1250 |
| | | | | | | | | | | | | | | | | | 0 | 0 | 1500 |
| Glazing | 6 | 240 | 300 | | | | | | | | | | | | | | | | |
| External doors | 4 | 1600 | 1800 | | | | | | | | | | | | | | | | |
| Roof covering | 5 | 1000 | 1200 | | | | | | | | | | | | | | 0 | 0 | 0 |
| | | | | | | | | | | | | | | | | | 0 | 0 | 0 |
| External plumbing | 6 | 180 | 210 | | | | | | | | | | | | | | 0 | 0 | 0 |
| | | | | | | | | | | | | | | | | | 0 | 0 | 0 |
| Block partitions first floor | 7 | 2100 | 2450 | | | | | | | | | | | | | | | 0 | 0 |
| | | | | | | | | | | | | | | | | | | 0 | 0 |
| Plumbing first fix | 5 | 900 | 1050 | | | | | | | | | | | | | | | | 300 |
| | | | | | | | | | | | | | | | | | | | 350 |
| Electrical first fix | 8 | 400 | 480 | | | | | | | | | | | | | | | | |
| External painting | 12 | 240 | 300 | | | | | | | | | | | | | | | | 0 |
| | | | | | | | | | | | | | | | | | | | 0 |
| External works | 20 | 7564 | 9000 | | | | | | | | | | | | | | | | 0 |
| | | | | | | | | | | | | | | | | | | | 0 |
| Joinery first fix | 8 | 800 | 880 | | | | | | | | | | | | | | | | |
| Plastering | 18 | 1800 | 2160 | | | | | | | | | | | | | | | | |
| Floor screeding | 10 | 600 | 700 | | | | | | | | | | | | | | | | |
| Quarry tiling | 1 | 100 | 140 | | | | | | | | | | | | | | | | |
| Plumbing second fix | 10 | 1000 | 1200 | | | | | | | | | | | | | | | | |
| Joinery second fix | 10 | 1000 | 1200 | | | | | | | | | | | | | | | | |
| Internal painting | 14 | 420 | 504 | | | | | | | | | | | | | | | | |
| Vinyl tiling | 6 | 240 | 300 | | | | | | | | | | | | | | | | |
| Electrical second fix | 6 | 360 | 540 | | | | | | | | | | | | | | | | |
| Cleaning and handover | 1 | 50 | 60 | | | | | | | | | | | | | | | | |
| Preliminaries | | 13175 | 13950 | 425 | 425 | 425 | 425 | 425 | 425 | 425 | 425 | 425 | 425 | 425 | 425 | 425 | 425 | 425 | 425 |
| | | | | 450 | 450 | 450 | 450 | 450 | 450 | 450 | 450 | 450 | 450 | 450 | 450 | 450 | 450 | 450 | 450 |
| Planned expenditure | | 78719 | | 795 | 925 | 1125 | 1925 | 1925 | 1005 | 775 | 5425 | 5425 | 6075 | 5575 | 4165 | 3325 | 4125 | 4285 | 3475 |
| Planned income | | | 98574 | 890 | 1050 | 1330 | 2450 | 2450 | 1170 | 850 | 7950 | 7950 | 8730 | 8130 | 5400 | 3700 | 4600 | 4810 | 4150 |
| Planned expenditure monthly | | | | | | | 4770 | | | | 9130 | | | | 21240 | | | | 15210 |
| Planned income monthly | | | | | | | 5720 | | | | 12420 | | | | 30210 | | | | 17260 |
| Retention on planned income | | | | | | | 172 | | | | 373 | | | | 906 | | | | 518 |
| Income less retention time adjusted | | | | | | | 0 | | | | 5548 | | | | 12047 | | | | 29304 |
| Accumulative expenditure | | | | | | | 4770 | | | | 13900 | | | | 35140 | | | | 50350 |
| Accumulative income time adjusted | | | | | | | 0 | | | | 5548 | | | | 17596 | | | | 46900 |
| Valuation number | | | | | | | 1 | 1 | | | 2 | 2 | | | 3 | 3 | | | 4 |
| Difference between expenditure and income | | | | | | | -4770 | -4770 | | | -13900 | -8352 | | -29592 | | -17544 | | | -32754 |

Graph plottings

| 0 | 1 | 1 | 2 | 2 | 3 | 3 | 4 | 4 | 5 | 5 | 6 | 6 | 7 | 7 | 7.75 | 8 | 8 | 9 | 9 |
|---|---|---|---|---|---|---|---|---|---|---|---|---|---|---|---|---|---|---|---|
| 0 | -4770 | -4770 | -13900 | -8352 | -29592 | -17544 | -32754 | -3451 | -13371 | 3372 | -888 | 10276 | 4165 | 8802 | 724 | 724 | 7597 | 7597 | 18376 |

**Fig. 6.15** Activities at latest times; 3% retention.

| 17 | 18 | 19 | 20 | 21 | 22 | 23 | 24 | 25 | 26 | 27 | 28 | 29 | 30 | 31 | 32 | 33 | 34 | 35 | 36 | 37 | 38 | 39 | 59 | 60 |
|---|---|---|---|---|---|---|---|---|---|---|---|---|---|---|---|---|---|---|---|---|---|---|---|---|
| 200 | | | | | | | | | | | | | | | | | | | | | | | | |
| 240 | | | | | | | | | | | | | | | | | | | | | | | | |
| 1000 | | | | | | | | | | | | | | | | | | | | | | | | |
| 1250 | | | | | | | | | | | | | | | | | | | | | | | | |
| 500 | | | | | | | | | | | | | | | | | | | | | | | | |
| 600 | | | | | | | | | | | | | | | | | | | | | | | | |
| 120 | 120 | | | | | | | | | | | | | | | | | | | | | | | |
| 150 | 150 | | | | | | | | | | | | | | | | | | | | | | | |
| 1200 | 400 | | | | | | | | | | | | | | | | | | | | | | | |
| 1350 | 450 | | | | | | | | | | | | | | | | | | | | | | | |
| 600 | 400 | | | | | | | | | | | | | | | | | | | | | | | |
| 720 | 480 | | | | | | | | | | | | | | | | | | | | | | | |
| 90 | 90 | | | | | | | | | | | | | | | | | | | | | | | |
| 105 | 105 | | | | | | | | | | | | | | | | | | | | | | | |
| 900 | 1200 | | | | | | | | | | | | | | | | | | | | | | | |
| 1050 | 1400 | | | | | | | | | | | | | | | | | | | | | | | |
| 600 | | | | | | | | | | | | | | | | | | | | | | | | |
| 700 | | | | | | | | | | | | | | | | | | | | | | | | |
| | 100 | 250 | 50 | | | | | | | | | | | | | | | | | | | | | |
| | 120 | 300 | 60 | | | | | | | | | | | | | | | | | | | | | |
| 0 | 0 | 0 | 0 | 0 | 0 | 0 | 0 | 0 | 0 | 0 | 0 | 40 | 100 | 100 | | | | | | | | | | |
| 0 | 0 | 0 | 0 | 0 | 0 | 0 | 0 | 0 | 0 | 0 | 0 | 50 | 125 | 125 | | | | | | | | | | |
| 0 | 0 | 0 | 0 | 0 | 0 | 0 | 0 | 0 | 0 | 0 | 1891 | 1891 | 1891 | 1891 | | | | | | | | | | |
| 0 | 0 | 0 | 0 | 0 | 0 | 0 | 0 | 0 | 0 | 0 | 2250 | 2250 | 2250 | 2250 | | | | | | | | | | |
| | | | 0 | 0 | 300 | 500 | | | | | | | | | | | | | | | | | | |
| | | | 0 | 0 | 330 | 550 | | | | | | | | | | | | | | | | | | |
| | | | 400 | 500 | 500 | 400 | | | | | | | | | | | | | | | | | | |
| | | | 480 | 600 | 600 | 480 | | | | | | | | | | | | | | | | | | |
| | | | | | | 60 | 300 | 240 | | | | | | | | | | | | | | | | |
| | | | | | | 70 | 350 | 280 | | | | | | | | | | | | | | | | |
| | | | | | | | | 0 | 0 | 100 | | | | | | | | | | | | | | |
| | | | | | | | | 0 | 0 | 140 | | | | | | | | | | | | | | |
| | | | | | | | | 100 | 500 | 400 | | | | | | | | | | | | | | |
| | | | | | | | | 120 | 600 | 480 | | | | | | | | | | | | | | |
| | | | | | | | | 100 | 500 | 400 | | | | | | | | | | | | | | |
| | | | | | | | | 120 | 600 | 480 | | | | | | | | | | | | | | |
| | | | | | | | | | | 30 | 150 | 150 | 90 | | | | | | | | | | | |
| | | | | | | | | | | 36 | 180 | 180 | 108 | | | | | | | | | | | |
| | | | | | | | | | | | | | 80 | 160 | | | | | | | | | | |
| | | | | | | | | | | | | | 100 | 200 | | | | | | | | | | |
| | | | | | | | | | | | | | 120 | 240 | | | | | | | | | | |
| | | | | | | | | | | | | | 180 | 360 | | | | | | | | | | |
| | | | | | | | | | | | | | | 50 | | | | | | | | | | |
| | | | | | | | | | | | | | | 60 | | | | | | | | | | |
| 425 | 425 | 425 | 425 | 425 | 425 | 425 | 425 | 425 | 425 | 425 | 425 | 425 | 425 | 425 | | | | | | | | | | |
| 450 | 450 | 450 | 450 | 450 | 450 | 450 | 450 | 450 | 450 | 450 | 450 | 450 | 450 | 450 | | | | | | | | | | |
| 5635 | 2735 | 675 | 875 | 925 | 1225 | 1385 | 725 | 865 | 1425 | 1355 | 2466 | 2506 | 2706 | 2866 | | | | | | | | | | |
| 6615 | 3155 | 750 | 990 | 1050 | 1380 | 1550 | 800 | 970 | 1650 | 1586 | 2880 | 2930 | 3213 | 3445 | | | | | | | | | | |
| | | | 9920 | | | | 4260 | | | | 6111 | | | 8078 | | | | | | | | | | |
| | | | 11510 | | | | 4780 | | | | 7086 | | | 9588 | | | | | | | | | | |
| | | | 345 | | | | 143 | | | | 213 | | | 288 | | | | | | | | | | |
| | | | 16742 | | | | 11165 | | | | 4637 | | | | 6873 | | | | 9300 | | | | | |
| | | | 60270 | | | | 64530 | | | | 70641 | | | 78719 | | | | | | | | | | |
| | | | 63642 | | | | 74806 | | | | 79443 | | | | 86316 | | | | 97095 | | | | | |
| 4 | | | 5 | 5 | | | 6 | 6 | | | 7 | 7 | | 7.75 | 8 | 8 | | | 9 | 9 | | | 10 | 10 |
| -3451 | | | -13371 | 3372 | | | -888 | 10276 | | | 4165 | 8802 | | 724 | 724 | 7597 | | | 7597 | 18376 | | | 18376 | 19855 |
| 10 | 10 | | | | | | | | | | | | | | | | | | | | | | | |
| 18376 | 19855 | | | | | | | | | | | | | | | | | | | | | | | |

Programme for small office building
Expenditure and income at earliest time 1 Retention 3%
Expenditure and income at latest time

| Operation | Duration (days) | Planned expenditure | Planned income | Weeks 1 | 2 | 3 | 4 | 5 | 6 | 7 | 8 | 9 | 10 | 11 | 12 | 13 | 14 | 15 | 16 |
|---|---|---|---|---|---|---|---|---|---|---|---|---|---|---|---|---|---|---|---|
| Site set up | 3 | 210 | 240 | 210 | | | | | | | | | | | | | | | |
| | | | | 240 | | | | | | | | | | | | | | | |
| Setting out | 4 | 60 | 80 | 60 | | | | | | | | | | | | | | | |
| | | | | 80 | | | | | | | | | | | | | | | |
| Excavate to reduce level | 10 | 1000 | 1200 | 100 | 500 | 400 | | | | | | | | | | | | | |
| | | | | 120 | 600 | 480 | | | | | | | | | | | | | |
| Piling | 12 | 3600 | 4800 | | | 300 | 1500 | 1500 | 300 | | | | | | | | | | |
| | | | | | | 400 | 2000 | 2000 | 400 | | | | | | | | | | |
| Drains | 6 | 300 | 360 | | | 50 | 250 | 0 | 0 | 0 | 0 | 0 | 0 | 0 | | | | | |
| | | | | | | 60 | 300 | 0 | 0 | 0 | 0 | 0 | 0 | 0 | | | | | |
| Pad foundation | 9 | 630 | 720 | | | | | | 280 | 350 | | | | | | | | | |
| | | | | | | | | | 320 | 400 | | | | | | | | | |
| Strip foundation | 5 | 500 | 600 | | | | | | 400 | 100 | 0 | 0 | 0 | 0 | | | | | |
| | | | | | | | | | 480 | 120 | 0 | 0 | 0 | 0 | | | | | |
| In situ concrete frame | 22 | 22000 | 33000 | | | | | | | | 5000 | 5000 | 5000 | 5000 | 2000 | | | | |
| | | | | | | | | | | | 7500 | 7500 | 7500 | 7500 | 3000 | | | | |
| Ground floor slab | 5 | 800 | 900 | | | | | | | | | | | | 480 | 320 | 0 | | |
| | | | | | | | | | | | | | | | 540 | 360 | 0 | | |
| Brick infill panels | 15 | 8700 | 9750 | | | | | | | | | | | | 1740 | 2900 | 2900 | 1160 | |
| | | | | | | | | | | | | | | | 1950 | 3250 | 3250 | 1300 | |
| Precast stairs and floors | 3 | 2400 | 2700 | | | | | | | | | | | | | 2400 | 0 | 0 | |
| | | | | | | | | | | | | | | | | 2700 | 0 | 0 | |
| Hardwood frames | 10 | 1000 | 1200 | | | | | | | | | | | | | | | 300 | 500 |
| | | | | | | | | | | | | | | | | | | 360 | 600 |
| Precast roof slab | 2 | 2000 | 2500 | | | | | | | | | | | | | | 2000 | 0 | 0 |
| | | | | | | | | | | | | | | | | | 2500 | 0 | 0 |
| Block partitions ground floor | 7 | 1750 | 2100 | | | | | | | | | | | | | | 1250 | 500 | 0 |
| | | | | | | | | | | | | | | | | | 1500 | 600 | 0 |
| Glazing | 6 | 240 | 300 | | | | | | | | | | | | | | | | |
| External doors | 4 | 1600 | 1800 | | | | | | | | | | | | | | | | |
| Roof covering | 5 | 1000 | 1200 | | | | | | | | | | | | | | 600 | 400 | 0 |
| | | | | | | | | | | | | | | | | | 720 | 480 | 0 |
| External plumbing | 6 | 180 | 210 | | | | | | | | | | | | | | 90 | 90 | 0 |
| | | | | | | | | | | | | | | | | | 105 | 105 | 0 |
| Block partitions first floor | 7 | 2100 | 2450 | | | | | | | | | | | | | | | 900 | 1200 |
| | | | | | | | | | | | | | | | | | | 1050 | 1400 |
| Plumbing first fix | 5 | 900 | 1050 | | | | | | | | | | | | | | | | 300 |
| | | | | | | | | | | | | | | | | | | | 350 |
| Electrical first fix | 8 | 400 | 480 | | | | | | | | | | | | | | | | |
| External painting | 12 | 240 | 300 | | | | | | | | | | | | | | | | 40 |
| | | | | | | | | | | | | | | | | | | | 50 |
| External works | 20 | 7564 | 9000 | | | | | | | | | | | | | | | | 756 |
| | | | | | | | | | | | | | | | | | | | 900 |
| Joinery first fix | 8 | 800 | 880 | | | | | | | | | | | | | | | | |
| Plastering | 18 | 1800 | 2160 | | | | | | | | | | | | | | | | |
| Floor screeding | 10 | 600 | 700 | | | | | | | | | | | | | | | | |
| Quarry tiling | 1 | 100 | 140 | | | | | | | | | | | | | | | | |
| Plumbing second fix | 10 | 1000 | 1200 | | | | | | | | | | | | | | | | |
| Joinery second fix | 10 | 1000 | 1200 | | | | | | | | | | | | | | | | |
| Internal painting | 14 | 420 | 504 | | | | | | | | | | | | | | | | |
| Vinyl tiling | 6 | 240 | 300 | | | | | | | | | | | | | | | | |
| Electrical second fix | 6 | 360 | 540 | | | | | | | | | | | | | | | | |
| Cleaning and handover | 1 | 50 | 60 | | | | | | | | | | | | | | | | |
| Preliminaries | | 13175 | 13950 | 425 | 425 | 425 | 425 | 425 | 425 | 425 | 425 | 425 | 425 | 425 | 425 | 425 | 425 | 425 | 425 |
| | | | | 450 | 450 | 450 | 450 | 450 | 450 | 450 | 450 | 450 | 450 | 450 | 450 | 450 | 450 | 450 | 450 |
| Planned expenditure | | 78719 | | 795 | 925 | 1175 | 2175 | 1925 | 1405 | 875 | 5425 | 5425 | 5425 | 5425 | 4645 | 6045 | 7265 | 3773 | 3221 |
| Planned income | | | 98574 | 890 | 1050 | 1390 | 2750 | 2450 | 1650 | 970 | 7950 | 7950 | 7950 | 7950 | 5940 | 6760 | 8525 | 4345 | 3750 |
| Planned expenditure monthly | | | | | | | 5070 | | | | 9630 | | | | 20920 | | | | 20306 |
| Planned income monthly | | | | | | | 6080 | | | | 13020 | | | | 29790 | | | | 23380 |
| Retention on planned income | | | | | | | 182 | | | | 391 | | | | 894 | | | | 701 |
| Income less retention time adjusted | | | | | | | 0 | | | | 5898 | | | | 12629 | | | | 28896 |
| Accumulative expenditure | | | | | | | 5070 | | | | 14700 | | | | 35620 | | | | 55926 |
| Accumulative income time adjusted | | | | | | | 0 | | | | 5898 | | | | 18527 | | | | 47423 |
| Valuation number | | | | | | | 1 | 1 | | | 2 | 2 | | | 3 | 3 | | | 4 |
| Difference between expenditure and income | | | | | | | -5070 | -5070 | | | -14700 | -8802 | | | -29722 | -17093 | | | -37399 |

Graph plottings

| | | | | | | | | | | | | | | | | | | | |
|---|---|---|---|---|---|---|---|---|---|---|---|---|---|---|---|---|---|---|---|
| 0 | 1 | 1 | 2 | 2 | 3 | 3 | 4 | 4 | 5 | 5 | 6 | 6 | 7 | 7 | 7.75 | 8 | 8 | 9 | 9 |
| 0 | -5070 | -5070 | -14700 | -8802 | -29722 | -17093 | -37399 | -8503 | -21051 | 1628 | -2232 | 11920 | 7700 | 11910 | 9745 | 9745 | 14436 | 14436 | 18376 |

**Fig. 6.16** Activities at earliest times; 3% retention.

| 17 | 18 | 19 | 20 | 21 | 22 | 23 | 24 | 25 | 26 | 27 | 28 | 29 | 30 | 31 | 32 | 33 | 34 | 35 | 36 | 37 | 38 | 39 | 59 | 60 |
|---|---|---|---|---|---|---|---|---|---|---|---|---|---|---|---|---|---|---|---|---|---|---|---|---|
| 200 | | | | | | | | | | | | | | | | | | | | | | | | |
| 240 | | | | | | | | | | | | | | | | | | | | | | | | |
| 0 | | | | | | | | | | | | | | | | | | | | | | | | |
| 0 | | | | | | | | | | | | | | | | | | | | | | | | |
| 0 | | | | | | | | | | | | | | | | | | | | | | | | |
| 0 | | | | | | | | | | | | | | | | | | | | | | | | |
| 120 | 120 | | | | | | | | | | | | | | | | | | | | | | | |
| 150 | 150 | | | | | | | | | | | | | | | | | | | | | | | |
| 1200 | 400 | | | | | | | | | | | | | | | | | | | | | | | |
| 1350 | 450 | | | | | | | | | | | | | | | | | | | | | | | |
| 0 | 0 | | | | | | | | | | | | | | | | | | | | | | | |
| 0 | 0 | | | | | | | | | | | | | | | | | | | | | | | |
| 0 | 0 | | | | | | | | | | | | | | | | | | | | | | | |
| 0 | 0 | | | | | | | | | | | | | | | | | | | | | | | |
| 0 | 0 | | | | | | | | | | | | | | | | | | | | | | | |
| 0 | 0 | | | | | | | | | | | | | | | | | | | | | | | |
| 600 | | | | | | | | | | | | | | | | | | | | | | | | |
| 700 | | | | | | | | | | | | | | | | | | | | | | | | |
| | 100 | 250 | 50 | | | | | | | | | | | | | | | | | | | | | |
| | 120 | 300 | 60 | | | | | | | | | | | | | | | | | | | | | |
| 100 | 100 | 0 | 0 | 0 | 0 | 0 | 0 | 0 | 0 | 0 | 0 | 0 | 0 | 0 | | | | | | | | | | |
| 125 | 125 | 0 | 0 | 0 | 0 | 0 | 0 | 0 | 0 | 0 | 0 | 0 | 0 | 0 | | | | | | | | | | |
| 1891 | 1891 | 1891 | 1135 | 0 | 0 | 0 | 0 | 0 | 0 | 0 | 0 | 0 | 0 | 0 | | | | | | | | | | |
| 2250 | 2250 | 2250 | 1350 | 0 | 0 | 0 | 0 | 0 | 0 | 0 | 0 | 0 | 0 | 0 | | | | | | | | | | |
| | | | 400 | 400 | 0 | 0 | | | | | | | | | | | | | | | | | | |
| | | | 440 | 440 | 0 | 0 | | | | | | | | | | | | | | | | | | |
| | | | 400 | 500 | 500 | 400 | | | | | | | | | | | | | | | | | | |
| | | | 480 | 600 | 600 | 480 | | | | | | | | | | | | | | | | | | |
| | | | | | | 60 | 300 | 240 | | | | | | | | | | | | | | | | |
| | | | | | | 70 | 350 | 280 | | | | | | | | | | | | | | | | |
| | | | | | | | | 100 | 0 | 0 | | | | | | | | | | | | | | |
| | | | | | | | | 140 | 0 | 0 | | | | | | | | | | | | | | |
| | | | | | | | | 100 | 500 | 400 | | | | | | | | | | | | | | |
| | | | | | | | | 120 | 600 | 480 | | | | | | | | | | | | | | |
| | | | | | | | | 100 | 500 | 400 | | | | | | | | | | | | | | |
| | | | | | | | | 120 | 600 | 480 | | | | | | | | | | | | | | |
| | | | | | | | | | | 30 | 150 | 150 | 90 | | | | | | | | | | | |
| | | | | | | | | | | 36 | 180 | 180 | 108 | | | | | | | | | | | |
| | | | | | | | | | | | | | 80 | 160 | | | | | | | | | | |
| | | | | | | | | | | | | | 100 | 200 | | | | | | | | | | |
| | | | | | | | | | | | | | 120 | 240 | | | | | | | | | | |
| | | | | | | | | | | | | | 180 | 360 | | | | | | | | | | |
| | | | | | | | | | | | | | | 50 | | | | | | | | | | |
| | | | | | | | | | | | | | | 60 | | | | | | | | | | |
| 425 | 425 | 425 | 425 | 425 | 425 | 425 | 425 | 425 | 425 | 425 | 425 | 425 | 425 | 425 | | | | | | | | | | |
| 450 | 450 | 450 | 450 | 450 | 450 | 450 | 450 | 450 | 450 | 450 | 450 | 450 | 450 | 450 | | | | | | | | | | |
| 4536 | 3036 | 2566 | 2410 | 1325 | 925 | 885 | 725 | 965 | 1425 | 1255 | 575 | 575 | 715 | 875 | | | | | | | | | | |
| 5265 | 3545 | 3000 | 2780 | 1490 | 1050 | 1000 | 800 | 1110 | 1650 | 1446 | 630 | 630 | 838 | 1070 | | | | | | | | | | |
| | | | 12548 | | | | 3860 | | | | 4220 | | | 2165 | | | | | | | | | | |
| | | | 14590 | | | | 4340 | | | | 4836 | | | 2538 | | | | | | | | | | |
| | | | 438 | | | | 130 | | | | 145 | | | 76 | | | | | | | | | | |
| | | | 22679 | | | | 14152 | | | | 4210 | | | | 4691 | | | | 2462 | | | | | |
| | | | 68474 | | | | 72334 | | | | 76554 | | | 78719 | | | | | | | | | | |
| | | | 70102 | | | | 84254 | | | | 88464 | | | | 93155 | | | | 97095 | | | | | |
| 4 | | | 5 | 5 | | | 6 | 6 | | | 7 | 7 | | 7.75 | 8 | 8 | | | 9 | 9 | | | 10 | 10 |
| -8503 | | | -21051 | 1628 | | | -2232 | 11920 | | | 7700 | 11910 | | 9745 | 9745 | 14436 | | | 14436 | 18376 | | | 18376 | 19855 |

| 10 | 10 |
|---|---|
| 18376 | 19855 |

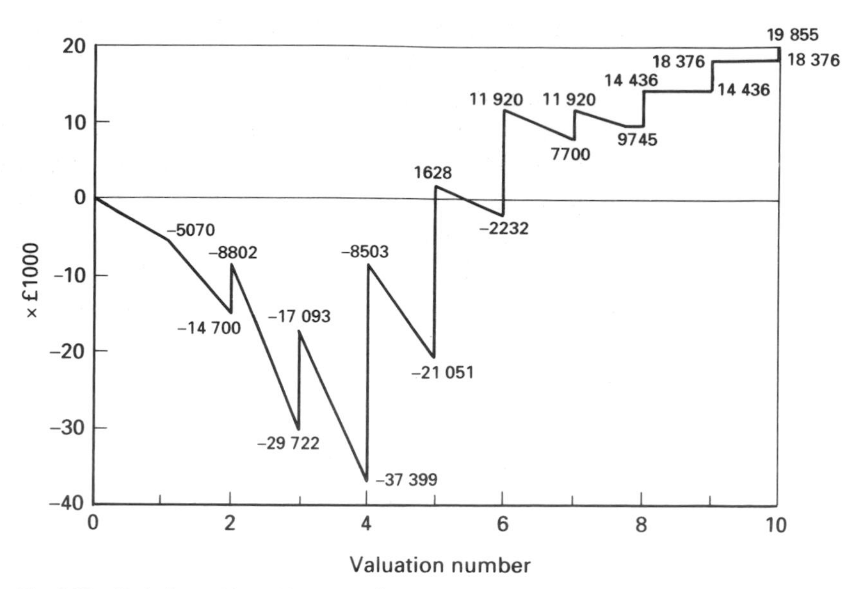


**Fig. 6.17** Cash flow with activities at earliest times; 3% retention.

everything that can be done on a typewriter and much more. Some advantages over a standard electric typewriter are the simple quick manipulation of text, as follows:

- ❑ Margins, spacing between lines, and length of lines can be altered in seconds.
- ❑ Whole documents, paragraphs or blocks of text can be reformed, saved, moved, copied, deleted, altered or called from disk.
- ❑ Spelling can be corrected automatically.
- ❑ Simple mailing facilities can be implemented.
- ❑ Line height and character width can be changed and many other facilities exploited.

### *6.5.2 The use of word processing to the construction manager*

Whilst the construction manager may not physically use word processing (although some do) it has much to offer because of the above capabilities. Typical uses could be as follows:

- ❑ Setting up a library of standard descriptions for building up work packages based on method statements for targeting purposes.
- ❑ In the case of small firms, setting up a standard specification form to be sent out with quotations. This could then be edited to suit the particular project in question.

- Setting up standard documents, e.g. agendas and minutes for meetings, and the many other forms that may simply need editing when being reused.
- Again in the case of small firms, setting up a library of descriptions for standard quotations or standard paragraphs for repetitive work.
- Keeping employee data up to date. Simple amendments can be made as circumstances change.
- Standard headings and endings for letters and other documents.
- Sending out general information to clients and subcontractors using 'mail-merge'.
- Producing draft reports.

There are many more uses for word processing, and its use can save much time for secretaries and typists. The text produced by word processing can result in extremely professional-looking documents.

Chapter 7

# Case Studies

## 7.1 Case study 1: Refurbishment project

### 7.1.1 *Purpose*

The purpose of this case study is to show the integration of the production process from the tender stage through to the planning and control stages.

Much has been said of the need to integrate the management processes, and this should be the aim of all construction firms.

### 7.1.2 *Vehicle for carrying out the work*

In order to illustrate the integration of the processes, the MASTER estimating and production system, which is described in Chapter 6 was used, but it is of course possible to carry out the procedures manually.

### 7.1.3 *Description of the project*

The project consisted of the complete refurbishment of a number of terrace houses, which were very similar in construction.

To help in understanding the procedure, *one* of the terrace houses has been illustrated. The extent of the work can be seen in Figs 7.1 and 7.2.

Typical of many terraced houses, there was no front garden, the property fronting onto the road. There was a rear garden area.

A photograph of this project is shown in Fig. 7.3.

### 7.1.4 *Architect and main contractor*

- *Architect*: Ritchie & Rennie Architects
- *Main contractor*: O & P Construction Services Limited.

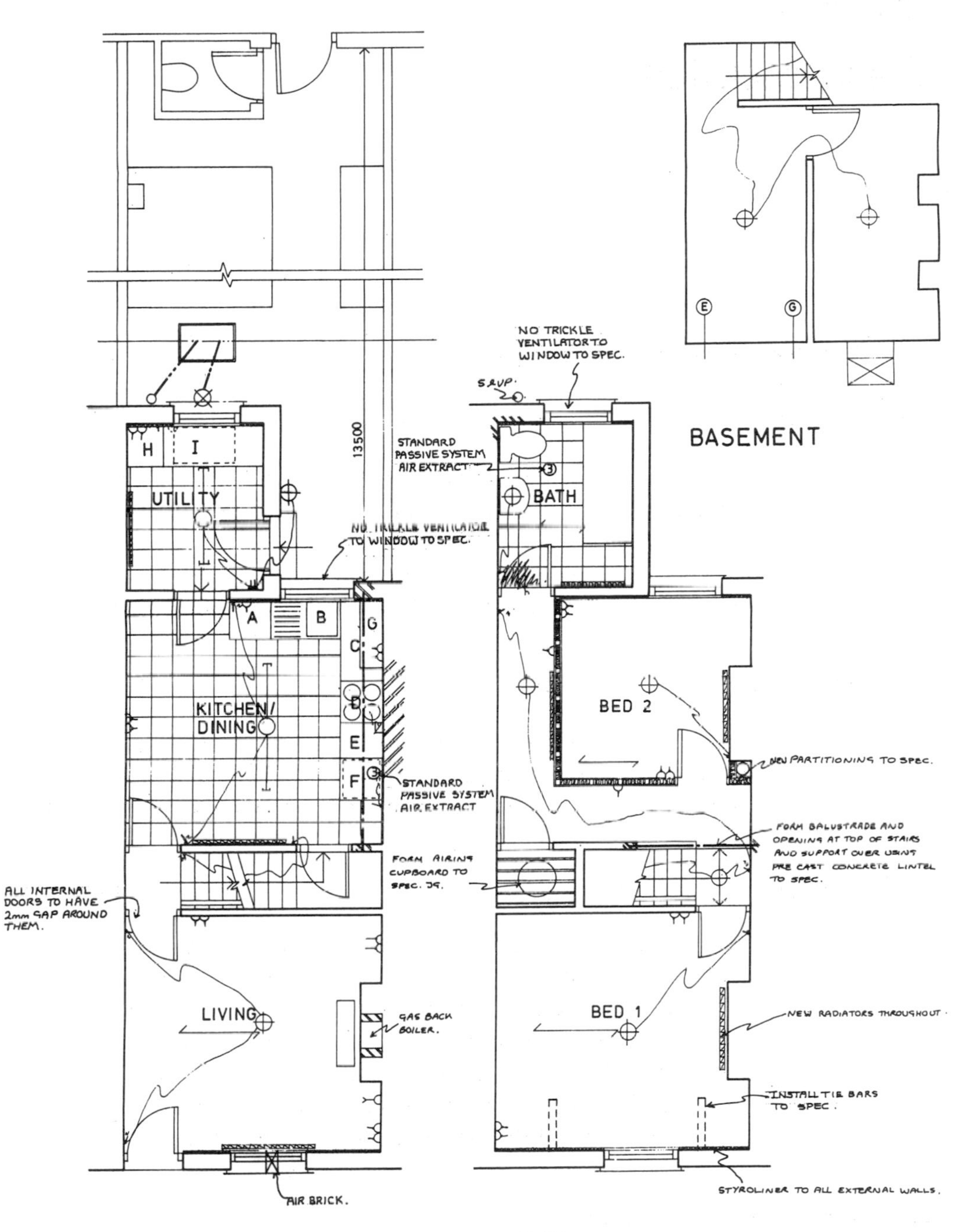


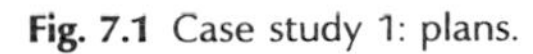

**Fig. 7.1** Case study 1: plans.

81 NORTH STREET

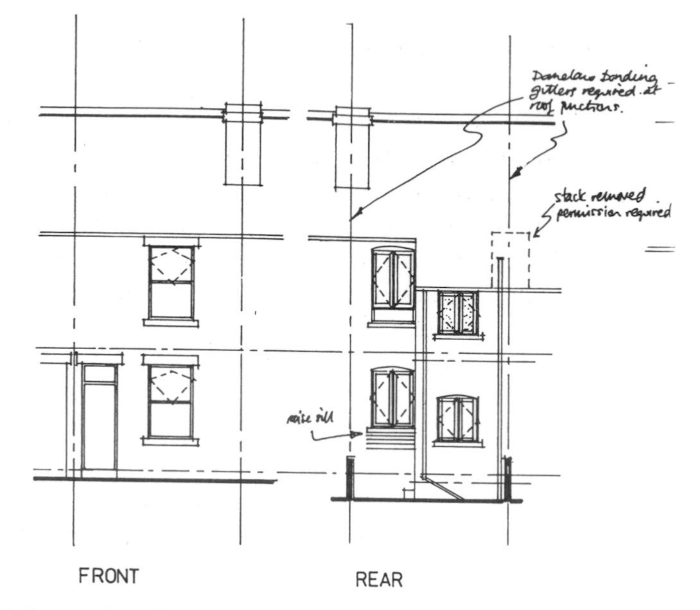


**Fig. 7.2** Case study 1: elevations.

**Fig. 7.3** Case study 1: 81 North Street, Doncaster.

### 7.1.5 Documents provided at tender stage

The drawings (Figs 7.1 and 7.2) were accompanied by a Schedule of Work, which included quantities. A section of the schedule can be seen in Figs 7.4 and 7.5. In addition to the schedule a comprehensive detailed specification was provided, the clauses of which were referred to in the Schedule of Work.

As an example, on page 146 in the Schedule of Work Clause 1.01 refers to specification CE.

The relevant specification clause for this item was as follows:

CE NEW CEILING – VAPOUR CHECKED

PREPARATION: In the case of existing ceilings to the first floor strip out and cart away existing ceiling plaster, laths or boards, nails and mouldings if present, prepare for new ceiling by packing out timbers so as to give a true and level surface. The Architect's attention should be brought to cases where excessive packing is required; install noggins at max 450 mm c/c and at perimeter (for setting out reasons the use of 2.4 × 1.2 m boards should be avoided, and it is preferred that 900 mm boards be employed).
CEILING: Provide and fix new ceiling as 12.5 mm Gyproc Duplex board with grey paper face using 40 mm galvanised nails @ 150 mm c/c. Scrim all joints and skim, minimum thickness 3 mm. Make good to walls.

### 7.1.6 Estimating and tendering procedure

It can be seen from the extract from the Schedule of Work that the items are scheduled room by room, and while they could be priced as shown, it was decided to group operations to save time at the estimating stage. As an example, 'Strip out ceiling and replace with vapour check ceiling to specification CE' appears in a number of rooms. These were grouped together under one item. When the items had been grouped they were then entered into the computer.

When considering each item the following data were entered:

- ❑ description of operation
- ❑ quantity and unit ($m^3$, $m^2$, m, etc.)
- ❑ all-in labour rate
- ❑ all-in material rate

File 50.82 81 North Street, Doncaster Page 146

| | | | B.F. | |
|---|---|---|---|---|
| | ***Bedroom 1*** | | | |
| | *Ceiling* | | | |
| 1.01 | Strip out existing ceiling and replace with vapour check ceiling to spec | CE | 12 m$^2$ | |
| | *Walls* | | | |
| 1.02 | Strip off existing finish and supply and fix wall insulation to spec | PE | 8 m$^2$ | |
| 1.03 | Replaster to spec | PB | 28 m$^2$ | |
| 1.04 | Remove hearthstone and vent flue to spec | STJ | 1 no | |
| 1.05 | Supply and fix skirtings to spec | JF | 14 m | |
| | *Window* | | | |
| 1.06 | Replace window to spec<br>Note: top hung with Cotswold reflex hinges | WB2 | 1 no | |
| | *Door* | | | |
| 1.07 | Supply and fix internal door and frame to spec | DC | 1 no | |
| 1.08 | Supply and fix new ironmongery to spec | DA | 1 no | |
| | *Floor* | | | |
| 1.09 | Supply and install tie bars as detail sheet and to spec | STT | 2 no | |

**Fig. 7.4** Case study 1: part of the schedule of work.

File 50.82 81 North Street, Doncaster Page 147

| | | | B.F. | |
|---|---|---|---|---|
| | ***Bedroom 2 and corridor*** | | | |
| | *Ceilings* | | | |
| 2.01 | Strip out existing ceiling and replace with vapour check ceiling to spec | CE | 13.5 m$^2$ | |
| | *Walls* | | | |
| 2.02 | Strip off existing finish and supply and fix wall insulation to spec | PE | 6 m$^2$ | |
| 2.03 | Replaster to spec | PB | 35 m$^2$ | |
| 2.04 | Construct partition filled with insulation between studs and spec | JA | | |
| 2.05 | Remove hearthstone and vent flue to spec | STJ | 1 no | |
| 2.06 | Construct airing cupboard shelves to spec | JG | 2 no | |
| 2.07 | Supply and fix skirtings to spec | JF | 15m | |
| | *Window* | | | |
| 2.08 | Replace window to spec | WB8 | 1 no | |
| | *Doors* | | | |
| 2.09 | Supply and fix internal door and frame to spec | DC | 2 no | |
| 2.10 | Supply and fix new ironmongery to spec | DA | 2 no | |

**Fig. 7.5** Case study 1: part of the schedule of work.

- all-in plant rate
- subcontractor rate or subcontractor quotation
- target description*
- location*.

The data shown * could have been left until the project was won, but *all* data were entered at the estimating stage. The target description, e.g. Car 1, was given to all items that were included in that operation.

### *Estimate for management*

When all items had been entered a report was generated, as shown in Fig. 7.6.

This estimate was then used as a basis for converting into a tender. It set out the labour, materials and plant and subcontractor costs and the total costs.

### *Preliminaries*

Based on the above figures and a site survey, which had already been carried out for the purpose of gathering information relating to access,

81 North Street, Doncaster - estimate for management Page 1

| Ref | Trad | No | Targ | Output | Quantity | Unit | Description | Lab cost | Mat cost | Plt cost | Scn cost | Totals |
|---|---|---|---|---|---|---|---|---|---|---|---|---|
| 5 | Titl | 9000 | | | | | *Ceilings* | | | | | |
| 10 | Lab | 9001 | Prep | 0.45 | 78.50 | $m^2$ | Strip down ceiling, denail joists and remove to skip | 167.09 | 0.00 | 17.27 | 0.00 | 184.36 |
| 15 | Plas | 9002 | Plas | 0.63 | 33.50 | $m^2$ | 12.5 mm foil backed plasterboard and skim (CE) | 115.44 | 70.35 | 0.00 | 0.00 | 185.79 |
| 20 | Plas | 9003 | Plas | 0.63 | 33.00 | $m^2$ | 12.5 plaster board and skim to ceilings (CA) | 113.72 | 61.05 | 0.00 | 0.00 | 174.77 |
| 100 | Titl | 9004 | | | | | *Wall plastering* | | | | | |
| 105 | Lab | 9005 | Prep | 0.45 | 157.00 | $m^2$ | Hack off wall plaster and remove to skip | 334.17 | 0.00 | 34.54 | 0.00 | 368.71 |
| 110 | Plas | 9006 | Plas | 1.02 | 53.00 | $m^2$ | 39.5 mm styroliner and skim to walls (PE) | 295.71 | 530.00 | 0.00 | 0.00 | 825.71 |
| 115 | Plas | 9007 | Plas | 0.52 | 69.00 | $m^2$ | Replast walls with additive (PA) | 196.26 | 165.60 | 0.00 | 0.00 | 361.86 |
| 120 | Plas | 9008 | Plas | 0.52 | 86.00 | $m^2$ | Replaster walls (PB) | 244.62 | 133.30 | 0.00 | 0.00 | 377.92 |
| 200 | Titl | 9009 | | | | | *Partitions* | | | | | |
| 205 | Carp | 9010 | Car1 | 0.50 | 2.00 | no | Short length of joist bolted to wall as trimmer | 5.47 | 4.00 | 0.00 | 0.00 | 9.47 |
| 210 | Carp | 9011 | Car1 | 0.50 | 2.00 | no | Joist hangers | 5.47 | 1.60 | 0.00 | 0.00 | 7.07 |
| 215 | Carp | 9012 | Car1 | 0.33 | 3.80 | m | Floor joist under partition | 6.86 | 6.65 | 0.00 | 0.00 | 13.51 |
| 220 | Carp | 9013 | Car1 | 0.25 | 79.00 | m | Construct stud walling | 108.03 | 59.25 | 0.00 | 0.00 | 167.28 |
| 225 | Plas | 9014 | Plas | 0.23 | 16.00 | $m^2$ | Insulation to stud walling | 20.13 | 32.00 | 0.00 | 0.00 | 52.13 |
| 230 | Plas | 9015 | Plas | 0.63 | 32.00 | $m^2$ | Plaster board and skim to stud partitions | 110.28 | 67.20 | 0.00 | 0.00 | 177.48 |
| 235 | Carp | 9016 | Car2 | 0.20 | 9.00 | m | Skirting to partitions | 9.85 | 8.55 | 0.00 | 0.00 | 18.40 |
| 300 | Titl | 9017 | | | | | *Rem h/stone and vent flue (STJ)* | | | | | |
| 305 | Lab | 9018 | Prep | 1.00 | 2.00 | no | Remove hearthstone and deposit in skip | 9.46 | 0.00 | 2.60 | 0.00 | 12.06 |
| | | | | | | | Page totals | 1742.56 | 1139.55 | 54.41 | 0.00 | 2936.52 |

**Fig. 7.6** Case study 1: estimate for management.

81 North Street, Doncaster - estimate for management Page 2

| Ref | Trad | No | Targ | Output | Quantity | Unit | Description | Lab cost | Mat cost | Plt cost | Scn cost | Totals |
|---|---|---|---|---|---|---|---|---|---|---|---|---|
| 310 | Carp | 9019 | Car1 | 0.50 | 4.00 | m | New joists in opening to match existing | 10.94 | 6.00 | 0.00 | 0.00 | 16.94 |
| 315 | Carp | 9020 | Car1 | 0.25 | 8.00 | no | Joist hangers to hearth joists | 10.94 | 6.40 | 0.00 | 0.00 | 17.34 |
| 320 | Blr | 9021 | Blr1 | 1.00 | 2.00 | no | Form opening at high level in flue for plaster vent | 10.94 | 0.00 | 0.00 | 0.00 | 10.94 |
| 325 | Plas | 9022 | Plas | 0.50 | 2.00 | no | 225x150 plaster vent in flue | 5.47 | 1.50 | 0.00 | 0.00 | 6.97 |
| 330 | Carp | 9023 | Car1 | 1.50 | 1.00 | $m^2$ | Floor boards in hearth area to match existing | 8.21 | 10.00 | 0.00 | 0.00 | 18.21 |
| 335 | Subc | 9024 | Subc | 0.00 | 2.00 | no | Sweep flues | 0.00 | 0.00 | 0.00 | 28.00 | 28.00 |
| 400 | Carp | 9025 | Car2 | 0.20 | 67.00 | m | 125x19 torus skirting (JF) | 73.30 | 63.65 | 0.00 | 0.00 | 136.95 |
| 500 | Titl | 9026 | | | | | *Windows* | | | | | |
| 505 | Supp | 9027 | Supp | 0.00 | 1.00 | item | Supply only windows | 0.00 | 712.76 | 0.00 | 0.00 | 712.76 |
| 510 | Car1 | 9028 | Car1 | 2.50 | 6.00 | no | Take out window remove to skip, fix new window | 82.05 | 0.00 | 12.00 | 0.00 | 94.05 |
| 515 | Blr | 9029 | Blr1 | 0.66 | 18.00 | m | Brick up window recesses | 64.98 | 9.00 | 0.00 | 0.00 | 73.98 |
| 520 | Carp | 9030 | Car1 | 0.20 | 24.00 | m | Window linings | 26.26 | 21.60 | 0.00 | 0.00 | 47.86 |
| 525 | Carp | 9031 | Car3 | 0.15 | 30.00 | m | Mastic pointing around windows | 24.62 | 10.50 | 0.00 | 0.00 | 35.11 |
| 530 | Carp | 9032 | Car2 | 0.20 | 24.00 | m | Architrave around windows | 26.26 | 10.80 | 0.00 | 0.00 | 37.06 |
| 535 | Carp | 9033 | Car1 | 1.50 | 6.20 | $m^2$ | Double glazing | 50.87 | 184.14 | 0.00 | 0.00 | 235.01 |
| 540 | Carp | 9219 | Car1 | 1.00 | 6.00 | m | Window boards | 32.82 | 21.00 | 0.00 | 0.00 | 53.82 |
| 600 | Titl | 9035 | | | | | *Doors and frames* | | | | | |
| 605 | Carp | 9036 | Car1 | 3.10 | 1.00 | no | T/out door and fr and CA fit new door fr inc fanlight and temp door | 16.96 | 29.36 | 3.00 | 0.00 | 49.32 |
| | | | | | | | Page totals | 444.60 | 1086.71 | 15.00 | 28.00 | 1574.31 |

**Fig. 7.6** (continued).

81 North Street, Doncaster - estimate for management Page 3

| Ref | Trad | No | Targ | Output | Quantity | Unit | Description | Lab cost | Mat cost | Plt cost | Scn cost | Totals |
|---|---|---|---|---|---|---|---|---|---|---|---|---|
| 610 | Carp | 9037 | Car1 | 3.10 | 1.00 | no | T/out door and frame and CA fit new frame and temporary door | 16.96 | 23.40 | 3.00 | 0.00 | 43.36 |
| 612 | Carp | 9040 | Car1 | 1.40 | 0.40 | $m^2$ | Double glazing to fanlights | 3.06 | 12.00 | 0.00 | 0.00 | 15.06 |
| 615 | Carp | 9038 | Car2 | 1.75 | 1.00 | no | Take off temporary door and store for reuse, fix new door | 9.57 | 112.00 | 0.00 | 0.00 | 121.57 |
| 620 | Carp | 9220 | Car2 | 1.75 | 1.00 | no | Take off temporary door and store for reuse, fix new door | 9.57 | 33.50 | 0.00 | 0.00 | 43.07 |
| 625 | Carp | 9041 | Car2 | 1.40 | 1.10 | $m^2$ | Toughened double glazing to doors | 8.42 | 59.40 | 0.00 | 0.00 | 67.82 |
| 630 | Carp | 9042 | Car3 | 0.30 | 10.80 | m | Mastic pointing around doors | 17.72 | 3.78 | 0.00 | 0.00 | 21.50 |
| 635 | Carp | 9043 | Car1 | 0.20 | 10.80 | m | Linings to external door openings | 11.82 | 9.72 | 0.00 | 0.00 | 21.54 |
| 640 | Carp | 9044 | Car2 | 0.15 | 10.80 | m | Architraves to external linings | 8.86 | 5.72 | 0.00 | 0.00 | 14.59 |
| 645 | Blr | 9045 | Blr1 | 0.66 | 11.00 | m | Brick up recesses to external doors | 39.71 | 5.50 | 0.00 | 0.00 | 45.21 |
| 650 | Lab | 9046 | Prep | 0.25 | 9.00 | no | Take out door linings and remove to skip | 10.64 | 0.00 | 1.80 | 0.00 | 12.44 |
| 655 | Carp | 9047 | Car1 | 1.00 | 8.00 | no | Assemble and fix door linings (DC) | 43.76 | 109.76 | 0.00 | 0.00 | 153.52 |
| 660 | Carp | 9048 | Car2 | 2.67 | 8.00 | no | Fix internal door and architraves (DC) | 116.84 | 186.40 | 0.00 | 0.00 | 303.24 |
| 665 | Carp | 9049 | Car2 | 0.33 | 8.00 | no | 35 x 19 strip screwed to bottom of doors | 14.44 | 2.40 | 0.00 | 0.00 | 16.84 |
| 700 | Titl | 9050 | | | | | *Ironmongery* | | | | | |
| 705 | Supp | 9051 | Supp | 0.00 | 1.00 | item | Supply only all ironmongery | 0.00 | 397.00 | 0.00 | 0.00 | 397.00 |
| 710 | Carp | 9052 | Car2 | 1.50 | 7.00 | no | Internal door ironmongery (DA) | 57.44 | 0.00 | 0.00 | 0.00 | 57.44 |
| 715 | Carp | 9053 | Car2 | 1.50 | 1.00 | no | Bathroom door ironmongery (DA) | 8.21 | 0.00 | 0.00 | 0.00 | 8.21 |
| 720 | Carp | 9054 | Car2 | 9.30 | 1.00 | no | Front door ironmongery (DG) | 50.87 | 0.00 | 0.00 | 0.00 | 50.87 |
| | | | | | | | Page totals | 427.89 | 960.58 | 4.80 | 0.00 | 1393.28 |

**Fig. 7.6** (continued).

81 North Street, Doncaster - estimate for management    Page 4

| Ref | Trad | No | Targ | Output | Quantity | Unit | Description | Lab cost | Mat cost | Plt cost | Scn cost | Totals |
|---|---|---|---|---|---|---|---|---|---|---|---|---|
| 725 | Carp | 9055 | Car2 | 4.76 | 1.00 | no | Back door ironmongery (DG) | 26.04 | 0.00 | 0.00 | 0.00 | 26.04 |
| 800 | Titl | 9056 | | | | | *Tie bars* | | | | | |
| 805 | Blr | 9057 | Blr1 | 1.50 | 2.00 | no | Form holes in external walls for tie bars | 16.41 | 0.00 | 0.00 | 0.00 | 16.41 |
| 810 | Blr | 9226 | Blr1 | 2.00 | 2.00 | no | Cut and lay brick slips to outside and MG at ends of tie bar | 21.88 | 8.00 | 0.00 | 0.00 | 29.88 |
| 815 | Carp | 9059 | Car1 | 2.25 | 1.00 | item | Carefully take up floor boards for tie bars (2 no areas) | 12.31 | 0.00 | 0.00 | 0.00 | 12.31 |
| 820 | Carp | 9060 | Car1 | 3.00 | 1.00 | item | Notch joists and screw fit 50×6 galv steel bars (2 no) | 16.41 | 0.00 | 0.00 | 0.00 | 16.41 |
| 825 | Blr | 9061 | Blr1 | 1.50 | 2.00 | no | Formwork and concrete to ends of bars in wall | 16.41 | 6.00 | 0.00 | 0.00 | 22.41 |
| 830 | Carp | 9062 | Car1 | 3.00 | 1.00 | item | Replace floor boards over ties (2 no ties) | 16.41 | 1.00 | 0.00 | 0.00 | 17.41 |
| 835 | Carp | 9227 | Car1 | 0.20 | 12.00 | m | 50×50 noggins to sides of ties | 13.13 | 6.00 | 0.00 | 0.00 | 19.13 |
| 840 | Carp | 9064 | Car1 | 0.16 | 4.00 | no | Packings at wall at ends of ties | 3.50 | 0.40 | 0.00 | 0.00 | 3.90 |
| 900 | Titl | 9065 | | | | | *Corner flue, stack removal (STR)* | | | | | |
| 905 | Lab | 9066 | Prep | 10.66 | 0.30 | $m^3$ | Take down corner flue and remove to skip | 15.13 | 0.00 | 3.60 | 0.00 | 18.73 |
| 910 | Lab | 9067 | Prep | 7.33 | 0.81 | $m^3$ | Take down stack and remove to skip | 28.08 | 0.00 | 9.72 | 0.00 | 37.80 |
| 915 | Blr | 9068 | Blr1 | 6.00 | 1.00 | item | Brick up where flue removed and generally make good | 32.82 | 12.00 | 0.00 | 0.00 | 44.82 |
| 920 | Carp | 9069 | Car1 | 2.00 | 1.00 | item | Remove short lengths of roof timber and replace (as STR) | 10.94 | 6.00 | 0.00 | 0.00 | 16.94 |
| 970 | Blr | 9070 | Blr1 | 2.00 | 1.00 | no | Take out timber lintel and replace with PC lintel (STV) | 10.94 | 4.00 | 0.00 | 0.00 | 14.94 |
| 1000 | Titl | 9071 | | | | | *Raise sill (STA)* | | | | | |
| 1005 | Blr | 9072 | Blr1 | 0.75 | 1.00 | no | Carefully take out sill and store for re-use (STA) | 3.55 | 0.00 | 0.00 | 0.00 | 3.55 |
| 1010 | Blr | 9073 | Blr1 | 3.30 | 0.40 | $m^2$ | Raise sill to 1050 above fl using 2nd hand bks inc cleaning | 7.22 | 4.80 | 0.00 | 0.00 | 12.02 |
| | | | | | | | Page totals | 251.17 | 48.20 | 13.32 | 0.00 | 312.69 |

**Fig. 7.6** (continued).

81 North Street, Doncaster - estimate for management    Page 5

| Ref | Trad | No | Targ | Output | Quantity | Unit | Description | Lab cost | Mat cost | Plt cost | Scn cost | Totals |
|---|---|---|---|---|---|---|---|---|---|---|---|---|
| 1015 | Blr | 9074 | Blr1 | 0.75 | 1.00 | no | Refix sill | 4.10 | 0.50 | 0.00 | 0.00 | 4.60 |
| 1020 | Blr | 9075 | Blr1 | 1.00 | 0.70 | m | Cut tooth and bond | 3.83 | 0.14 | 0.00 | 0.00 | 3.97 |
| 1030 | Titl | 9211 | | | | | *Cylinder cupboard* | | | | | |
| 1060 | Carp | 9204 | Car2 | 2.00 | 1.00 | no | Construct HW cylinder support stand (JG) | 10.94 | 4.00 | 0.00 | 0.00 | 14.94 |
| 1065 | Carp | 9205 | Car2 | 2.00 | 2.00 | no | Construct slatted shelves (JG) | 21.88 | 8.00 | 0.00 | 0.00 | 29.88 |
| 1100 | Titl | 9076 | | | | | *Bathroom fittings, etc. (BA)* | | | | | |
| 1105 | Subc | 9077 | Subc | 0.00 | 1.00 | | Supply and fix bathroom suite (BA) | 0.00 | 0.00 | 0.00 | 415.00 | 415.00 |
| 1110 | Tilg | 9078 | Tile | 0.25 | 3.00 | m | Proprietary plastic tiling strip to bath and WHB | 4.10 | 3.00 | 0.00 | 0.00 | 7.10 |
| 1115 | Tile | 9221 | Tile | 2.00 | 1.00 | $m^2$ | Three courses of 100×100 tiles above bath and washbasin | 10.94 | 10.00 | 0.00 | 0.00 | 20.94 |
| 1120 | Carp | 9080 | Car1 | 0.25 | 10.00 | M | Fix pipe boxing back boards (BA) | 13.68 | 15.00 | 0.00 | 0.00 | 28.68 |
| 1125 | Carp | 9081 | Car3 | 2.00 | 1.00 | item | Towel rail, toilet roll holder and mirror | 10.94 | 33.00 | 0.00 | 0.00 | 43.94 |
| 1130 | Carp | 9082 | Car3 | 0.25 | 10.00 | m | Pipe boxing front and sides | 13.68 | 10.00 | 0.00 | 0.00 | 23.68 |
| 1135 | Carp | 9083 | Car3 | 1.50 | 2.00 | no | Fix bath panels | 16.41 | 1.00 | 0.00 | 0.00 | 17.41 |
| 1140 | Subc | 9207 | Subc | 0.00 | 1.00 | item | Supply and fix above ground drainage (BB) | 0.00 | 0.00 | 0.00 | 90.00 | 90.00 |
| 1200 | Titl | 9084 | | | | | *Floor coverings inc preparation* | | | | | |
| 1205 | Carp | 9085 | Car3 | 1.00 | 4.00 | $m^2$ | Knock down nails and supply and lay ply for floor coverings | 21.88 | 8.00 | 0.00 | 0.00 | 29.88 |
| 1210 | Subc | 9086 | Subc | 0.00 | 4.00 | $m^2$ | Supply and lay anti-slip flooring (FE4) | 0.00 | 0.00 | 0.00 | 81.64 | 81.64 |
| 1215 | Subc | 9087 | Subc | 0.00 | 15.00 | $m^2$ | Anti-slip sheetg to conc floor inc latex (FE5) | 0.00 | 0.00 | 0.00 | 504.00 | 504.00 |
| 1300 | Titl | 9088 | | | | | *Form opening and balustrade* | | | | | |
| | | | | | | | Page totals | 132.37 | 92.64 | 0.00 | 1090.64 | 1315.65 |

**Fig. 7.6** (continued).

81 North Street, Doncaster - estimate for management Page 6

| Ref | Trad | No | Targ | Output | Quantity | Unit | Description | Lab cost | Mat cost | Plt cost | Scn cost | Totals |
|---|---|---|---|---|---|---|---|---|---|---|---|---|
| 1305 | Blr | 9089 | Blr1 | 0.75 | 1.70 | $m^2$ | Take down half brick wall | 6.97 | 0.00 | 0.00 | 0.00 | 6.97 |
| 1310 | Blr | 9090 | Blr1 | 2.66 | 0.26 | $m^3$ | Remove wall to skip | 3.78 | 0.00 | 3.12 | 0.00 | 6.90 |
| 1315 | Blr | 9091 | Blr1 | 0.50 | 2.20 | m | Quoin up jambs | 6.02 | 1.10 | 0.00 | 0.00 | 7.12 |
| 1320 | Blr | 9092 | Blr1 | 0.75 | 2.00 | m | Insert lintel and make good brickwork over | 8.21 | 8.00 | 0.00 | 0.00 | 16.21 |
| 1325 | Carp | 9093 | Car1 | 1.00 | 1.00 | m | 160 x 25 capping with bullnosed arrises | 5.47 | 1.50 | 0.00 | 0.00 | 6.97 |
| 1400 | Titl | 9094 | | | | | *Handrail* | | | | | |
| 1405 | Carp | 9095 | Prep | 0.40 | 4.00 | m | Take down handrail and make good | 8.75 | 0.00 | 0.00 | 0.00 | 8.75 |
| 1410 | Carp | 9096 | Car1 | 0.33 | 4.50 | m | 150 x 25 sw backboard fixed with Rawlbolts @ 1200 cts | 8.12 | 6.98 | 0.00 | 0.00 | 15.10 |
| 1415 | Carp | 9097 | Car2 | 0.33 | 4.50 | m | Moulded handrail with SAA brackets @ 1200 cts | 8.12 | 21.15 | 0.00 | 0.00 | 29.27 |
| 1470 | Carp | 9098 | Car1 | 0.03 | 28.00 | no | Unscrew stair clips and remove to skip | 4.59 | 0.00 | 0.00 | 0.00 | 4.59 |
| 1500 | Titl | 9099 | | | | | *Rem FP form opng for GB boiler* | | | | | |
| 1505 | Lab | 9100 | Prep | 2.00 | 1.00 | no | Take out fireplace fireback and hearthstone and remove to skip | 10.94 | 0.00 | 3.00 | 0.00 | 13.94 |
| 1510 | Blr | 9101 | Blr1 | 3.00 | 1.00 | $m^2$ | HB in brking up incorp opg to take gas fire bld in DPC (STG1) | 16.41 | 6.00 | 0.00 | 0.00 | 22.41 |
| 1515 | Blr | 9102 | Blr1 | 0.50 | 2.00 | m | Cut tooth and bond (STG1) | 5.47 | 1.20 | 0.00 | 0.00 | 6.67 |
| 1520 | Subc | 9103 | Subc | 0.00 | 1.00 | no | Sweep flues | 0.00 | 0.00 | 0.00 | 14.00 | 14.00 |
| 1525 | Carp | 9104 | Car1 | 2.00 | 1.00 | item | Jsts on hangers, T and G boarding in trimming opg and MG (STG1) | 10.94 | 10.00 | 0.00 | 0.00 | 20.94 |
| 1600 | Titl | 9105 | | | | | *Air brick (STE)* | | | | | |
| 1605 | Blr | 9106 | Blr1 | 2.00 | 5.00 | no | Chop opg in wall and insert 220 x 220 redbank 374 AB sleeve-MG | 54.70 | 15.00 | 0.00 | 0.00 | 69.70 |
| 1610 | Plas | 9107 | Plas | 0.50 | 5.00 | no | Plaster vents (STE) | 13.68 | 3.75 | 0.00 | 0.00 | 17.43 |
| | | | | | | | Page totals | 172.18 | 74.68 | 6.12 | 14.00 | 266.97 |

**Fig. 7.6** (continued).

81 North Street, Doncaster - estimate for management Page 7

| Ref | Trad | No | Targ | Output | Quantity | Unit | Description | Lab cost | Mat cost | Plt cost | Scn cost | Totals |
|---|---|---|---|---|---|---|---|---|---|---|---|---|
| 1690 | Titl | 9212 | | | | | *Matwell* | | | | | |
| 1700 | Carp | 9108 | Car1 | 2.00 | 1.00 | item | Remove and CA Matwell and replace with floorboards inc bearers | 10.94 | 11.00 | 1.00 | 0.00 | 22.94 |
| 1800 | Titl | 9109 | | | | | *Replace concrete floors (FD)* | | | | | |
| 1805 | Lab | 9110 | Prep | 1.20 | 17.00 | $m^2$ | Hack up existing concrete floor (FD) | 96.49 | 0.00 | 0.00 | 0.00 | 96.49 |
| 1810 | Lab | 9111 | Prep | 1.33 | 2.60 | $m^3$ | CA concrete floor | 16.36 | 0.00 | 31.20 | 0.00 | 47.56 |
| 1820 | Lab | 9113 | Prep | 0.08 | 20.00 | $m^2$ | DPM turned up 150 mm at walls | 7.57 | 10.00 | 0.00 | 0.00 | 17.57 |
| 1825 | Lab | 9114 | Prep | 0.08 | 17.00 | m | Secure DPM to wall with treated batten | 6.43 | 5.10 | 0.00 | 0.00 | 11.53 |
| 1830 | Lab | 9115 | Prep | 5.33 | 2.60 | $m^3$ | Concrete floor slab | 65.55 | 130.00 | 0.00 | 0.00 | 195.55 |
| 1835 | Lab | 9116 | Prep | 0.25 | 18.00 | $m^2$ | Insulation under solid floors | 21.29 | 36.00 | 0.00 | 0.00 | 57.29 |
| 1840 | Lab | 9117 | Prep | 0.10 | 17.00 | m | Insulation around edges of new solid floors | 8.04 | 6.80 | 0.00 | 0.00 | 14.84 |
| 1845 | Plas | 9118 | Plas | 0.46 | 17.00 | $m^2$ | Screed | 42.78 | 83.30 | 0.00 | 0.00 | 126.08 |
| 1900 | Titl | 9119 | | | | | *Rem chim brst, fix RSJ over (ST)* | | | | | |
| 1905 | Lab | 9120 | Prep | 4.33 | 1.75 | $m^3$ | Take down chimney breast and remove to skip | 35.84 | 0.00 | 21.00 | 0.00 | 56.84 |
| 1910 | Blr | 9121 | Blr1 | 1.00 | 4.60 | $m^2$ | Make good wall where breast removed | 25.16 | 27.60 | 0.00 | 0.00 | 52.76 |
| 1915 | Blr | 9122 | Blr1 | 1.00 | 2.00 | no | Form holes for and insert padstones for RSJs | 10.94 | 6.00 | 0.00 | 0.00 | 16.94 |
| 1920 | Blr | 9123 | Blr1 | 8.00 | 1.00 | no | 203 x 150 RSJ inc pockets at ends and all making good | 43.76 | 0.00 | 0.00 | 0.00 | 43.76 |
| 1925 | Carp | 9124 | Car1 | 0.25 | 23.00 | m | Bracketing around RSJs to take plasterboard | 31.45 | 10.35 | 0.00 | 0.00 | 41.80 |
| 1930 | Plas | 9125 | Plas | 1.26 | 3.00 | $m^2$ | Double plasterboard and skim to beams | 20.68 | 25.20 | 0.00 | 0.00 | 45.88 |
| 1935 | Supp | 9126 | Supp | 0.00 | 1.00 | item | Supply only all steel work (2 RSJs and ties) | 0.00 | 156.00 | 0.00 | 0.00 | 156.00 |
| | | | | | | | Page totals | 443.27 | 507.35 | 53.20 | 0.00 | 1003.82 |

**Fig. 7.6** (continued).

81 North Street, Doncaster - estimate for management Page 8

| Ref | Trad | No | Targ | Output | Quantity | Unit | Description | Lab cost | Mat cost | Plt cost | Scn cost | Totals |
|---|---|---|---|---|---|---|---|---|---|---|---|---|
| 1940 | Blr | 9127 | Blr1 | 3.00 | 1.00 | no | Take out cellar door lintel and replace with RSJ | 16.41 | 0.00 | 0.00 | 0.00 | 16.41 |
| 2000 | Titl | 9128 | | | | | *Vents* | | | | | |
| 2005 | Supp | 9222 | Supp | 0.00 | 1.00 | item | Delivery charge for vents | 0.00 | 18.00 | 0.00 | 0.00 | 18.00 |
| 2010 | Blr | 9130 | Blr1 | 1.00 | 2.00 | no | 100mm dia PVC vent tubes to cellar | 10.94 | 10.00 | 0.00 | 0.00 | 20.94 |
| 2015 | Blr | 9131 | Blr1 | 1.00 | 2.00 | no | Periscope vents in rear wall | 10.94 | 7.50 | 0.00 | 0.00 | 18.44 |
| 2100 | Titl | 9132 | | | | | *Kitchen fittings, etc.* | | | | | |
| 2105 | Supp | 9133 | Supp | 0.00 | 1.00 | item | Supply only kitchen fittings and accessories | 0.00 | 678.00 | 0.00 | 0.00 | 678.00 |
| 2110 | Carp | 9134 | Car2 | 0.20 | 2.60 | m | 50 x 25 softwood bearers | 2.84 | 1.04 | 0.00 | 0.00 | 3.88 |
| 2115 | Carp | 9136 | Car2 | 1.00 | 6.00 | m | Worktops | 32.82 | 0.00 | 0.00 | 0.00 | 32.82 |
| 2120 | Carp | 9135 | Car2 | 1.00 | 2.40 | m | Aluminium trims | 13.13 | 0.00 | 0.00 | 0.00 | 13.13 |
| 2125 | Carp | 9137 | Car2 | 0.20 | 1.00 | no | Chrome legs | 1.09 | 0.00 | 0.00 | 0.00 | 1.09 |
| 2130 | Carp | 9138 | Car2 | 0.90 | 3.00 | no | Single base unit | 14.77 | 0.00 | 0.00 | 0.00 | 14.77 |
| 2135 | Carp | 9139 | Car2 | 1.45 | 1.00 | no | Double base unit | 7.93 | 0.00 | 0.00 | 0.00 | 7.93 |
| 2140 | Carp | 9140 | Car2 | 1.50 | 1.00 | no | Double sink base unit | 8.21 | 0.00 | 0.00 | 0.00 | 8.21 |
| 2145 | Carp | 9141 | Car2 | 2.00 | 1.00 | no | Double wall unit | 10.94 | 0.00 | 0.00 | 0.00 | 10.94 |
| 2150 | Tilg | 9142 | Tile | 2.00 | 5.00 | $m^2$ | Ceramic wall tiling | 54.70 | 60.00 | 0.00 | 0.00 | 114.70 |
| 2155 | Tilg | 9143 | Tile | 0.35 | 8.40 | m | Sealant | 16.08 | 2.94 | 0.00 | 0.00 | 19.02 |
| 2160 | Subc | 9206 | Subc | 0.00 | 1.00 | item | Plumbing to kitchen fittings (FG1) | 0.00 | 0.00 | 0.00 | 130.00 | 130.00 |
| | | | | | | | Page totals | 200.80 | 777.48 | 0.00 | 130.00 | 1108.28 |

**Fig. 7.6** (continued)

81 North Street, Doncaster - estimate for management Page 9

| Ref | Trad | No | Targ | Output | Quantity | Unit | Description | Lab cost | Mat cost | Plt cost | Scn cost | Totals |
|---|---|---|---|---|---|---|---|---|---|---|---|---|
| 2200 | Blr | 9144 | Fire | 1.50 | 20.00 | $m^2$ | Const fire stop walls in roof space of 100mm lightwt blks | 164.10 | 80.00 | 0.00 | 0.00 | 244.10 |
| 2300 | Titl | 9145 | | | | | *Insulation* | | | | | |
| 2305 | Lab | 9146 | Insl | 0.10 | 34.00 | $m^2$ | Supply and lay 200mm Rockwool insulation to roof space (INS1) | 16.08 | 88.40 | 0.00 | 0.00 | 104.48 |
| 2310 | Lab | 9147 | Insl | 0.33 | 13.00 | $m^2$ | 200mm Rockwool ins to cellar fixed with netting (INS2) | 20.29 | 39.00 | 0.00 | 0.00 | 59.29 |
| 2400 | Titl | 9148 | | | | | *Roof hatch* | | | | | |
| 2405 | Carp | 9149 | Car1 | 2.00 | 1.00 | item | Cut out clg joists, insert 75 x 50 trimmers and 125 x 25 lining | 10.94 | 7.00 | 0.00 | 0.00 | 17.94 |
| 2410 | Carp | 9150 | Car2 | 3.00 | 1.00 | no | Roof hatch stops, door fxd with brass C&S, insul and architrave | 16.41 | 8.40 | 0.00 | 0.00 | 24.81 |
| 2490 | Titl | 9213 | | | | | *Floorboards* | | | | | |
| 2500 | Carp | 9151 | Car3 | 0.25 | 15.00 | m | Replace floor boards in small quantities (FA) | 20.51 | 18.00 | 0.00 | 0.00 | 38.51 |
| 2590 | Titl | 9214 | | | | | *Clear out* | | | | | |
| 2600 | Lab | 9152 | Prep | 12.00 | 1.00 | item | Remove and CA all rubbish internally and externally (RUB) | 65.64 | 0.00 | 95.00 | 0.00 | 160.64 |
| 2605 | Lab | 9210 | Prep | 0.00 | 1.00 | item | Take up floor coverings and remove to skip | 0.00 | 0.00 | 0.00 | 0.00 | 0.00 |
| 2690 | Titl | 9215 | | | | | *Services* | | | | | |
| 2700 | Carp | 9153 | Car1 | 3.35 | 1.00 | no | Construct tank support and walkway in roof (SA) | 18.32 | 25.00 | 0.00 | 0.00 | 43.32 |
| 2705 | Subc | 9154 | Subc | 0.00 | 1.00 | no | Supply and fix cold water tank (SA) | 0.00 | 0.00 | 0.00 | 95.00 | 95.00 |
| 2800 | Subc | 9155 | Subc | 0.00 | 1.00 | no | Supply and fix HW cylinder (SB) | 0.00 | 0.00 | 0.00 | 90.00 | 90.00 |
| 2900 | Subc | 9156 | Subc | 0.00 | 1.00 | no | Water plumbing installation (SD) | 0.00 | 0.00 | 0.00 | 280.00 | 280.00 |
| 3000 | Subc | 9157 | Subc | 0.00 | 1.00 | no | Gas plumbing installation (SK) | 0.00 | 0.00 | 0.00 | 60.00 | 60.00 |
| 3100 | Subc | 9158 | Subc | 0.00 | 1.00 | no | Electrical rewiring installation (ELE) | 0.00 | 0.00 | 0.00 | 967.00 | 967.00 |
| | | | | | | | Page totals | 332.30 | 265.80 | 95.00 | 1492.00 | 2185.10 |

**Fig. 7.6** (continued).

81 North Street, Doncaster - estimate for management Page 10

| Ref | Trad | No | Targ | Output | Quantity | Unit | Description | Lab cost | Mat cost | Plt cost | Scn cost | Totals |
|---|---|---|---|---|---|---|---|---|---|---|---|---|
| 3200 | Subc | 9159 | Subc | 0.00 | 1.00 | no | Supply and fix CH gas boiler (SF) | 0.00 | 0.00 | 0.00 | 775.00 | 775.00 |
| 3300 | Subc | 9160 | Subc | 0.00 | 1.00 | item | Heating installation including radiators | 0.00 | 0.00 | 0.00 | 705.00 | 705.00 |
| 3400 | Carp | 9161 | Car1 | 14.00 | 1.00 | item | Passive ventilation system as spec | 76.58 | 173.86 | 0.00 | 0.00 | 250.44 |
| 3500 | Titl | 9162 | | | | | *Roof* | | | | | |
| 3505 | Subc | 9223 | Subc | 0.00 | 1.00 | item | Relace roof covering, bonding gutters, flashings and soakers | 0.00 | 0.00 | 0.00 | 1645.00 | 1645.00 |
| 3525 | Blr | 9167 | Blr2 | 8.00 | 1.00 | no | Scaffold for pointing stacks | 43.76 | 40.00 | 0.00 | 0.00 | 83.76 |
| 3530 | Blr | 9168 | Bwk1 | 1.00 | 4.00 | $m^2$ | Repoint stacks (EA3) | 21.88 | 0.88 | 0.00 | 0.00 | 22.76 |
| 3535 | Blr | 9169 | Blr2 | 1.00 | 4.00 | no | Remove chimney pots and return to yard | 21.88 | 0.00 | 0.00 | 0.00 | 21.88 |
| 3540 | Blr | 9170 | Blr2 | 2.50 | 1.00 | no | Fix gas cowls (EG) | 13.68 | 28.00 | 0.00 | 0.00 | 41.68 |
| 3545 | Blr | 9171 | Blr2 | 1.50 | 3.00 | no | Fix mushroom caps (EH) | 24.62 | 36.00 | 0.00 | 0.00 | 60.62 |
| 3600 | Titl | 9178 | | | | | *Rainwater goods* | | | | | |
| 3605 | Rwg | 9209 | Rwg | 1.25 | 8.00 | m | Replace fascia and gutter inc 2 coats preservative | 54.70 | 36.00 | 0.00 | 0.00 | 90.70 |
| 3610 | Carp | 9174 | Rwg | 0.25 | 6.00 | no | Stop ends or joints to existing | 8.21 | 3.00 | 0.00 | 0.00 | 11.21 |
| 3615 | Carp | 9175 | Rwg | 0.50 | 2.00 | no | Outlets | 5.47 | 2.60 | 0.00 | 0.00 | 8.07 |
| 3625 | Carp | 9176 | Rwg | 0.60 | 12.00 | m | Take down and replace RWPS | 39.38 | 22.80 | 0.00 | 0.00 | 62.18 |
| 3630 | Carp | 9177 | Rwg | 0.25 | 2.00 | no | Shoes | 2.74 | 4.50 | 0.00 | 0.00 | 7.23 |
| 3690 | Titl | 9216 | | | | | *External brickwork* | | | | | |
| 3700 | Blr | 9179 | Blr2 | 1.00 | 29.00 | $m^2$ | Repoint elevations (EA1) | 158.63 | 6.38 | 0.00 | 0.00 | 165.01 |
| 3705 | Blr | 9180 | Blr2 | 0.50 | 15.00 | no | Chop indiv bricks and replace with second hand brks (EL) | 41.03 | 3.75 | 0.00 | 0.00 | 44.78 |
| | | | | | | | Page totals | 512.54 | 357.77 | 0.00 | 3125.00 | 3995.31 |

**Fig. 7.6** (continued).

81 North Street, Doncaster - estimate for management Page 11

| Ref | Trad | No | Targ | Output | Quantity | Unit | Description | Lab cost | Mat cost | Plt cost | Scn cost | Totals |
|---|---|---|---|---|---|---|---|---|---|---|---|---|
| 3800 | Titl | 9181 | | | | | *Paving* | | | | | |
| 3805 | Lab | 9182 | Ext | 0.75 | 26.00 | $m^2$ | Take up paving and exc for new paving flags and rem to skip | 106.67 | 0.00 | 31.20 | 0.00 | 137.87 |
| 3810 | Lab | 9183 | Ext | 1.00 | 21.00 | m | Exc, lay bed, lay 152 x 52 edging with haunching (GG1) | 99.33 | 56.70 | 0.00 | 0.00 | 156.03 |
| 3815 | Lab | 9184 | Ext | 0.75 | 26.00 | $m^2$ | Precast concrete paving flags as (GE1) | 92.24 | 130.00 | 0.00 | 0.00 | 222.24 |
| 3820 | Blr | 9185 | Blr2 | 0.50 | 1.00 | no | Remove and CA door steps | 2.74 | 0.00 | 0.60 | 0.00 | 3.34 |
| 3830 | Blr | 9186 | Blr2 | 1.00 | 1.00 | no | Supply and fix external door step (GD2) | 5.47 | 15.00 | 0.00 | 0.00 | 20.47 |
| 3840 | Blr | 9224 | Blr2 | 5.00 | 1.00 | item | Rebuild top course of bdy wall as necessary and soldier course | 27.35 | 5.00 | 0.00 | 0.00 | 32.35 |
| 3900 | Titl | 9187 | | | | | *Yard door* | | | | | |
| 3905 | Carp | 9188 | Car2 | 1.20 | 1.00 | no | Yard door and frame including sealant (DH) | 6.56 | 22.00 | 0.00 | 0.00 | 28.56 |
| 3910 | Carp | 9189 | Car2 | 2.00 | 1.00 | no | Yard door hung on tee hinges including latch | 10.94 | 40.00 | 0.00 | 0.00 | 50.94 |
| 3990 | Titl | 9208 | | | | | *Drainage* | | | | | |
| 4000 | Blr | 9190 | Drn | 14.60 | 1.00 | no | Excavate for and construct manhole (GL) | 79.86 | 120.00 | 0.00 | 0.00 | 199.86 |
| 4005 | Blr | 9191 | Drn | 2.50 | 1.50 | m | Excavate for and lay drains | 20.51 | 6.00 | 0.00 | 0.00 | 26.51 |
| 4010 | Blr | 9192 | Drn | 0.50 | 4.00 | no | Bends or rest bends | 10.94 | 11.00 | 0.00 | 0.00 | 21.94 |
| 4015 | Blr | 9193 | Drn | 1.33 | 0.70 | $m^3$ | Backfill drains | 4.40 | 0.00 | 0.00 | 0.00 | 4.40 |
| 4020 | Blr | 9194 | Drn | 2.00 | 1.00 | no | Back inlet gulley | 10.94 | 16.00 | 0.00 | 0.00 | 26.94 |
| 4100 | Lab | 9195 | Ext | 2.86 | 1.00 | no | Exc, lay found and fix cloths post and hook (GC) | 13.53 | 16.27 | 0.00 | 0.00 | 29.80 |
| 4200 | Titl | 9196 | | | | | *Landscaping* | | | | | |
| 4205 | Lab | 9197 | Ext | 0.33 | 20.00 | $m^2$ | Dig over garden 150 mm deep | 31.22 | 0.00 | 0.00 | 0.00 | 31.22 |
| | | | | | | | Page totals | 522.69 | 437.97 | 31.80 | 0.00 | 992.46 |

**Fig. 7.6** (continued).

81 North Street, Doncaster - estimate for management Page 12

| Ref | Trad | No | Targ | Output | Quantity | Unit | Description | Lab cost | Mat cost | Plt cost | Scn cost | Totals |
|---|---|---|---|---|---|---|---|---|---|---|---|---|
| 4210 | Lab | 9198 | Ext | 0.05 | 20.00 | m$^2$ | Rake level | 4.73 | 0.00 | 0.00 | 0.00 | 4.73 |
| 4215 | Lab | 9199 | Ext | 4.00 | 1.40 | m$^3$ | Lay topsoil 70 mm thick | 26.49 | 28.00 | 0.00 | 0.00 | 54.49 |
| 4290 | Titl | 9217 | | | | | *Decoration* | | | | | |
| 4300 | Subc | 9225 | Subc | 0.00 | 1.00 | item | Internal and external decoration | 0.00 | 0.00 | 0.00 | 874.00 | 874.00 |
| 4340 | Titl | 9218 | | | | | *Specialist subcontractors* | | | | | |
| 4350 | Subc | 9202 | Subc | 0.00 | 1.00 | item | DPC installation (P&A) | 0.00 | 0.00 | 0.00 | 350.00 | 350.00 |
| | | | | | | | **Totals** | **5213.60** | **5776.73** | **273.65** | **7103.64** | **18367.62** |

**Fig. 7.6** (continued).

site difficulties, security risk, general condition of property, etc., the preliminaries were then priced, and included:

- ❑ insurance
- ❑ plant
- ❑ general clearing up
- ❑ final clean
- ❑ scaffolding (for pointing)
- ❑ travel to site
- ❑ allowance for collecting materials, etc.

These were added to the costs in the Estimate for Management to give the total estimated cost.

*Conversion to a tender and submission of tender*

Overheads and mark-up were then added to the above total giving the tender sum. This was submitted to the architect and was successful.

### *7.1.7 Priced schedule for the architect (Fig. 7.7)*

As there was no provision in the schedule supplied for pricing preliminaries separately, these were added back into the rates for the items together with the mark-up.

Provision is made within MASTER for adding different percentages to labour, plant, materials and subcontractors. The percentages added were as follows:

- ❑ labour 39%
- ❑ plant 10%
- ❑ materials 10%
- ❑ subcontractors 10%.

81 North Street, Doncaster - priced schedule for architect Page 1

| Ref | Description | Quantity | Unit | Rate | Totals | Location |
|---|---|---|---|---|---|---|
| 5 | *Ceilings* | | | | | |
| 10 | Strip down ceiling, denail joists and remove to skip | 78.50 | $m^2$ | 3.20 | 251.20 | |
| 15 | 12.5mm foil backed plasterboard and skim (CE) | 33.50 | $m^2$ | 7.10 | 237.85 | |
| 20 | 12.5 Plasterboard and skim to ceilings (CA) | 33.00 | $m^2$ | 6.83 | 225.39 | |
| 100 | *Wall plastering* | | | | | |
| 105 | Hack off wall plaster and remove to skip | 157.00 | $m^2$ | 3.20 | 502.40 | |
| 110 | 39.5mm Styroliner and skim to walls (PE) | 53.00 | $m^2$ | 18.76 | 994.28 | Ext wall |
| 115 | Replast walls with additive (PA) | 69.00 | $m^2$ | 6.59 | 454.71 | Grd flr |
| 120 | Replaster walls (PB) | 86.00 | $m^2$ | 5.66 | 486.76 | 1st flr |
| 200 | *Partitions* | | | | | |
| 205 | Short length of joist bolted to wall as trimmer | 2.00 | no | 6.00 | 12.00 | Bed2/cor |
| 210 | Joist hangers | 2.00 | no | 4.68 | 9.36 | Bed2/cor |
| 215 | Floor joist under partition | 3.80 | m | 4.43 | 16.83 | Bed2/cor |
| 220 | Construct stud walling | 79.00 | m | 2.73 | 215.67 | Bed2/cor |
| 225 | Insulation to stud walling | 16.00 | $m^2$ | 3.95 | 63.20 | Bed2/cor |
| 230 | Plasterboard and skim to stud partitions | 32.00 | $m^2$ | 7.10 | 227.20 | Bed2/cor |
| 235 | Skirting to partitions | 9.00 | m | 2.57 | 23.13 | Bed2/cor |
| 300 | *Rem H-stone and vent flue (STJ)* | | | | | |
| 305 | Remove hearthstone and deposit in skip | 2.00 | no | 8.01 | 16.02 | Bedrooms |
| 310 | New joists in opening to match existing | 4.00 | m | 5.45 | 21.80 | Bedrooms |
| 315 | Joist hangers to hearth joists | 8.00 | no | 2.78 | 22.24 | Bedrooms |
| 320 | Form opening at high level in flue for plaster vent | 2.00 | no | 7.61 | 15.22 | Bedrooms |
| 325 | 225 x 150 plaster vent in flue | 2.00 | no | 4.63 | 9.26 | Bedrooms |
| 330 | Floorboards in hearth area to match existing | 1.00 | $m^2$ | 22.40 | 22.40 | Bedrooms |
| 335 | Sweep flues | 2.00 | no | 15.40 | 30.80 | Bedrooms |
| 400 | 125 x 19 torus skirting (JF) | 67.00 | m | 2.57 | 172.19 | |
| 500 | *Windows* | | | | | |
| 505 | Supply only windows | 1.00 | item | 784.04 | 784.04 | |
| Page total | | | | | 4813.23 | |

**Fig. 7.7** Case study 1: priced schedule for the architect.

81 North Street, Doncaster - priced schedule for architect Page 2

| Ref | Description | Quantity | Unit | Rate | Totals | Location |
|---|---|---|---|---|---|---|
| 510 | Take out window remove to skip, fix new window | 6.00 | no | 21.21 | 127.26 | |
| 515 | Brick up window recesses | 18.00 | m | 5.57 | 100.26 | |
| 520 | Window linings | 24.00 | m | 2.51 | 60.24 | |
| 525 | Mastic pointing around windows | 30.00 | m | 1.53 | 45.90 | |
| 530 | Architrave around windows | 24.00 | m | 2.02 | 48.48 | |
| 535 | Double glazing | 6.20 | $m^2$ | 44.07 | 273.23 | |
| 540 | Window boards | 6.00 | m | 11.45 | 68.70 | |
| 600 | *Doors and frames* | | | | | |
| 605 | T/out door and fr and CA, fit new door fr inc fanlight and temp door | 1.00 | no | 59.17 | 59.17 | Front dr |
| 610 | T/out door and frame and CA, fit new frame and temporary door | 1.00 | no | 52.61 | 52.61 | Back dr |
| 612 | Double glazing to fanlights | 0.40 | $m^2$ | 43.65 | 17.46 | Front dr |
| 615 | Take off temporary door and store for reuse, fix new door | 1.00 | no | 136.51 | 136.51 | Front dr |
| 620 | Take off temporary door and store for reuse, fix new door | 1.00 | no | 50.16 | 50.16 | |
| 625 | Toughened double glazing to doors | 1.10 | $m^2$ | 70.05 | 77.06 | Back dr |
| 630 | Mastic pointing around doors | 10.80 | m | 2.67 | 28.84 | External |
| 635 | Linings to external door openings | 10.80 | m | 2.51 | 27.11 | |
| 640 | Architraves to external linings | 10.80 | m | 1.72 | 18.58 | |
| 645 | Brick up recesses to external doors | 11.00 | m | 5.57 | 61.27 | |
| 650 | Take out door linings and remove to skip | 9.00 | no | 1.86 | 16.74 | Internal |
| 655 | Assemble and fix door linings (DC) | 8.00 | no | 22.70 | 181.60 | Internal |
| 660 | Fix internal door and architraves (DC) | 8.00 | no | 45.93 | 367.44 | |
| 665 | 35 x 19 strip screwed to bottom of doors | 8.00 | no | 2.84 | 22.72 | Internal |
| 700 | *Ironmongery* | | | | | |
| 705 | Supply only all ironmongery | 1.00 | item | 436.70 | 436.70 | |
| 710 | Internal door ironmongery (DA) | 7.00 | no | 11.40 | 79.80 | |
| 715 | Bathroom door ironmongery (DA) | 1.00 | no | 11.40 | 11.40 | |
| 720 | Front door ironmongery (DG) | 1.00 | no | 70.71 | 70.71 | |
| 725 | Back door ironmongery (DG) | 1.00 | no | 36.19 | 36.19 | |
| Page total | | | | | 2475.93 | |

**Fig. 7.7** (continued).

81 North Street, Doncaster - priced schedule for architect Page 3

| Ref | Description | Quantity | Unit | Rate | Totals | Location |
|---|---|---|---|---|---|---|
| 800 | *Tie bars* | | | | | |
| 805 | Form holes in external walls for tie bars | 2.00 | no | 11.41 | 22.82 | Bedroom 1 |
| 810 | Cut and lay brick slips to outside and MG at ends of tie bar | 2.00 | no | 19.61 | 39.22 | Bedroom 1 |
| 815 | Carefully take up floorboards for tie bars (2 no areas) | 1.00 | item | 17.11 | 17.11 | Bedroom 1 |
| 820 | Notch joists and screw fit 50 x6 galv steel bars (2 no) | 1.00 | item | 22.81 | 22.81 | Bedroom 1 |
| 825 | Formwork and concrete to ends of bars in wall | 2.00 | no | 14.70 | 29.40 | Bedroom 1 |
| 830 | Replace floorboards over ties (2 no ties) | 1.00 | item | 23.91 | 23.91 | Bedroom 1 |
| 835 | 50 x 50 noggins to sides of ties | 12.00 | m | 2.07 | 24.84 | Bedroom 1 |
| 840 | Packings at wall at ends of ties | 4.00 | no | 1.33 | 5.32 | Bedroom 1 |
| 900 | *Corner flue, stack removal (STR)* | | | | | |
| 905 | Take down corner flue and remove to skip | 0.30 | $m^3$ | 83.30 | 24.99 | Bathroom |
| 910 | Take down stack and remove to skip | 0.81 | $m^3$ | 61.41 | 49.74 | Bathroom |
| 915 | Brick up where flue removed and generally make good | 1.00 | item | 58.82 | 58.82 | Bathroom |
| 920 | Remove short lengths of roof timber and replace (as STR) | 1.00 | item | 21.81 | 21.81 | Bathroom |
| 970 | Take out timber lintel and replace with PC lintel (STV) | 1.00 | no | 19.61 | 19.61 | Bathroom |
| 1000 | *Raise sill (STA)* | | | | | |
| 1005 | Carefully take out sill and store for re-use (STA) | 1.00 | no | 4.93 | 4.93 | Kitchen |
| 1010 | Raise sill to 1050 above fl using 2nd hand bks inc cleaning | 0.40 | $m^2$ | 38.30 | 15.32 | Kitchen |
| 1015 | Refix sill | 1.00 | no | 6.25 | 6.25 | Kitchen |
| 1020 | Cut tooth and bond | 0.70 | m | 7.81 | 5.47 | Kitchen |
| 1030 | *Cylinder cupboard* | | | | | |
| 1060 | Construct HW cylinder support stand (JG) | 1.00 | no | 19.61 | 19.61 | Corridor |
| 1065 | Construct slatted shelves (JG) | 2.00 | no | 19.61 | 39.22 | Corridor |
| 1100 | *Bathroom fittings, etc. (BA)* | | | | | |
| 1105 | Supply and fix bathroom suite (BA) | 1.00 | item | 456.50 | 456.50 | Bathroom |
| 1110 | Proprietary plastic tiling strip to bath and WHB | 3.00 | m | 3.00 | 9.00 | Bathroom |
| 1115 | Three courses of 100 x 100 tiles above bath and washbasin | 1.00 | $m^2$ | 26.21 | 26.21 | Bathroom |
| 1120 | Fix pipe boxing back boards (BA) | 10.00 | m | 3.55 | 35.50 | Bathroom |
| Page total | | | | | 978.40 | |

**Fig. 7.7** (continued).

81 North Street, Doncaster - priced schedule for architect Page 4

| Ref | Description | Quantity | Unit | Rate | Totals | Location |
|---|---|---|---|---|---|---|
| 1125 | Towel rail, toilet roll holder and mirror | 1.00 | item | 51.51 | 51.51 | Bathroom |
| 1130 | Pipe boxing front and sides | 10.00 | m | 3.00 | 30.00 | Bathroom |
| 1135 | Fix bath panels | 2.00 | no | 11.96 | 23.92 | Bathroom |
| 1140 | Supply and fix above ground drainage (BB) | 1.00 | item | 99.00 | 99.00 | See drg |
| 1200 | *Floor coverings inc preparation* | | | | | |
| 1205 | Knock down nails and supply and lay ply for floor coverings | 4.00 | $m^2$ | 9.80 | 39.20 | Bathroom |
| 1210 | Supply and lay anti-slip flooring (FE4) | 4.00 | $m^2$ | 22.45 | 89.80 | Bathroom |
| 1215 | Anti-slip sheetg to conc floor inc latex (FE5) | 15.00 | $m^2$ | 36.96 | 554.40 | Di/ki/ut |
| 1300 | *Form opening and balustrade* | | | | | |
| 1305 | Take down half brick wall | 1.70 | $m^2$ | 5.70 | 9.69 | Stair |
| 1310 | Remove wall to skip | 0.26 | $m^3$ | 33.46 | 8.70 | Stair |
| 1315 | Quoin up jambs | 2.20 | m | 4.35 | 9.57 | Stair |
| 1320 | Insert lintel and make good brickwork over | 2.00 | m | 10.10 | 20.20 | Stair |
| 1325 | 160 x 25 capping with bullnosed arrises | 1.00 | m | 9.25 | 9.25 | Stair |
| 1400 | *Handrail* | | | | | |
| 1405 | Take down handrail and make good | 4.00 | m | 3.04 | 12.16 | Stair |
| 1410 | 150 x 25 SW backboard fixed with rawbolts @ 1200 cts | 4.50 | m | 4.21 | 18.95 | Stair |
| 1415 | Moulded handrail with SAA brackets @ 1200 cts | 4.50 | m | 7.68 | 34.56 | Stair |
| 1470 | Unscrew stair clips and remove to skip | 28.00 | no | 0.23 | 6.44 | Stair |
| 1500 | *Rem FP, form opng for GB boiler (STG1)* | | | | | |
| 1505 | Take out fireplace, fireback and hearthstone and remove to skip | 1.00 | no | 18.51 | 18.51 | Living |
| 1510 | HB in brking up incorp opg to take gas fire, bld in DPC (STG1) | 1.00 | $m^2$ | 29.41 | 29.41 | Living |
| 1515 | Cut tooth and bond (STG1) | 2.00 | m | 4.46 | 8.92 | Living |
| 1520 | Sweep flues | 1.00 | no | 15.40 | 15.40 | Living |
| 1525 | Jsts on hangers, T and G boarding in trimming opg and MG (STG1) | 1.00 | item | 26.21 | 26.21 | Living |
| 1600 | *Air brick (STE)* | | | | | |
| 1605 | Chop opg in wall and insert 220 x 220 Redbank 374 AB sleeve-MG | 5.00 | no | 18.51 | 92.55 | Liv/cell |
| 1610 | Plaster vents (STE) | 5.00 | no | 4.63 | 23.15 | Liv/cell |
| Page total | | | | | 1231.45 | |

**Fig. 7.7** (continued).

81 North Street, Doncaster - priced schedule for architect Page 5

| Ref | Description | Quantity | Unit | Rate | Totals | Location |
|---|---|---|---|---|---|---|
| 1690 | *Matwell* | | | | | |
| 1700 | Remove and CA Matwell and replace with floorboards inc bearers | 1.00 | item | 28.41 | 28.41 | Living |
| 1800 | *Replace concrete floors (FD)* | | | | | |
| 1805 | Hack up existing concrete floor (FD) | 17.00 | $m^2$ | 7.89 | 134.13 | Ki/di/ut |
| 1810 | CA concrete floor | 2.60 | $m^3$ | 21.96 | 57.10 | Ki/di/ut |
| 1820 | DPM turned up 150 mm at walls | 20.00 | $m^2$ | 1.08 | 21.60 | Ki/di/ut |
| 1825 | Secure DPM to wall with treated batten | 17.00 | m | 0.86 | 14.62 | Ki/di/ut |
| 1830 | Concrete floor slab | 2.60 | $m^3$ | 90.04 | 234.10 | Ki/di/ut |
| 1835 | Insulation under solid floors | 18.00 | $m^2$ | 3.84 | 69.12 | Ki/di/ut |
| 1840 | Insulation around edges of new solid floors | 17.00 | m | 1.10 | 18.70 | Ki/di/ut |
| 1845 | Screed | 17.00 | $m^2$ | 8.89 | 151.13 | Ki/di/ut |
| 1900 | *Rem chim brst, fix RSJ over (STL1)* | | | | | |
| 1905 | Take down chimney breast and remove to skip | 1.75 | $m^3$ | 41.68 | 72.94 | Kit/din |
| 1910 | Make good wall where breast removed | 4.60 | $m^2$ | 14.20 | 65.32 | Kit/din |
| 1915 | Form holes for and insert padstones for RSJs | 2.00 | no | 10.91 | 21.82 | Kit/din |
| 1920 | 203 x 150 RSJ inc pockets at ends and all making good | 1.00 | no | 60.83 | 60.83 | Kit/din |
| 1925 | Bracketing around RSJs to take plasterboard | 23.00 | m | 2.40 | 55.20 | Kit/din |
| 1930 | Double plasterboard and skim to beams | 3.00 | $m^2$ | 18.82 | 56.46 | Kit/din |
| 1935 | Supply only all steel work (2 RSJs and ties) | 1.00 | item | 171.60 | 171.60 | Kit/din |
| 1940 | Take out cellar door lintel and relace with RSJ | 1.00 | no | 22.81 | 22.81 | Kit/din |
| 2000 | *Vents* | | | | | |
| 2005 | Delivery charge for vents | 1.00 | item | 19.80 | 19.80 | |
| 2010 | 100 m dia PVC vent tubes to cellar | 2.00 | no | 13.11 | 26.22 | Kit/din |
| 2015 | Periscope vents in rear wall | 2.00 | no | 11.73 | 23.46 | Kit/din |
| 2100 | *Kitchen fittings, etc.* | | | | | |
| 2105 | Supply only kitchen fittings and accessories | 1.00 | item | 745.80 | 745.80 | |
| 2110 | 50 x 25 softwood bearers | 2.60 | m | 1.96 | 5.10 | Ki/di/ut |
| 2115 | Worktops | 6.00 | m | 7.60 | 45.60 | Ki/di/ut |
| Page total | | | | | 2121.62 | |

**Fig. 7.7** (continued).

81 North Street, Doncaster - priced schedule for architect Page 6

| Ref | Description | Quantity | Unit | Rate | Totals | Location |
|---|---|---|---|---|---|---|
| 2120 | Aluminium trims | 2.40 | m | 7.60 | 18.24 | Kit/din |
| 2125 | Chrome legs | 1.00 | no | 1.52 | 1.52 | Kit/din |
| 2130 | Single base unit | 3.00 | no | 6.84 | 20.52 | Kit/di/ut |
| 2135 | Double base unit | 1.00 | no | 11.02 | 11.02 | Kit/din |
| 2140 | Double sink base unit | 1.00 | no | 11.40 | 11.40 | Kit/din |
| 2145 | Double wall unit | 1.00 | no | 15.21 | 15.21 | Kit/din |
| 2150 | Ceramic wall tiling | 5.00 | $m^2$ | 28.41 | 142.05 | Ki/di/ut |
| 2155 | Sealant | 8.40 | m | 3.05 | 25.62 | Ki/di/ut |
| 2160 | Plumbing to kitchen fittings (FG1) | 1.00 | item | 143.00 | 143.00 | Ki/di/ut |
| 2200 | Const fire stop walls in roof space of 100 mm lightwt blks | 20.00 | $m^2$ | 15.81 | 316.20 | Roof |
| 2300 | *Insulation* | | | | | |
| 2305 | Supply and lay 200 mm Rockwool insulation to roof space (INS1) | 34.00 | $m^2$ | 3.52 | 119.68 | Roof |
| 2310 | 200 mm Rockwool ins to cellar fixed with netting (INS2) | 13.00 | $m^2$ | 5.47 | 71.11 | Cellar |
| 2400 | *Roof hatch* | | | | | |
| 2405 | Cut out clg joists, insert 75 x 50 trimmers and 125 x 25 lining | 1.00 | item | 22.91 | 22.91 | See arch |
| 2410 | Roof hatch stops, door fxd with brass C&S, insul and architrave | 1.00 | no | 32.05 | 32.05 | See arch |
| 2490 | *Floorboards* | | | | | |
| 2500 | Replace floorboards in small quantities (FA) | 15.00 | m | 3.22 | 48.30 | General |
| 2590 | *Clear out* | | | | | |
| 2600 | Remove and CA all rubbish internally and externally (RUB) | 1.00 | item | 195.84 | 195.84 | |
| 2605 | Take up floor coverings and remove to skip | 1.00 | item | 0.00 | 0.00 | |
| 2690 | *Services* | | | | | |
| 2700 | Construct tank support and walkway in roof (SA) | 1.00 | no | 52.97 | 52.97 | Roof |
| 2705 | Supply and fix cold water tank (SA) | 1.00 | no | 104.50 | 104.50 | Roof |
| 2800 | Supply and fix HW cylinder (SB) | 1.00 | no | 99.00 | 99.00 | Corridor |
| 2900 | Water plumbing installation (SD) | 1.00 | no | 308.00 | 308.00 | |
| 3000 | Gas plumbing installation (SK) | 1.00 | no | 66.00 | 66.00 | |
| 3100 | Electrical rewiring installation (ELE) | 1.00 | no | 1063.70 | 1063.70 | |
| Page total | | | | | 2888.62 | |

**Fig. 7.7** (continued).

81 North Street, Doncaster - priced schedule for architect Page 7

| Ref | Description | Quantity | Unit | Rate | Totals | Location |
|---|---|---|---|---|---|---|
| 3200 | Supply and fix CH gas boiler (SF) | 1.00 | no | 852.50 | 852.50 | Living |
| 3300 | Heating installation including radiators | 1.00 | item | 775.50 | 775.50 | |
| 3400 | Passive ventilation system as spec | 1.00 | item | 297.70 | 297.70 | See drg |
| 3500 | *Roof* | | | | | |
| 3505 | Replace roof covering, bonding gutters, flashings and soakers | 1.00 | item | 1809.50 | 1809.50 | |
| 3525 | Scaffold for pointing stacks | 1.00 | no | 104.83 | 104.83 | Roof |
| 3530 | Repoint stacks (EA3) | 4.00 | $m^2$ | 7.85 | 31.40 | Roof |
| 3535 | Remove chimney pots and return to yard | 4.00 | no | 7.60 | 30.40 | Roof |
| 3540 | Fix gas cowls (EG) | 1.00 | no | 49.81 | 49.81 | Roof |
| 3545 | Fix mushroom caps (EH) | 3.00 | no | 24.60 | 73.80 | Roof |
| 3600 | *Rainwater goods* | | | | | |
| 3605 | Replace fascia and gutter inc 2 coats preservative | 8.00 | m | 14.45 | 115.60 | All |
| 3610 | Stop ends or joints to existing | 6.00 | no | 2.45 | 14.70 | |
| 3615 | Outlets | 2.00 | no | 5.23 | 10.46 | See drg |
| 3625 | Take down and replace RWPS | 12.00 | m | 6.65 | 79.80 | All |
| 3630 | Shoes | 2.00 | no | 4.38 | 8.76 | To RWP |
| 3690 | *External brickwork* | | | | | |
| 3700 | Repoint elevations (EA1) | 29.00 | $m^2$ | 7.85 | 227.65 | All |
| 3705 | Chop out indiv bricks and replace with second hand brks (EL) | 15.00 | no | 4.08 | 61.20 | See arch |
| 3800 | *Paving* | | | | | |
| 3805 | Take up paving and exc for new paving flags and rem to skip | 26.00 | $m^2$ | 7.02 | 182.52 | See drg |
| 3810 | Exc, lay bed, lay 152 x 52 edging with haunching (GG1) | 21.00 | m | 9.54 | 200.34 | See drg |
| 3815 | Precast concrete paving flags as (GE1) | 26.00 | $m^2$ | 10.43 | 271.18 | See drg |
| 3820 | Remove and CA door steps | 1.00 | no | 4.46 | 4.46 | Rear |
| 3830 | Supply and fix external door step (GD2) | 1.00 | no | 24.10 | 24.10 | Rear |
| 3840 | Rebuild top course of bdy wall as necessary and soldier course | 1.00 | item | 43.52 | 43.52 | |
| 3900 | *Yard door* | | | | | |
| 3905 | Yard door and frame including sealant (DH) | 1.00 | no | 33.32 | 33.32 | Rear |
| Page total | | | | | 5303.13 | |

Fig. 7.7 (continued).

81 North Street, Doncaster - priced schedule for architect Page 8

| Ref | Description | Quantity | Unit | Rate | Totals | Location |
|---|---|---|---|---|---|---|
| 3910 | Yard door hung on tee hinges including latch | 1.00 | no | 59.21 | 59.21 | Rear |
| 3990 | *Drainage* | | | | | |
| 4000 | Excavate for and construct manhole (GL) | 1.00 | no | 243.01 | 243.01 | As drg |
| 4005 | Excavate for and lay drains | 1.50 | m | 23.41 | 35.12 | As drg |
| 4010 | Bends or rest bends | 4.00 | no | 6.83 | 27.32 | As drg |
| 4015 | Backfill drains | 0.70 | $m^3$ | 8.74 | 6.12 | As drg |
| 4020 | Back inlet gulley | 1.00 | no | 32.81 | 32.81 | As drg |
| 4100 | Exc lay found and fix cloths post and hook (GC) | 1.00 | no | 36.70 | 36.70 | As drg |
| 4200 | *Landscaping* | | | | | |
| 4205 | Dig over garden 150 mm deep | 20.00 | $m^2$ | 2.17 | 43.40 | As drg |
| 4210 | Rake level | 20.00 | $m^2$ | 0.33 | 6.60 | As drg |
| 4215 | Lay topsoil 70 mm thick | 1.40 | $m^3$ | 48.30 | 67.62 | As drg |
| 4290 | *Decoration* | | | | | |
| 4300 | Internal and external decoration | 1.00 | item | 961.40 | 961.40 | |
| 4340 | *Specialist subcontractors* | | | | | |
| 4350 | DPC installation (+P & A) | 1.00 | item | 385.00 | 385.00 | |
| **Summary** | | | | | **21716.63** | |

Fig. 7.7 (continued).

These percentages were entered into the computer, and a report in the form of Priced Schedule for Architect was generated (Fig. 7.7).

*Note*: The amount of time taken to generate this report is minimal, because the only input required of the estimator is entering the percentages.

In this case, the schedule provided by the architect had to be priced, as this was to be a contract document.

The rates were ascertained from the above-mentioned priced schedule (Fig. 7.7).

The schedule submitted was accepted and the contract was signed.

### 7.1.8 Production information

The information entered into the computer at the estimating stage was used in generating reports for use in the office and on sites as follows.

*(a) Full information for management*

A complete set of reports is shown in Fig. 7.8. This includes a subcontractor report.

81 North Street, Doncaster - Prep work - management info - targ prep - management Page 1

| Ref | Description | Hrs | Targ | Sub cost | Plt cost | Mat cost |
|---|---|---|---|---|---|---|
| 10 | Strip down ceiling, denail joists and remove to skip<br>78.50 $m^2$ 0.45 $h/m^2$ | 35.33 | 33.56 | | 17.27 | |
| 105 | Hack off wall plaster and remove to skip<br>157.00 $m^2$ 0.45 $h/m^2$ | 70.65 | 67.12 | | 34.54 | |
| 305 | Remove hearthstone and deposit in skip<br>2.00 no 1.00 h/no | 2.00 | 1.90 | | 2.60 | |
| 650 | Take out door linings and remove to skip<br>9.00 no 0.25 h/no | 2.25 | 2.14 | | 1.80 | |
| 905 | Take down corner flue and remove to skip<br>0.30 $m^3$ 10.66 $h/m^3$ | 3.20 | 3.04 | | 3.60 | |
| 910 | Take down stack and remove to skip<br>0.81 $m^3$ 7.33 $h/m^3$ | 5.94 | 5.64 | | 9.72 | |
| 1405 | Take down handrail and make good<br>4.00 m 0.40 h/m | 1.60 | 1.52 | | | |
| 1505 | Take out fireplace, fireback and hearthstone and remove to skip<br>1.00 no 2.00 h/no | 2.00 | 1.90 | | 3.00 | |
| 1805 | Hack up existing concrete floor (FD)<br>17.00 $m^2$ 1.20 $h/m^2$ | 20.40 | 19.38 | | | |
| 1810 | CA concrete floor<br>2.60 $m^3$ 1.33 $h/m^3$ | 3.46 | 3.29 | | 31.20 | |
| 1820 | DPM turned up 150 mm at walls<br>20.00 $m^2$ 0.08 $h/m^2$ | 1.60 | 1.52 | | | 10.00 |
| 1825 | Secure DPM to wall with treated batten<br>17.00 m 0.08 h/m | 1.36 | 1.29 | | | 5.10 |
| 1830 | Concrete floor slab<br>2.60 $m^3$ 5.33 $h/m^3$ | 13.86 | 13.17 | | | 130.00 |
| 1835 | Insulation under solid floors<br>18.00 $m^2$ 0.25 $h/m^2$ | 4.50 | 4.27 | | | 36.00 |
| 1840 | Insulation around edges of new solid floors<br>17.00 m 0.10 h/m | 1.70 | 1.62 | | | 6.80 |
| 1905 | Take down chimney breast and remove to skip<br>1.75 $m^3$ 4.33 $h/m^3$ | 7.58 | 7.20 | | 21.00 | |
| 2600 | Remove and CA all rubbish internally and externally (RUB)<br>1.00 item 12.00 h/item | 12.00 | 11.40 | | 95.00 | |
| 2605 | Take up floor coverings and remove to skip<br>1.00 item 0.00 h/item | 0.00 | 0.00 | | | |
| | Labour hours | | 189.41 | | Tot mat cost | 187.90 |
| | Factor 0.95 Factor total | | 179.94 | | Tot lab cost | 907.47 |
| | | | | | Tot sub cost | 0.00 |
| | | | | | Tot plt cost | 219.73 |

**Fig. 7.8** Case study 1: complete set of reports for management.

81 North Street, Doncaster - internal brickwork - management info - targ blr1 - management Page 1

| Ref | Description | Hrs | Targ | Sub cost | Plt cost | Mat cost |
|---|---|---|---|---|---|---|
| 320 | Form opening at high level in flue for plaster vent<br>2.00 no 1.00 h/no | 2.00 | 1.90 | | | |
| 515 | Brick up window recesses<br>18.00 m 0.66 h/m | 11.88 | 11.29 | | | 9.00 |
| 645 | Brick up recesses to external doors<br>11.00 m 0.66 h/m | 7.26 | 6.90 | | | 5.50 |
| 805 | Form holes in external walls for tie bars<br>2.00 no 1.50 h/no | 3.00 | 2.85 | | | |
| 810 | Cut and lay brick slips to outside and MG at ends of tie bar<br>2.00 no 2.00 h/no | 4.00 | 3.80 | | | 8.00 |
| 825 | Formwork and concrete to ends of bars in wall<br>2.00 no 1.50 h/no | 3.00 | 2.85 | | | 6.00 |
| 915 | Brick up where flue removed and generally make good<br>1.00 item 6.00 h/item | 6.00 | 5.70 | | | 12.00 |
| 970 | Take out timber lintel and relace with PC lintel (STV)<br>1.00 no 2.00 h/no | 2.00 | 1.90 | | | 4.00 |
| 1005 | Carefully take out sill and store for re-use (STA)<br>1.00 no 0.75 h/no | 0.75 | 0.71 | | | |
| 1010 | Raise sill to 1050 above fl using 2nd hand bks inc cleaning<br>0.40 $m^2$ 3.30 h/$m^2$ | 1.32 | 1.25 | | | 4.80 |
| 1015 | Refix sill<br>1.00 no 0.75 h/no | 0.75 | 0.71 | | | 0.50 |
| 1020 | Cut tooth and bond<br>0.70 m 1.00 h/m | 0.70 | 0.66 | | | 0.14 |
| 1305 | Take down half brick wall<br>1.70 $m^2$ 0.75 h/$m^2$ | 1.28 | 1.21 | | | |
| 1310 | Remove wall to skip<br>0.26 $m^3$ 2.66 h/$m^3$ | 0.69 | 0.66 | | 3.12 | |
| 1315 | Quoin up jambs<br>2.20 m 0.50 h/m | 1.10 | 1.05 | | | 1.10 |
| 1320 | Insert lintel and make good brickwork over<br>2.00 m 0.75 h/m | 1.50 | 1.43 | | | 8.00 |
| 1510 | HB in brking up incorp opg to take gas fire bld in DPC (STG1)<br>1.00 $m^2$ 3.00 h/$m^2$ | 3.00 | 2.85 | | | 6.00 |
| 1515 | Cut tooth and bond (STG1)<br>2.00 m 0.50 h/m | 1.00 | 0.95 | | | 1.20 |
| 1605 | Chop opg in wall and insert 220 × 220 Redbank 374 AB sleeve-MG<br>5.00 no 2.00 h/no | 10.00 | 9.50 | | | 15.00 |
| 1910 | Make good wall where breast removed<br>4.60 $m^2$ 1.00 h/$m^2$ | 4.60 | 4.37 | | | 27.60 |
| 1915 | Form holes for and insert padstones for RSJs<br>2.00 no 1.00 h/no | 2.00 | 1.90 | | | 6.00 |
| 1920 | 203 × 150 RSJ inc pockets at ends and all making good<br>1.00 no 8.00 h/no | 8.00 | 7.60 | | | |
| 1940 | Take out cellar door lintel and replace with RSJ<br>1.00 no 3.00 h/no | 3.00 | 2.85 | | | |
| 2010 | 100 mm dia PVC vent tubes to cellar<br>2.00 no 1.00 h/no | 2.00 | 1.90 | | | 10.00 |
| 2015 | Periscope vents in rear wall<br>2.00 no 1.00 h/no | 2.00 | 1.90 | | | 7.50 |

| | | | | |
|---|---|---|---|---|
| Factor 0.95 | Labour hours | 82.83 | Tot mat cst | 132.34 |
| | Factor total | 78.69 | Tot lab cst | 452.51 |
| | | | Tot sub cst | 0.00 |
| | | | Tot plt cst | 3.12 |

**Fig. 7.8** (continued).

81 North Street, Doncaster - firestop walls - management info - targ fire - management Page 1

| Ref | Description | Hrs | Targ | Sub cost | Plt cost | Mat cost |
|---|---|---|---|---|---|---|
| 2200 | Const fire stop walls in roof space of 100 mm lightwt blks<br>20.00 $m^2$ 1.50 h/$m^2$ | 30.00 | 28.50 | | | 80.00 |

| | | | | |
|---|---|---|---|---|
| Factor 0.95 | Labour hours | 30.00 | Tot mat cst | 80.00 |
| | Factor total | 28.50 | Tot lab cst | 164.10 |
| | | | Tot sub cst | 0.00 |
| | | | Tot plt cst | 0.00 |

**Fig. 7.8** (continued).

81 North Street, Doncaster - rainwater goods - management info - targ RWG - management Page 1

| Ref | Description | Hrs | Targ | Sub cost | Plt cost | Mat cost |
|---|---|---|---|---|---|---|
| 3605 | Relace fascia and gutter inc 2 coats preservative | 10.00 | | | | |
| | 8.00 m 1.25 h/m | | 9.50 | | | 36.00 |
| 3610 | Stop ends or joints to existing | 1.50 | | | | |
| | 6.00 no 0.25 h/no | | 1.42 | | | 3.00 |
| 3615 | Outlets | 1.00 | | | | |
| | 2.00 no 0.50 h/no | | 0.95 | | | 2.60 |
| 3625 | Take down and replace RWPS | 7.20 | | | | |
| | 12.00 m 0.60 h/m | | 6.84 | | | 22.80 |
| 3630 | Shoes | 0.50 | | | | |
| | 2.00 no 0.25 h/no | | 0.48 | | | 4.50 |
| | Labour hours | | 20.20 | | Tot mat cst | 68.90 |
| | Factor 0.95 Factor total | | 19.19 | | Tot lab cst | 110.49 |
| | | | | | Tot sub cst | 0.00 |
| | | | | | Tot plt cst | 0.00 |

Fig. 7.8 (continued).

8 North Street, Doncaster - carpentry 1st fix - management info - targ car1 - management Page 1

| Ref | Description | Hrs | Targ | Sub cost | Plt cost | Mat cost |
|---|---|---|---|---|---|---|
| 205 | Short length of joist bolted to wall as trimmer | 1.00 | | | | |
| | 2.00 no 0.50 h/no | | 0.95 | | | 4.00 |
| 210 | Joist hangers | 1.00 | | | | |
| | 2.00 no 0.50 h/no | | 0.95 | | | 1.60 |
| 215 | Floor joist under partition | 1.25 | | | | |
| | 3.80 m 0.33 h/m | | 1.19 | | | 6.65 |
| 220 | Construct stud walling | 19.75 | | | | |
| | 79.00 m 0.25 h/m | | 18.76 | | | 59.25 |
| 310 | New joists in opening to match existing | 2.00 | | | | |
| | 4.00 m 0.50 h/m | | 1.90 | | | 6.00 |
| 315 | Joist hangers to hearth joists | 2.00 | | | | |
| | 8.00 no 0.25 h/no | | 1.90 | | | 6.40 |
| 330 | Floorboards in hearth area to match existing | 1.50 | | | | |
| | 1.00 m² 1.50 h/m² | | 1.43 | | | 10.00 |
| 510 | Take out window remove to skip, fix new window | 15.00 | | | | |
| | 6.00 no 2.50 h/no | | 14.25 | | 12.00 | |
| 520 | Window linings | 4.80 | | | | |
| | 24.00 m 0.20 h/m | | 4.56 | | | 21.60 |
| 535 | Double glazing | 9.30 | | | | |
| | 6.20 m² 1.50 h/m² | | 8.84 | | | 184.14 |
| 540 | Window boards | 6.00 | | | | |
| | 6.00 m 1.00 h/m | | 5.70 | | | 21.00 |
| 605 | T/out door and fr and CA, fit new door fr inc fanlight and temp door | 3.10 | | | | |
| | 1.00 no 3.10 h/no | | 2.95 | | 3.00 | 29.36 |
| 610 | T/out door and frame and CA, fit new frame and temporary door | 3.10 | | | | |
| | 1.00 no 3.10 h/no | | 2.95 | | 3.00 | 23.40 |
| 612 | Double glazing to fanlights | 0.56 | | | | |
| | 0.40 m² 1.40 h/m² | | 0.53 | | | 12.00 |
| 635 | Linings to external door openings | 2.16 | | | | |
| | 10.80 m 0.20 h/m | | 2.05 | | | 9.72 |
| 655 | Assemble and fix door linings (DC) | 8.00 | | | | |
| | 8.00 no 1.00 h/no | | 7.60 | | | 109.76 |
| 815 | Carefully take up floor boards for tie bars (2 no areas) | 2.25 | | | | |
| | 1.00 item 2.25 h/item | | 2.14 | | | |
| 820 | Notch joists and screw fit 50 x 6 galv steel bars (2 no) | 3.00 | | | | |
| | 1.00 item 3.00 h/item | | 2.85 | | | |
| 830 | Replace floorboards over ties (2 no ties) | 3.00 | | | | |
| | 1.00 item 3.00 h/item | | 2.85 | | | 1.00 |

Fig. 7.8 (continued).

8 North Street, Doncaster - carpentry 1st fix - management info - targ car1 - management Page 2

| Ref | Description | Hrs | Targ | Sub cost | Plt cost | Mat cost |
|---|---|---|---|---|---|---|
| 835 | 50 x 50 noggins to sides of ties<br>12.00 m 0.20 h/m | 2.40 | 2.28 | | | 6.00 |
| 840 | Packings at wall at ends of ties<br>4.00 no 0.16 h/no | 0.64 | 0.61 | | | 0.40 |
| 920 | Remove short lengths of roof timber and replace (as STR)<br>1.00 item 2.00 h/item | 2.00 | 1.90 | | | 6.00 |
| 1120 | Fix pipe boxing back boards (BA)<br>10.00 m 0.25 h/m | 2.50 | 2.38 | | | 15.00 |
| 1325 | 160 x 25 capping with bullnosed arrises<br>1.00 m 1.00 h/m | 1.00 | 0.95 | | | 1.50 |
| 1410 | 150 x 25 SW backboard fixed with Rawlbolts @ 1200 cts<br>4.50 m 0.33 h/m | 1.49 | 1.41 | | | 6.98 |
| 1470 | Unscrew stair clips and remove to skip<br>28.00 no 0.03 h/no | 0.84 | 0.80 | | | |
| 1525 | Jsts on hangers, T and G boarding in trimming opg and MG (STG1)<br>1.00 item 2.00 h/item | 2.00 | 1.90 | | | 10.00 |
| 1700 | Remove and CA matwell and replace with floorboards inc bearers<br>1.00 item 2.00 h/item | 2.00 | 1.90 | | 1.00 | 11.00 |
| 1925 | Bracketing around RSJs to take plasterboard<br>23.00 m 0.25 h/m | 5.75 | 5.46 | | | 10.35 |
| 2405 | Cut out clg joists, insert 75 x 50 trimmers and 125 x 25 lining<br>1.00 item 2.00 h/item | 2.00 | 1.90 | | | 7.00 |
| 2700 | Construct tank support and walkway in roof (SA)<br>1.00 no 3.35 h/no | 3.35 | 3.18 | | | 25.00 |
| 3400 | Passive ventilation system as spec<br>1.00 item 14.00 h/item | 14.00 | 13.30 | | | 173.86 |
| | Factor 0.95 | Labour hours<br>Factor total | 128.74<br>122.30 | | Tot mat cst<br>Tot lab cst<br>Tot sub cst<br>Tot plt cst | 778.97<br>704.20<br>0.00<br>19.00 |

**Fig. 7.8** (continued).

81 North Street, Doncaster - Plastering - management info - targ plas - management Page 1

| Ref | Description | Hrs | Targ | Sub cost | Plt cost | Mat cost |
|---|---|---|---|---|---|---|
| 15 | 12.5 mm foil backed plasterboard and skim (CE)<br>33.50 m² 0.63 h/m² | 21.11 | 20.05 | | | 70.35 |
| 20 | 12.5 plasterboard and skim to ceilings (CA)<br>33.00 m² 0.63 h/m² | 20.79 | 19.75 | | | 61.05 |
| 110 | 39.5 mm Styroliner and skim to walls (PE)<br>53.00 m² 1.02 h/m² | 54.06 | 51.36 | | | 530.00 |
| 115 | Replast walls with additive (PA)<br>69.00 m² 0.52 h/m² | 35.88 | 34.09 | | | 165.60 |
| 120 | Replaster walls (PB)<br>86.00 m² 0.52 h/m² | 44.72 | 42.48 | | | 133.30 |
| 225 | Insulation to stud walling<br>16.00 m² 0.23 h/m² | 3.68 | 3.50 | | | 32.00 |
| 230 | Plasterboard and skim to stud partitions<br>32.00 m² 0.63 h/m² | 20.16 | 19.15 | | | 67.20 |
| 325 | 225x150 plaster vent in flue<br>2.00 no 0.50 h/no | 1.00 | 0.95 | | | 1.50 |
| 1610 | Plaster vents (STE)<br>5.00 no 0.50 h/no | 2.50 | 2.38 | | | 3.75 |
| 1845 | Screed<br>17.00 m² 0.46 h/m² | 7.82 | 7.43 | | | 83.30 |
| 1930 | Double plasterboard and skim to beams<br>3.00 m² 1.26 h/m² | 3.78 | 3.59 | | | 25.20 |
| | Factor 0.95 | Labour hours<br>Factor total | 215.50<br>204.72 | | Tot mat cst<br>Tot lab cst<br>Tot sub cst<br>Tot plt cst | 1173.25<br>1178.76<br>0.00<br>0.00 |

**Fig. 7.8** (continued).

81 North Street, Doncaster - carpentry 2nd fix - management info - targ car2 - management Page 1

| Ref | Description | Hrs | Targ | Sub cost | Plt cost | Mat cost |
|---|---|---|---|---|---|---|
| 235 | Skirting to partitions | 1.80 | | | | |
| | 9.00 m 0.20 h/m | | 1.71 | | | 8.55 |
| 400 | 125x19 torus skirting (JF) | 13.40 | | | | |
| | 67.00 m 0.20 h/m | | 12.73 | | | 63.65 |
| 530 | Architrave around windows | 4.80 | | | | |
| | 24.00 m 0.20 h/m | | 4.56 | | | 10.80 |
| 615 | Take off temporary door and store for reuse, fix new door | 1.75 | | | | |
| | 1.00 no 1.75 h/no | | 1.66 | | | 112.00 |
| 620 | Take off temporary door and store for reuse, fix new door | 1.75 | | | | |
| | 1.00 no 1.75 h/no | | 1.66 | | | 33.50 |
| 625 | Toughened double glazing to doors | 1.54 | | | | |
| | 1.10 $m^2$ 1.40 h/$m^2$ | | 1.46 | | | 59.40 |
| 640 | Architraves to external linings | 1.62 | | | | |
| | 10.80 m 0.15 h/m | | 1.54 | | | 5.72 |
| 660 | Fix internal door and architraves (DC) | 21.36 | | | | |
| | 8.00 no 2.67 h/no | | 20.29 | | | 186.40 |
| 665 | 35x19 strip screwed to bottom of doors | 2.64 | | | | |
| | 8.00 no 0.33 h/no | | 2.51 | | | 2.40 |
| 710 | Internal door ironmongery (DA) | 10.50 | | | | |
| | 7.00 no 1.50 h/no | | 9.98 | | | |
| 715 | Bathroom door ironmongery (DA) | 1.50 | | | | |
| | 1.00 no 1.50 h/no | | 1.43 | | | |
| 720 | Front door ironmongery (DG) | 9.30 | | | | |
| | 1.00 no 9.30 h/no | | 8.84 | | | |
| 725 | Back door ironmongery (DG) | 4.76 | | | | |
| | 1.00 no 4.76 h/no | | 4.52 | | | |
| 1060 | Construct HW cylinder support stand (JG) | 2.00 | | | | |
| | 1.00 no 2.00 h/no | | 1.90 | | | 4.00 |
| 1065 | Construct slatted shelves (JG) | 4.00 | | | | |
| | 2.00 no 2.00 h/no | | 3.80 | | | 8.00 |
| 1415 | Moulded handrail with SAA brackets @ 1200 cts | 1.49 | | | | |
| | 4.50 m 0.33 h/m | | 1.41 | | | 21.15 |
| 2110 | 50x25 softwood bearers | 0.52 | | | | |
| | 2.60 m 0.20 h/m | | 0.49 | | | 1.04 |
| 2115 | Worktops | 6.00 | | | | |
| | 6.00 m 1.00 h/m | | 5.70 | | | |
| 2120 | Aluminium trims | 2.40 | | | | |
| | 2.40 m 1.00 h/m | | 2.28 | | | |

**Fig. 7.8** (continued).

81 North Street, Doncaster - carpentry 2nd fix - management info - targ car2 - management Page 2

| Ref | Description | Hrs | Targ | Sub cost | Plt cost | Mat cost |
|---|---|---|---|---|---|---|
| 2125 | Chrome legs | 0.20 | | | | |
| | 1.00 no 0.20 h/no | | 0.19 | | | |
| 2130 | Single base unit | 2.70 | | | | |
| | 3.00 no 9.00 h/no | | 2.57 | | | |
| 2135 | Double base unit | 1.45 | | | | |
| | 1.00 no 1.45 h/no | | 1.38 | | | |
| 2140 | Double sink base unit | 1.50 | | | | |
| | 1.00 no 1.50 h/no | | 1.43 | | | |
| 2145 | Double wall unit | 2.00 | | | | |
| | 1.00 no 2.00 h/no | | 1.90 | | | |
| 2410 | Roof hatch stops, door fxd with brass c&s, insul and architrave | 3.00 | | | | |
| | 1.00 no 3.00 h/no | | 2.85 | | | 8.40 |
| 3905 | Yard door and frame including sealant (DH) | 1.20 | | | | |
| | 1.00 no 1.20 h/no | | 1.14 | | | 22.00 |
| 3910 | Yard door hung on tee hinges including latch | 2.00 | | | | |
| | 1.00 no 2.00 h/no | | 1.90 | | | 40.00 |
| | | Labour hours | 107.18 | | Tot mat cst | 587.01 |
| | Factor 0.95 | Factor total | 101.82 | | Tot lab cst | 586.25 |
| | | | | | Tot sub cst | 0.00 |
| | | | | | Tot plt cst | 0.00 |

**Fig. 7.8** (continued).

81 North Street, Doncaster - insulation - management info - targ insl - management Page 1

| Ref | Description | Hrs | Targ | Sub cost | Plt cost | Mat cost |
|---|---|---|---|---|---|---|
| 2305 | Supply and lay 200 mm Rockwool insulation to roof space (INS1) | 3.40 | | | | |
| | 34.00 m² 0.10 h/m² | | 3.23 | | | 88.40 |
| 2310 | 200 mm Rockwool ins to cellar fixed with netting (INS2) | 4.29 | | | | |
| | 13.00 m² 0.33 h/m² | | 4.08 | | | 39.00 |
| | | Labour hours | 7.69 | | Tot mat cst | 127.40 |
| | Factor 0.95 | Factor total | 7.31 | | Tot lab cst | 36.37 |
| | | | | | Tot sub cst | 0.00 |
| | | | | | Tot plt cst | 0.00 |

**Fig. 7.8** (continued).

81 North Street, Doncaster - tiling - management info - targ tile - management Page 1

| Ref | Description | Hrs | Targ | Sub cost | Plt cost | Mat cost |
|---|---|---|---|---|---|---|
| 1110 | Proprietary plastic tiling strip to bath and WHB | 0.75 | | | | |
| | 3.00 m 0.25 h/m | | 0.71 | | | 3.00 |
| 1115 | Three courses of 100 x 100 tiles above bath and washbasin | 2.00 | | | | |
| | 1.00 m² 2.00 h/m² | | 1.90 | | | 10.00 |
| 2150 | Ceramic wall tiling | 10.00 | | | | |
| | 5.00 m² 2.00 h/m² | | 9.50 | | | 60.00 |
| 2155 | Sealant | 2.94 | | | | |
| | 8.40 m 0.35 h/m | | 2.79 | | | 2.94 |
| | | Labour hours | 15.69 | | Tot mat cst | 75.94 |
| | Factor 0.95 | Factor total | 14.91 | | Tot lab cst | 85.82 |
| | | | | | Tot sub cst | 0.00 |
| | | | | | Tot plt cst | 0.00 |

**Fig. 7.8** (continued).

81 North Street, Doncaster - carpentry final fix, etc. - management info - targ car3 - management Page 1

| Ref | Description | Hrs | Targ | Sub cost | Plt cost | Mat cost |
|---|---|---|---|---|---|---|
| 525 | Mastic pointing around windows | 4.50 | | | | |
| | 30.00 m 0.15 h/m | | 4.27 | | | 10.50 |
| 630 | Mastic pointing around doors | 3.24 | | | | |
| | 10.80 m 0.30 h/m | | 3.08 | | | 3.78 |
| 1125 | Towel rail, toilet roll holder and mirror | 2.00 | | | | |
| | 1.00 item 2.00 h/item | | 1.90 | | | 33.00 |
| 1130 | Pipe boxing front and sides | 2.50 | | | | |
| | 10.00 m 0.25 h/m | | 2.38 | | | 10.00 |
| 1135 | Fix bath panels | 3.00 | | | | |
| | 2.00 no 1.50 h/no | | 2.85 | | | 1.00 |
| 1205 | Knock down nails and supply and lay ply for floor coverings | 4.00 | | | | |
| | 4.00 m² 1.00 h/m² | | 3.80 | | | 8.00 |
| 2500 | Replace floorboards in small quantities (FA) | 3.75 | | | | |
| | 15.00 m 0.25 h/m | | 3.56 | | | 18.00 |
| | | Labour hours | 22.99 | | Tot mat cst | 84.28 |
| | Factor 0.95 | Factor total | 21.84 | | Tot lab cst | 125.76 |
| | | | | | Tot sub cst | 0.00 |
| | | | | | Tot plt cst | 0.00 |

**Fig. 7.8** (continued).

81 North Street, Doncaster - external brickwork - management info - targ blr2 - management Page 1

| Ref | Description | Hrs | Targ | Sub cost | Plt cost | Mat cost |
|---|---|---|---|---|---|---|
| 3525 | Scaffold for pointing stacks | 8.00 | | | | |
| | 1.00 no 8.00 h/no | | 7.60 | | | 40.00 |
| 3535 | Remove chimney pots and return to yard | 4.00 | | | | |
| | 4.00 no 1.00 h/no | | 3.80 | | | |
| 3540 | Fix gas cowls (EG) | 2.50 | | | | |
| | 1.00 no 2.50 h/no | | 2.38 | | | 28.00 |
| 3545 | Fix mushroom caps (EH) | 4.50 | | | | |
| | 3.00 no 1.50 h/no | | 4.27 | | | 36.00 |
| 3700 | Repoint elevations (EA1) | 29.00 | | | | |
| | 29.00 m$^2$ 1.00 h/m$^2$ | | 27.55 | | | 6.38 |
| 3705 | Chop out indiv bricks and replace with second hand brks (EL) | 7.50 | | | | |
| | 15.00 no 0.50 h/no | | 7.13 | | | 3.75 |
| 3820 | Remove and CA door steps | 0.50 | | | | |
| | 1.00 no 0.50 h/no | | 0.48 | | 0.60 | |
| 3830 | Supply and fix external door step (GD2) | 1.00 | | | | |
| | 1.00 no 1.00 h/no | | 0.95 | | | 15.00 |
| 3840 | Rebuild top course of bdy wall as necessary and soldier course | 5.00 | | | | |
| | 1.00 item 5.00 h/item | | 4.75 | | | 5.00 |

| | | | | |
|---|---|---|---|---|
| | | Labour hours 62.00 | Tot mat cst | 134.13 |
| Factor | 0.95 | Factor total 58.90 | Tot lab cst | 339.14 |
| | | | Tot sub cst | 0.00 |
| | | | Tot plt cst | 0.60 |

**Fig. 7.8** (continued).

81 North Street, Doncaster - drainage - management info - targ drn - management Page 1

| Ref | Description | Hrs | Targ | Sub cost | Plt cost | Mat cost |
|---|---|---|---|---|---|---|
| 4000 | Excavate for and construct manhole (GL) | 14.60 | | | | |
| | 1.00 no 14.60 h/no | | 13.87 | | | 120.00 |
| 4005 | Excavate for and lay drains | 3.75 | | | | |
| | 1.50 m 2.50 h/m | | 3.56 | | | 6.00 |
| 4010 | Bends or rest bends | 2.00 | | | | |
| | 4.00 no 0.50 h/no | | 1.90 | | | 11.00 |
| 4015 | Backfill drains | 0.93 | | | | |
| | 0.70 m$^3$ 1.33 h/m$^3$ | | 0.88 | | | |
| 4020 | Back inlet gulley | 2.00 | | | | |
| | 1.00 no 2.00 h/no | | 1.90 | | | 16.00 |

| | | | | |
|---|---|---|---|---|
| | | Labour hours 23.28 | Tot mat cst | 153.00 |
| Factor | 0.95 | Factor total 22.12 | Tot lab cst | 126.66 |
| | | | Tot sub cst | 0.00 |
| | | | Tot plt cst | 0.00 |

**Fig. 7.8** (continued).

81 North Street, Doncaster - External works - management info - targ ext - management Page 1

| Ref | Description | Hrs | Targ | Sub cost | Plt cost | Mat cost |
|---|---|---|---|---|---|---|
| 3805 | Take up paving and exc for new paving flags and rem to skip | 19.50 | | | | |
| | 26.00 m$^2$ 0.75 h/m$^2$ | | 18.53 | | 31.20 | |
| 3810 | Exc, lay bed, lay 152×52 edging with haunching (GG1) | 21.00 | | | | |
| | 21.00 m 1.00 h/m | | 19.95 | | | 56.70 |
| 3815 | Precast concrete paving flags as (GE1) | 19.50 | | | | |
| | 26.00 m$^2$ 0.75 h/m$^2$ | | 18.53 | | | 130.00 |
| 4100 | Exc, lay found and fix cloths post and hook (GC) | 2.86 | | | | |
| | 1.00 no 2.86 h/no | | 2.72 | | | 16.27 |
| 4205 | Dig over garden 150 mm deep | 6.60 | | | | |
| | 20.00 m$^2$ 0.33 h/m$^2$ | | 6.27 | | | |
| 4210 | Rake level | 1.00 | | | | |
| | 20.00 m$^2$ 0.05 h/m$^2$ | | 0.95 | | | |
| 4215 | Lay topsoil 70 mm thick | 5.60 | | | | |
| | 1.40 m$^3$ 4.00 h/m$^3$ | | 5.32 | | | 28.00 |

| | | | | |
|---|---|---|---|---|
| | | Labour hours 76.06 | Tot mat cst | 230.97 |
| Factor | 0.95 | Factor total 72.26 | Tot lab cst | 374.19 |
| | | | Tot sub cst | 0.00 |
| | | | Tot plt cst | 31.20 |

**Fig. 7.8** (continued).

81 North Street, Doncaster - sub-contractor info for management - targ sub - management Page 1

| Ref | Description | Hrs | Targ | Sub cost | Plt cost | Mat cost |
|---|---|---|---|---|---|---|
| 335 | Sweep flues | 0.00 | | | | |
| | 2.00 no 0.00 h/no | | 0.00 | 28.00 | | |
| 1105 | Supply and fix bathroom suite (BA) | 0.00 | | | | |
| | 1.00 0.00 h | | 0.00 | 415.00 | | |
| 1140 | Supply and fix above ground drainage (BB) | 0.00 | | | | |
| | 1.00 item 0.00 h/item | | 0.00 | 90.00 | | |
| 1210 | Supply and lay anti-slip flooring (FE4) | 0.00 | | | | |
| | 4.00 $m^2$ 0.00 h/$m^2$ | | 0.00 | 81.64 | | |
| 1215 | Anti-slip sheetg to conc floor inc latex (FE5) | 0.00 | | | | |
| | 15.00 $m^2$ 0.00 h/$m^2$ | | 0.00 | 504.00 | | |
| 1520 | Sweep flues | 0.00 | | | | |
| | 1.00 no 0.00 h/no | | 0.00 | 14.00 | | |
| 2160 | Plumbing to kitchen fittings (FG1) | 0.00 | | | | |
| | 1.00 item 0.00 h/item | | 0.00 | 130.00 | | |
| 2705 | Supply and fix cold water tank (SA) | 0.00 | | | | |
| | 1.00 no 0.00 h/no | | 0.00 | 95.00 | | |
| 2800 | Supply and fix HW cylinder (SB) | 0.00 | | | | |
| | 1.00 no 0.00 h/no | | 0.00 | 90.00 | | |
| 2900 | Water plumbing installation (SD) | 0.00 | | | | |
| | 1.00 no 0.00 h/no | | 0.00 | 280.00 | | |
| 3000 | Gas plumbing installation (SK) | 0.00 | | | | |
| | 1.00 no 0.00 h/no | | 0.00 | 60.00 | | |
| 3100 | Electrical rewiring installation (ELE) | 0.00 | | | | |
| | 1.00 no 0.00 h/no | | 0.00 | 967.00 | | |
| 3200 | Supply and fix CH gas boiler (SF) | 0.00 | | | | |
| | 1.00 no 0.00 h/no | | 0.00 | 775.00 | | |
| 3300 | Heating installation including radiators | 0.00 | | | | |
| | 1.00 item 0.00 h/item | | 0.00 | 705.00 | | |
| 3505 | Replace roof covering, bonding gutters, flashings and soakers | 0.00 | | | | |
| | 1.00 item 0.00 h/item | | 0.00 | 1645.00 | | |
| 4300 | Internal and external decoration | 0.00 | | | | |
| | 1.00 item 0.00 h/item | | 0.00 | 874.00 | | |
| 4350 | DPC installation (+P & A) | 0.00 | | | | |
| | 1.00 item 0.00 h/item | | 0.00 | 350.00 | | |
| | Factor 1.00 | Labour hours | 0.00 | | Tot mat cst | 0.00 |
| | | Factor total | 0.00 | | Tot lab cst | 0.00 |
| | | | | | Tot sub cst | 7103.64 |
| | | | | | Tot plt cst | 0.00 |

**Fig. 7.8** (continued).

81 North Street, Doncaster - supplier info for management - targ supp - management Page 1

| Ref | Description | Hrs | Targ | Sub cost | Plt cost | Mat cost |
|---|---|---|---|---|---|---|
| 505 | Supply only windows | 0.00 | | | | |
| | 1.00 item 0.00 h/item | | 0.00 | | | 712.76 |
| 705 | Supply only all ironmongery | 0.00 | | | | |
| | 1.00 item 0.00 h/item | | 0.00 | | | 397.00 |
| 1935 | Supply only all steel work (2 RSJs and ties) | 0.00 | | | | |
| | 1.00 item 0.00 h/item | | 0.00 | | | 156.00 |
| 2005 | Delivery charge for vents | 0.00 | | | | |
| | 1.00 item 0.00 h/item | | 0.00 | | | 18.00 |
| | Factor 1.00 | Labour hours | 0.00 | | Tot mat cst | 1283.76 |
| | | Factor total | 0.00 | | Tot lab cst | 0.00 |
| | | | | | Tot sub cst | 0.00 |
| | | | | | Tot plt cst | 0.00 |

**Fig. 7.8** (continued).

A separate report was generated for supply only of materials. This covers materials that apply to more than one operation where all-inclusive prices were given by suppliers at the tender stage.

*(b)(i) Descriptions, quantities, location and target for operatives*

Typical examples of reports for carpentry first fix and carpentry second fix are shown in Figs 7.9 and 7.10. This type of report was used when quantities were necessary to clarify the extent of the work.

*(b)(ii) Descriptions, location and target for the operatives (i.e. excluding quantities)*

Typical examples of reports for preparation work and rainwater goods are shown in Figs 7.11 and 7.12.

### *7.1.9 Planning*

The target descriptions were used as programme operations together with subcontractor operations.

*Calculation Sheet (Fig. 7.13)*

The calculation sheet was generated using Lotus 1-2-3, but it could have been completed manually.

81 North Street, Doncaster - target for carpentry 1st fix - targ car1 - operatives Page 1

| Ref | Description | Quantity | Unit | Location |
|---|---|---|---|---|
| 205 | Short length of joist bolted to wall as trimmer | 2.00 | no | Bed2/cor |
| 210 | Joist hangers | 2.00 | no | Bed2/cor |
| 215 | Floor joist under partition | 3.80 | m | Bed2/cor |
| 220 | Construct stud walling | 79.00 | m | Bed2/cor |
| 310 | New joists in opening to match existing | 4.00 | m | Bedrooms |
| 315 | Joist hangers to hearth joists | 8.00 | no | Bedrooms |
| 330 | Floorboards in hearth area to match existing | 1.00 | $m^2$ | Bedrooms |
| 510 | Take out window remove to skip, fix new window | 6.00 | no | |
| 520 | Window linings | 24.00 | m | |
| 535 | Double glazing | 6.20 | $m^2$ | |
| 540 | Window boards | 6.00 | m | |
| 605 | T/out door and fr and CA, fit new door fr inc fanlight and temp door | 1.00 | no | Front dr |
| 610 | T/out door and frame and CA, fit new frame and temporary door | 1.00 | no | Back dr |
| 612 | Double glazing to fanlights | 0.40 | $m^2$ | Front dr |
| 635 | Linings to external door openings | 10.80 | m | |
| 655 | Assemble and fix door linings (DC) | 8.00 | no | Internal |
| 815 | Carefully take up floor boards for tie bars (2 no areas) | 1.00 | item | Bedroom 1 |
| 820 | Notch joists and screw fit 50 x 6 galv steel bars (2 no) | 1.00 | item | Bedroom 1 |
| 830 | Replace floor boards over ties (2 no ties) | 1.00 | item | Bedroom 1 |
| 835 | 50 x 50 noggins to sides of ties | 12.00 | m | Bedroom 1 |
| 840 | Packings at wall at ends of ties | 4.00 | no | Bedroom 1 |
| 920 | Remove short lengths of roof timber and replace (as STR) | 1.00 | item | Bathroom |
| 1120 | Fix pipe boxing back boards (BA) | 10.00 | m | Bathroom |
| 1325 | 160 x 25 capping with bullnosed arrises | 1.00 | m | Stair |
| 1410 | 150 x 25 SW backboard fixed with Rawlbolts @ 1200 cts | 4.50 | m | Stair |
| 1470 | Unscrew stair clips and remove to skip | 28.00 | no | Stair |
| 1525 | Jsts on hangers, T and G boarding in trimming opg and MG (STG1) | 1.00 | item | Living |
| 1700 | Remove and CA matwell and replace with floorboards inc bearers | 1.00 | item | Living |
| 1925 | Bracketing around RSJs to take plasterboard | 23.00 | m | Kit/din |
| 2405 | Cut out clg joists, insert 75 x 50 trimmers and 125 x 25 lining | 1.00 | item | See arch |
| 2700 | Construct tank support and walkway in roof (SA) | 1.00 | no | Roof |
| 3400 | Passive ventilation system as spec | 1.00 | item | See drg |
| | Target | | 122 | |

**Fig. 7.9** Case study 1: target for carpentry first fix.

81 North Street, Doncaster - target for carpentry 2nd fix - targ car2 - operatives Page 1

| Ref | Description | Quantity | Unit | Location |
|---|---|---|---|---|
| 235 | Skirting to partitions | 9.00 | m | Bed2/cor |
| 400 | 125 x 19 torus skirting (JF) | 67.00 | m | |
| 530 | Architrave around windows | 24.00 | m | |
| 615 | Take off temporary door and store for reuse, fix new door | 1.00 | no | Front dr |
| 620 | Take off temporary door and store for reuse, fix new door | 1.00 | no | Back dr |
| 625 | Toughened double glazing to doors | 1.10 | m$^2$ | Back dr |
| 640 | Architraves to external linings | 10.80 | m | |
| 660 | Fix internal door and architraves (DC) | 8.00 | no | |
| 665 | 35 x 19 strip screwed to bottom of doors | 8.00 | no | Internal |
| 710 | Internal door ironmongery (DA) | 7.00 | no | |
| 715 | Bathroom door ironmongery (DA) | 1.00 | no | |
| 720 | Front door ironmongery (DG) | 1.00 | no | |
| 725 | Back door ironmongery (DG) | 1.00 | no | |
| 1060 | Construct HW cylinder support stand (JG) | 1.00 | no | Corridor |
| 1065 | Construct slatted shelves (JG) | 2.00 | no | Corridor |
| 1415 | Moulded handrail with SAA brackets @ 1200 cts | 4.50 | m | Stair |
| 2110 | 50 x 25 softwood bearers | 2.60 | m | Ki/di/ut |
| 2115 | Worktops | 6.00 | m | Ki/di/ut |
| 2120 | Aluminium trims | 2.40 | m | Kit/din |
| 2125 | Chrome legs | 1.00 | no | Kit/din |
| 2130 | Single base unit | 3.00 | no | Ki/di/ut |
| 2135 | Double base unit | 1.00 | no | Kit/din |
| 2140 | Double sink base unit | 1.00 | no | Kit/din |
| 2145 | Double wall unit | 1.00 | no | Kit/din |
| 2410 | Roof hatch stops, door fxd with brass C&S, insul and architrave | 1.00 | no | See arch |
| 3905 | Yard door and frame including sealant (DH) | 1.00 | no | Rear |
| 3910 | Yard door hung on tee hinges including latch | 1.00 | no | Rear |
| | Total | 102 | | |

**Fig. 7.10** Case study 1: target for carpentry second fix.

81 North Street, Doncaster - target for preparation work - targ prep - operatives Page 1

| Ref | Description | Location |
|---|---|---|
| 10 | Strip down ceiling, denail joists and remove to skip | |
| 105 | Hack off wall plaster and remove to skip | |
| 305 | Remove hearthstones and deposit in skip | Bedrooms |
| 650 | Take out linings and remove to skip | Internal |
| 905 | Take down corner flue and remove to skip | Bathroom |
| 910 | Take down stack and remove to skip | Bathroom |
| 1405 | Take down handrail and make good | Stair |
| 1505 | Take out fireplace fireback and hearthstone and remove to skip | Living |
| 1805 | Hack up existing concrete floor (FD) | Ki/di/ut |
| 1810 | CA concrete floor | Ki/di/ut |
| 1820 | DPM turned up 150 mm at walls | Ki/di/ut |
| 1825 | Secure DPM to wall with treated batten | Ki/di/ut |
| 1830 | Concrete floor slab | Ki/di/ut |
| 1835 | Insulation under solid floors | Ki/di/ut |
| 1840 | Insulation around edges of new solid floors | Ki/di/ut |
| 1905 | Take down chimney breast and remove to skip | Kit/din |
| 2600 | Remove and CA all rubbish internally and externally (RUB) | |
| 2605 | Take up floor coverings and remove to skip | |
| | Target 180 | |

**Fig. 7.11** Case study 1: target for preparation work.

To calculate programme duration it was assumed that the operation would take 75% of the target hours. In fact the expected hours are approximately two-thirds of the target hours, but 75% was used to allow for flexibility.

81 North Street, Doncaster - target for rainwater goods - targ RWG - operatives Page 1

| Ref | Description | Location |
|---|---|---|
| 3605 | Replace fascia and gutter inc 2 coats preservative | |
| 3610 | Stop ends or joints to existing | |
| 3615 | Outlets | See drg |
| 3625 | Take down and replace RWPS | All |
| 3630 | Shoes | To RWP |
| | Target 19 | |

**Fig. 7.12** Case study 1: target for rainwater goods.

81 North Street - calculation sheet

| Ref | Operation | Target hours | Programme hours | Gang size | Gang hours | Duration (Days) |
|---|---|---|---|---|---|---|
| PREP | Preparation work | 180 | 135 | 2 | 68 | 8.4 |
| SUBC | Sweep flues | | | | | 1.0 |
| BWK1 | Internal brickwork | 79 | 59 | 2 | 30 | 3.7 |
| SUBC | DPC and timber treatment | | | | | 1.0 |
| SUBC | Roof covering | | | | | 3.0 |
| RWG | Rainwater goods | 19 | 14 | 2 | 7 | 0.9 |
| FIRE | Firestop walls | 29 | 22 | 2 | 11 | 1.4 |
| CAR1 | Carpentry 1st fix | 122 | 92 | 2 | 46 | 5.7 |
| SUBC | Plumbing and heating 1st fix | | | | | 2.0 |
| SUBC | Electrical work 1st fix | | | | | 2.0 |
| PLAS | Plastering | 205 | 154 | 2 | 77 | 9.6 |
| CAR2 | Carpentry 2nd fix | 102 | 77 | 2 | 38 | 4.8 |
| SUBC | Plumbing and heating 2nd fix | | | | | 8.0 |
| SUBC | Electrical work 2nd fix | | | | | 3.0 |
| INSL | Insulation | 7 | 5 | 1 | 5 | 0.7 |
| TILE | Tiling | 15 | 11 | 1 | 11 | 1.4 |
| CAR3 | Carpentry 3rd fix | 22 | 17 | 1 | 17 | 2.1 |
| SUBC | Decoration | | | | | 7.0 |
| SUBC | Floor covering | | | | | 2.0 |
| BLR2 | External brickwork | 59 | 44 | 2 | 22 | 2.8 |
| DRN | Drainage | 22 | 17 | 2 | 8 | 1.0 |
| EXT | External works | 72 | 54 | 2 | 27 | 3.4 |
| SUBC | Clear up | | | | | 1.0 |

**Fig. 7.13** Case study 1: calculation sheet.

*Programme (Fig. 7.14)*

The programme chart was drawn up, based on information from the calculation sheet, and a labour schedule was shown below the programme chart.

*Tabular subcontract schedule (Fig. 7.15)*

The tabular subcontractor schedule was drawn up, based on the programme.

*Materials schedule (Fig. 7.16)*

The materials schedules were drawn up, based on the Information for Management (Fig. 7.8). As this relates to targets, the timing of material call-up and delivery times, etc. could be ascertained.

Data for ordering materials were also available from this report. Clearly, the quantities need to be checked prior to ordering, as one

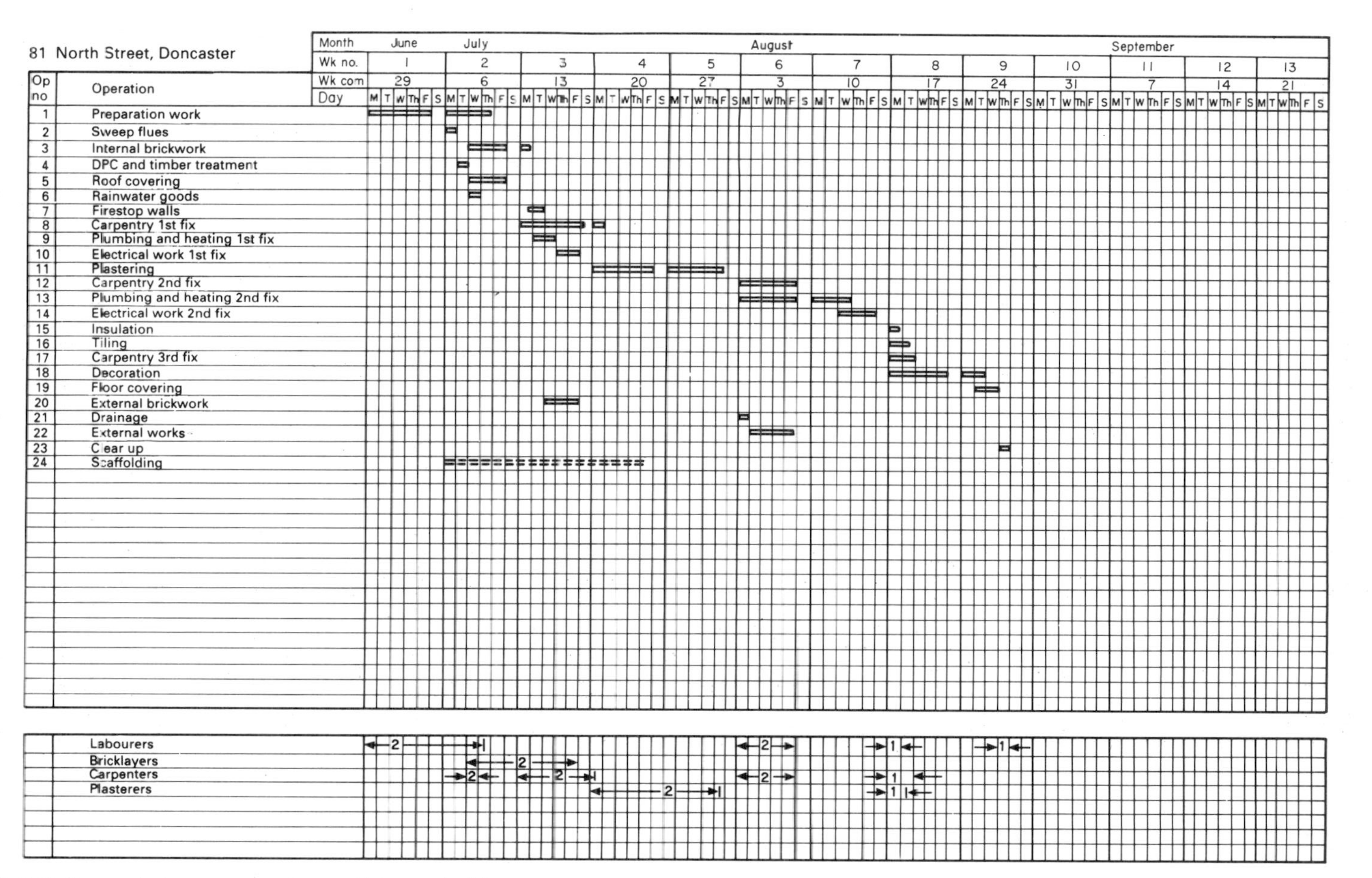


**Fig 7.14** Case study 1: programme chart and labour schedule.

81 North Street, Doncaster - subcontractor schedule

| Work subcontracted | Subcontractor | Date required | Notice | Date ordered | Order number |
|---|---|---|---|---|---|
| Sweep flues | R Bunting | 6 July | 3 days | 23 June | 145 |
| DPC and timber treatment | Multiskill | 7 July | 7 days | 23 June | 146 |
| Roofing, leadwork and scaffolding | K Sweeney | 8 July | 7 days | 23 June | 147 |
| Electrical work - 1st fix | A Caddick | 16 July | 7 days | 23 June | 148 |
| Electrical work - 2nd fix | A Caddick | 12 August | 7 days | 23 June | 148 |
| Plumbing and heating - 1st fix | J Burke | 14 July | 7 days | 23 June | 149 |
| Plumbing and heating - 2nd fix | J Burke | 3 August | 7 days | 23 June | 149 |
| Painting and decorating | W Bolden | 17 August | 14 days | | |
| Floor coverings | Shirtcliffes | 25 August | 14 days | | |
| Clean out on completion | Contract clean | 28 August | 7 days | | |

**Fig. 7.15** Case study 1: tabular subcontract schedule.

81 North Street, Doncaster - materials schedule

| Material | Supplier | Date required | Notice required | Date ordered | Order number | Date received | Comments |
|---|---|---|---|---|---|---|---|
| *Preparation* | | | | | | | |
| Jablite insulation | Aizlewoods | 30 June | 7 days | 23 June | 156 | 30 June | |
| DPM and DPC | Aizlewoods | 7 July | 7 days | 23 June | 156 | 30 June | |
| Concrete | Busybees | 2 July | 3 days | 23 June | 165 | 2 July | |
| *Brickwork - target 1* | | | | | | | |
| Bricks | Aizlewoods | 7 July | 7 days | 23 June | 156 | 7 July | |
| Precast lintels | Aizlewoods | 7 July | 7 days | 23 June | 156 | 7 July | |
| Padstones | Aizlewoods | 7 July | 7 days | 23 June | 156 | 7 July | |
| Catnic lintels | Aizlewoods | 7 July | 7 days | 23 June | 156 | 7 July | |
| Air grates and sleeves | Aizlewoods | 7 July | 7 days | 23 June | 156 | 7 July | |
| Vent tubes and periscope vents | Aizlewoods | 7 July | 7 days | 23 June | 156 | 7 July | |
| Steelwork | Daver Steels | 7 July | 7 days | 23 June | 157 | 8 July | |
| *Rainwater goods* | Aizlewoods | 6 July | 7 days | 23 June | 156 | 7 July | |
| *Carpentry first fix* | | | | | | | |
| Floor boards, noggins, joists, etc. | Malden Timber | 13 July | 7 days | 3 July | 178 | 13 July | |
| Studding | Malden Timber | 13 July | 7 days | 3 July | 178 | 13 July | |
| Window and door reveal linings | Malden Timber | 13 July | 7 days | 3 July | 178 | 13 July | |
| MDF window boards | Malden Timber | 13 July | 7 days | 3 July | 178 | 13 July | |
| Fascia boards | Malden Timber | 13 July | 7 days | 3 July | 178 | 13 July | |
| Roof timbers | Malden Timber | 13 July | 7 days | 3 July | 178 | 13 July | |
| Back boards for pipes | Malden Timber | 13 July | 7 days | 3 July | 178 | 13 July | |
| Backboard for handrail | Malden Timber | 13 July | 7 days | 3 July | 178 | 13 July | |
| Windows | GD Woodworking | 13 July | 7 days | 3 July | 179 | 13 July | |
| Door frames and internal linings | GD Woodworking | 13 July | 7 days | 3 July | 179 | 13 July | |
| Double and single glazing | Discount Glass | 13 July | 7 days | 3 July | 180 | 13 July | |
| Temporary doors | In Yard | | | | | | Old doors from previous jobs |

**Fig. 7.16** Case study 1: materials schedule.

should not rely entirely on the quantities stated on the schedule produced at the tender stage.

### *7.1.9 Control*

*Control of progress*

This was shown on the programme, based on analysis of progress on each target. Progress was shown at the end of each week. An illustration is given showing progress at the end of week 3 (Fig. 7.17).

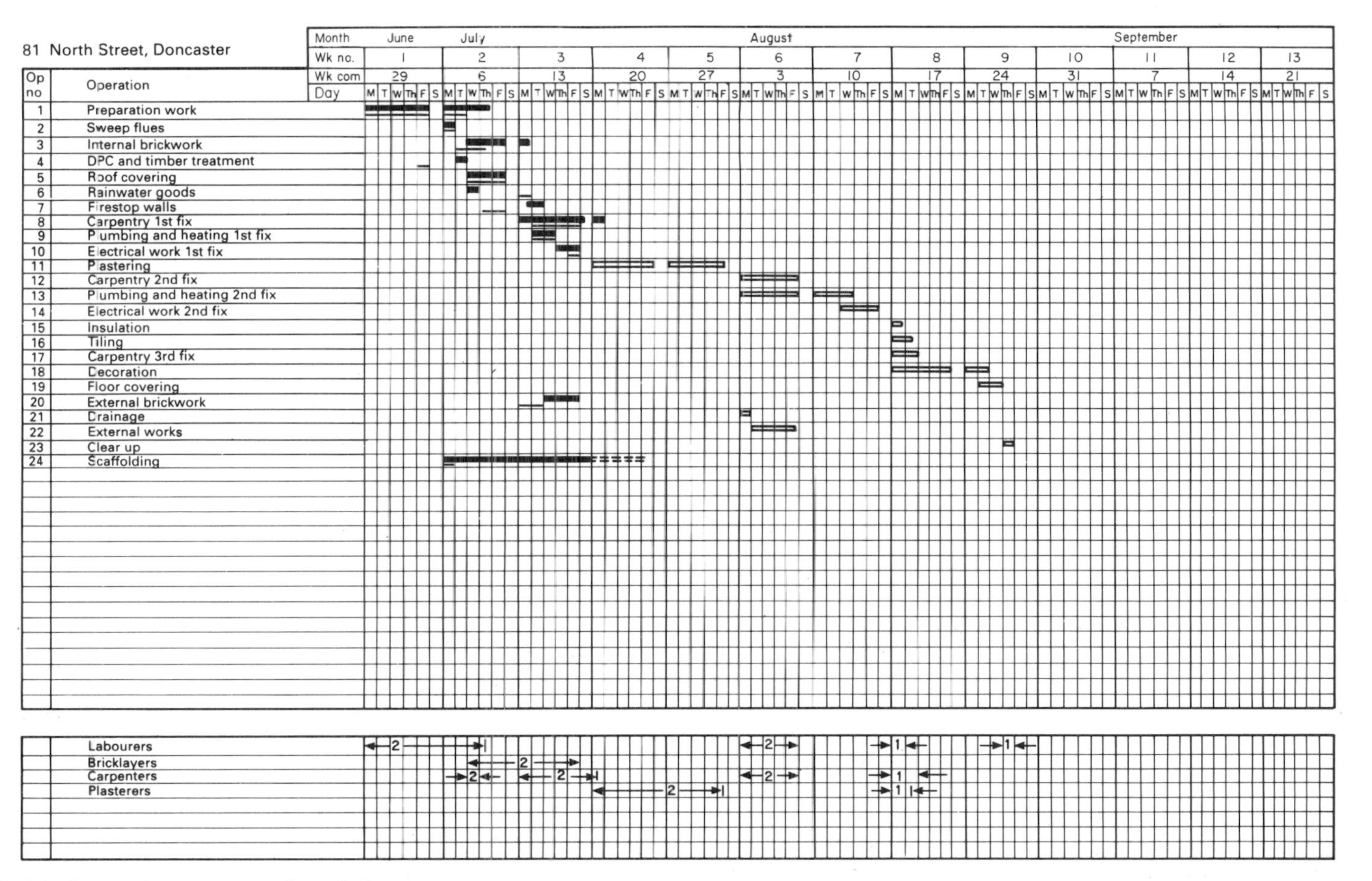

**Fig. 7.17** Case study 1: progress at the end of week 3.

A summary of progress is as follows:

- ❑ Preparation work was complete and the duration of the operation was one and a half days less than programmed. This operation is also covered in the cost control sheet (Fig. 7.20).
- ❑ Sweep flues was completed as programmed.

TIME SHEET

NAME: D. Charles

WEEK ENDING: 4th June

| JOB | DESCRIPTION OF WORK | Start and finish time | TOTAL OPERATIONAL TIME | | | | | | |
|---|---|---|---|---|---|---|---|---|---|
| | | | M | T | W | Th | F | Sa | Su |
| 81, North Street | Preparation work | 8.00<br>5.00 | 8½ | | | | | | |
| | — " — | 8.00<br>5.00 | | 8½ | | | | | |
| | — " — | 8.00<br>5.00 | | | 8½ | | | | |
| | — " — | 8.00<br>5.00 | | | | 8½ | | | |
| | — " — | 8.00<br>5.00 | | | | | 8 | | |
| | | | | | | | | | |

Fig. 7.18 Case study 1: typical example of a general operative's time sheet.

- ❑ Brickwork target 1 started one day early, and the duration of the operation was only three days. This operation is also covered by the cost control sheet (Fig. 7.20).
- ❑ DPC and timber treatment was completed earlier than programmed.
- ❑ Roof coverings proceeded exactly as programmed.
- ❑ Rainwater goods were carried out after the roof covering, and the duration of the operation was as programmed.
- ❑ Firestop walls were completed early, as they were carried out by the bricklayers from operation no. 3. The operation duration was as programmed.
- ❑ Carpentry first fix started one day late and was completed half a day early.
- ❑ Plumbing first fix was completed as programmed.
- ❑ Electrical first fix started one day late and has been completed.
- ❑ External brickwork target 2 started early as it was carried out by bricklayers from operation 7. The operation duration was half a day less than programmed.

As can be seen, overall progress was as programmed.

*Control of labour costs*

Timesheets were completed each week by all operatives. A typical example of a general operative's time sheet for week 1 of the contract is shown in Fig. 7.18.

Information from timesheets was recorded on the weekly summary sheet (Fig. 7.19). The information shown on the timesheet has been transferred onto this sheet together with the information from the timesheet for another operative working on the same operation.

Information from the weekly summary sheet was transferred onto the cost control sheet (Fig. 7.20). Cost control sheet entries are shown for the operations preparation work and brickwork target 1. The calculations are carried out as shown in section 4.2.11.

When this project was carried out, the all-in rate for general operatives was £4.80 per hour and for bricklayers was £5.47. All of the operatives were paid £2.58 per hour saved.

*Weekly Summary Sheet*
Week ending 4 June

| Project | Name | Trade | Operation | M | T | W | Th | F | Sa | Su | Travel time | Total | Remarks |
|---|---|---|---|---|---|---|---|---|---|---|---|---|---|
| 81 North Street | D. Charles | Gen op | Preparation work | 8.5 | 8.5 | 8.5 | 8.5 | 8 | | | | 42 | |
| | P. Rowbotham | Gen op | Preparation work | 8.5 | 8.5 | 8.5 | 8.5 | 8 | | | | 42 | |

**Fig. 7.19** Case study 1: weekly summary sheet.

*Cost Control Sheet*

| a | b | c | d | e | f | g | h | i | j | k | l | m | n | o |
|---|---|---|---|---|---|---|---|---|---|---|---|---|---|---|
| Week ending | Target no | Operation | Measure | Target rate | Target hours | Target value | Time taken | Cost | Time saved | Bonus | Adjustment | Total cost | Saving/ loss | Remarks |
| 4 June | | Preparation work | | | 180.00 | 907.47 | 84.00 | | | | | | | |
| 11 June | | | | | | | 26.00 | | | | | | | |
| | | | | | | | 110.00 | 528.00 | 70.00 | 180.60 | 48.00 | 756.60 | 150.87 | |
| 11 June | | Brickwork target 1 | | | 79.00 | 452.51 | 51.00 | 278.97 | 28.00 | 72.24 | 32.82 | 384.03 | 68.48 | |

**Fig. 7.20** Case study 1: cost control sheet.

The adjustments shown in the adjustments column are for work carried out by the operatives that was not included in the target set: in the brickwork target (item 1910 on the target sheet) 'making good walls where breast removed' was more extensive than expected as the adjacent wall was affected.

*Control of material costs*

At the estimating stage, quotations were obtained for all of the major materials. When ordering materials based on the items under each target, the estimator (who was also the buyer) ensured that all the costs were within original quotations wherever possible.

*Variations*

Variations were entered into the original MASTER estimate file as they occurred. Targets were generated immediately prior to an operation being carried out and consequently included all of the variations up to that date.

## 7.2 Case study 2: Line of balance applied to a refurbishment project

### *7.2.1 Purpose*

The purpose of this case study is to illustrate the method of programming and progressing repetitive construction when a fast handover rate is required. This is the only case study that was not actually built.

The refurbishment project shown in case study 1 has been used as a basis for this example.

### 7.2.2 *Description of project*

Assume a refurbishment project for a housing association consists of 50 units of similar construction, and is to be completed in 25 working weeks at a rate of ten units per week. Owing to the high handover rate it has been decided to use the Line of Balance Method of programming.

### 7.2.3 *Balancing gang*

In practice, gangs do not balance exactly and have to be rounded off, as shown in Table 7.1.

### 7.2.4 *Overlapping activities (Fig. 7.21)*

A number of activities have been omitted from the diagram, as they could overlap or run concurrently with the activities shown. If they were included the diagram would not be as clear. These activities are better shown on overlay sheets, and are listed below:

- Sweep flues carried out after preparation work and completed at approximately the same time as internal brickwork.
- DPC and timber treatment carried out after preparation work.
- Plumbing and heating first fix and electrical work first fix carried out at the same time as carpentry first fix.
- Plumbing and heating second fix and electrical work second fix carried out at the same time as carpentry second fix.
- Insulation is to be carried out after electrical second fix and before decoration. This is to be carried out in batches of 15 units, as it progresses faster than other activities proceeding at the same time.
- Tiling carried out before decoration and after units have been fitted. Again this is to be carried out in batches of 15 units, as this also progresses faster than other activities proceeding at the same time.
- External brickwork, drainage and external works to be commenced after roof covering and rainwater goods and completed after the scaffold has been dropped. These will progress at the same time as internal activities, as external and internal work have little influence on each other.

### 7.2.5 *Procedure for programming (Table 7.1)*

The procedure for programming is as set out in Chapter 1, but is repeated here for clarity.

**Table 7.1** Refurbishment project.

| Operation | Productive operative (hrs) | Min no. of productive operatives | Time req'd using min gang (hrs), $G$ | Total no. of gangs req'd, $g = G/t$ | No. of units per week, $R = 40g/G$ | No. of gangs to be used per unit, $P$ | Time req'd per unit (days) $T = G/8P$ | Total duration from start of first to start of last unit, $5(N-1)/R$ |
|---|---|---|---|---|---|---|---|---|
| Preparation | 135.00 | 2 | 67.50 | 8 | 4.74 | 2 | 4.22 | 51.68 |
| Sweep flues | subcontractor | | 8.00 | 1 | 5.00 | 1 | 1.00 | 49.00 |
| Internal brickwork (2 + 1) | 59.00 | 2 | 29.50 | 4 | 5.42 | 1 | 3.69 | 45.17 |
| DPC and timber treatment | subcontractor | | 8.00 | 1 | 5.00 | 1 | 1.00 | 49.00 |
| Roof covering | subcontractor | | 24.00 | 3 | 5.00 | 1 | 3.00 | 49.00 |
| Rainwater goods | 14.00 | 2 | 7.00 | 1 | 5.71 | 1 | 0.88 | 42.88 |
| Firestop wall | 22.00 | 2 | 11.00 | 1 | 3.64 | 1 | 1.38 | 67.38 |
| Carpentry first fix* | 92.00 | 2 | 46.00 | 6 | 5.22 | 1 | 5.75 | 46.96 |
| Plumbing and heating first fix | subcontractor | | 16.00 | 2 | 5.00 | 1 | 2.00 | 49.00 |
| Electrical first fix | subcontractor | | 16.00 | 2 | 5.00 | 1 | 2.00 | 49.00 |
| Plastering | 154.00 | 2 | 77.00 | 10 | 5.19 | 1 | 9.63 | 47.16 |
| Carpentry second fix | 77.00 | 2 | 38.50 | 5 | 5.19 | 1 | 4.81 | 47.16 |
| Plumbing and heating second fix | subcontractor | | 64.00 | 8 | 5.00 | 2 | 4.00 | 49.00 |
| Electrical second fix | subcontractor | | 4.00 | 1 | 10.00 | 1 | 0.50 | 24.50 |
| Insulation | 5.00 | 1 | 5.00 | 1 | 8.00 | 1 | 0.63 | 30.63 |
| Tiling | 11.00 | 1 | 11.00 | 2 | 7.27 | 1 | 1.38 | 33.69 |
| Carpentry third fix | 17.00 | 2 | 8.50 | 1 | 4.71 | 1 | 1.06 | 52.06 |
| Decoration | subcontractor | | 56.00 | 7 | 5.00 | 1 | 7.00 | 49.00 |
| Floor covering | subcontractor | | 16.00 | 2 | 5.00 | 1 | 2.00 | 49.00 |
| External brickwork (2 + 1) | 44.00 | 2 | 22.00 | 3 | 5.45 | 1 | 2.75 | 44.92 |
| Drainage | 17.00 | 2 | 8.50 | 1 | 4.71 | 1 | 1.06 | 52.06 |
| External works | 54.00 | 2 | 27.00 | 3 | 4.44 | 1 | 3.38 | 55.13 |
| Clean up | subcontractor | | 8.00 | 1 | 5.00 | 1 | 1.00 | 49.00 |

* Carpentry first fix includes floors, etc.

(1) Using the minimum optimum number of operatives, calculate the time required by each gang ($G$)
(2) The handover rate required is ten units per week. Based on a 40 hour week one unit must be handed over every 4 hours after completion of the first unit. Calculate the number of gangs required to give this handover rate, i.e. $G/4$. Round off the number of gangs to the nearest whole number.
(3) Calculate the number of units per week, i.e. $R=40g/G$.
(4) Determine the number of gangs per unit. i.e. $P$.
(5) Calculate the duration required per unit, i.e. $T=G/8P$.
(6) Calculate the overall duration from the start of the first unit to the start of the last unit, i.e. $5(N-1)/R$, when $N$ is the number of units.
(7) Plot the line of balance schedule based on the above information (Fig. 7.21). The criteria used in plotting this schedule were as follows:
    (a) All activities start at the beginning of the day.
    (b) Where an activity is faster than the one preceding it, no buffer is allowed on the last unit.
    (c) Where an activity is slower than the one preceding it, no buffer is allowed on the first unit.
    (d) Where activities are almost parallel, a two-day buffer is allowed.

### 7.2.6 *Factors governing handover rate*

The handover rate for housing association and for local authority work is usually determined by them. If the required handover rate does not fit in with the natural rhythm of the activities to be performed, either it must be adjusted or non-productive time will result, thereby increasing direct costs.

For private development, the forecast rate of selling will obviously have an influence on the rate of production, but again adjustments would be made to allow the natural rhythm of the activities to be achieved.

### 7.2.7 *Material schedules*

Material schedules can be obtained easily from the line of balance schedule. The amount of material required can be obtained by reading off the starts of activities at any particular time and allowing for such factors as call-up times, size of loads, etc.

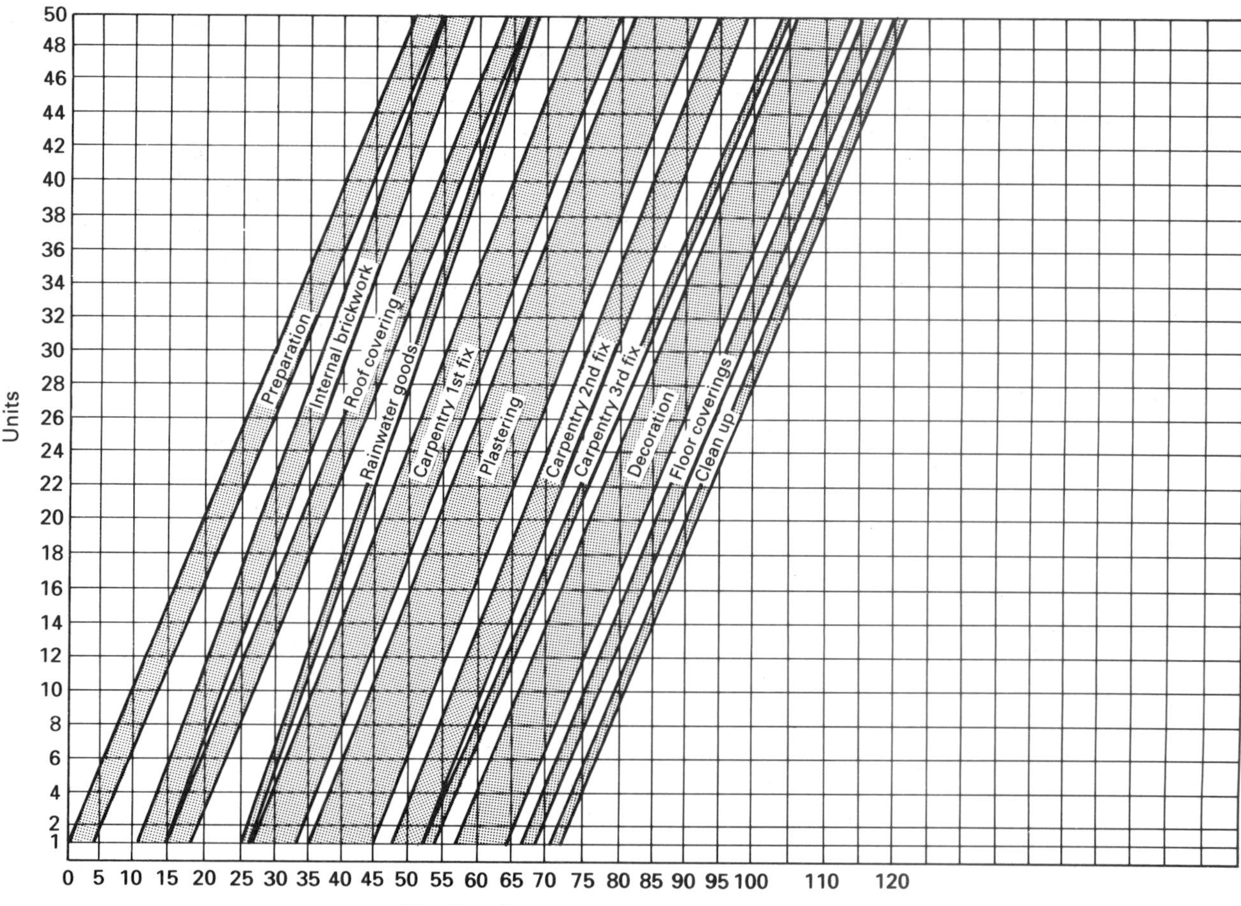

**Fig. 7.21** Case study 2: line of balance schedule.

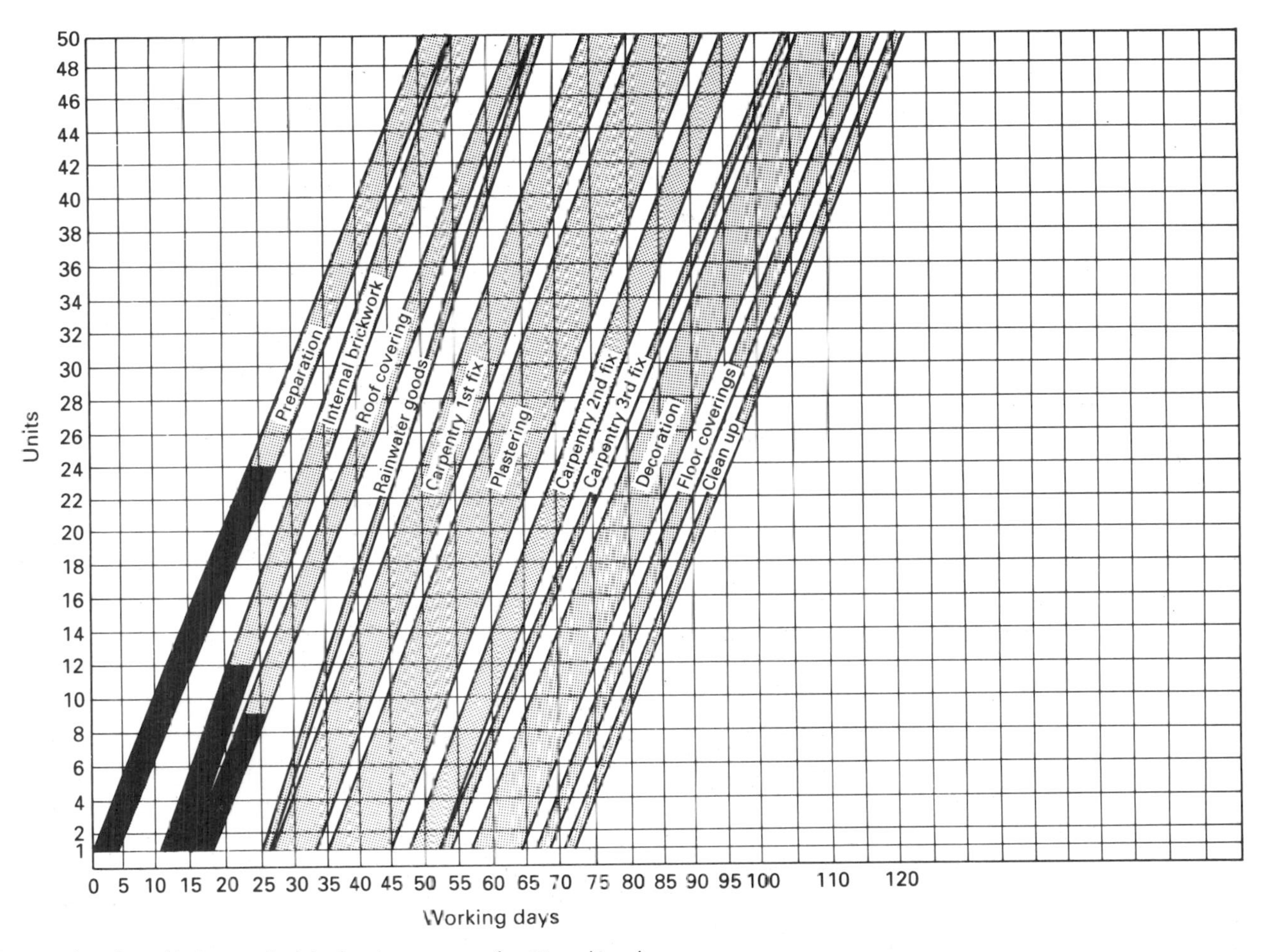


**Fig. 7.22** Case study 2: line of balance schedule showing progress after 25 working days.

### 7.2.8 *Labour and plant schedules*

Labour and plant schedules can be obtained simply by accumulating requirements as the starts occur and dropping them off when the first unit is complete.

### 7.2.9 *Progressing the schedule (Fig. 7.22)*

The activities on each unit are shaded in as they are completed. Progress is shown in the line of balance schedule after 25 working days and progress is as follows:

- ❑ *Preparation*: completed up to and including unit 24 and is therefore in front of programme.
- ❑ *Internal brickwork*: completed up to unit 12 and is therefore slightly behind programme.
- ❑ *Roof covering*: Complete up to unit 9 and is therefore slightly in front of programme.

Based on this information, no action is necessary at this stage.

## 7.3 Case study 3: Kimberworth Surgery

### 7.3.1 *Purpose*

The purpose of this case study is to illustrate the planning and control process on a medium-sized project.

### 7.3.2 *Description of project*

The project consisted of a two-storey building of traditional loadbearing walling at ground-floor level and loadbearing stud partitions at first-floor level. It contained a health centre mainly on the ground floor, with offices, etc. on the first floor (see Figs 7.23–7.25).

- ❑ *Architect*: Smith Rickard Partnership
- ❑ *Quantity surveyor*: Felton & Partners
- ❑ *Main contractor*: O & P Construction Services Limited.

This project was featured in the RIBA East Midlands Yearbook.

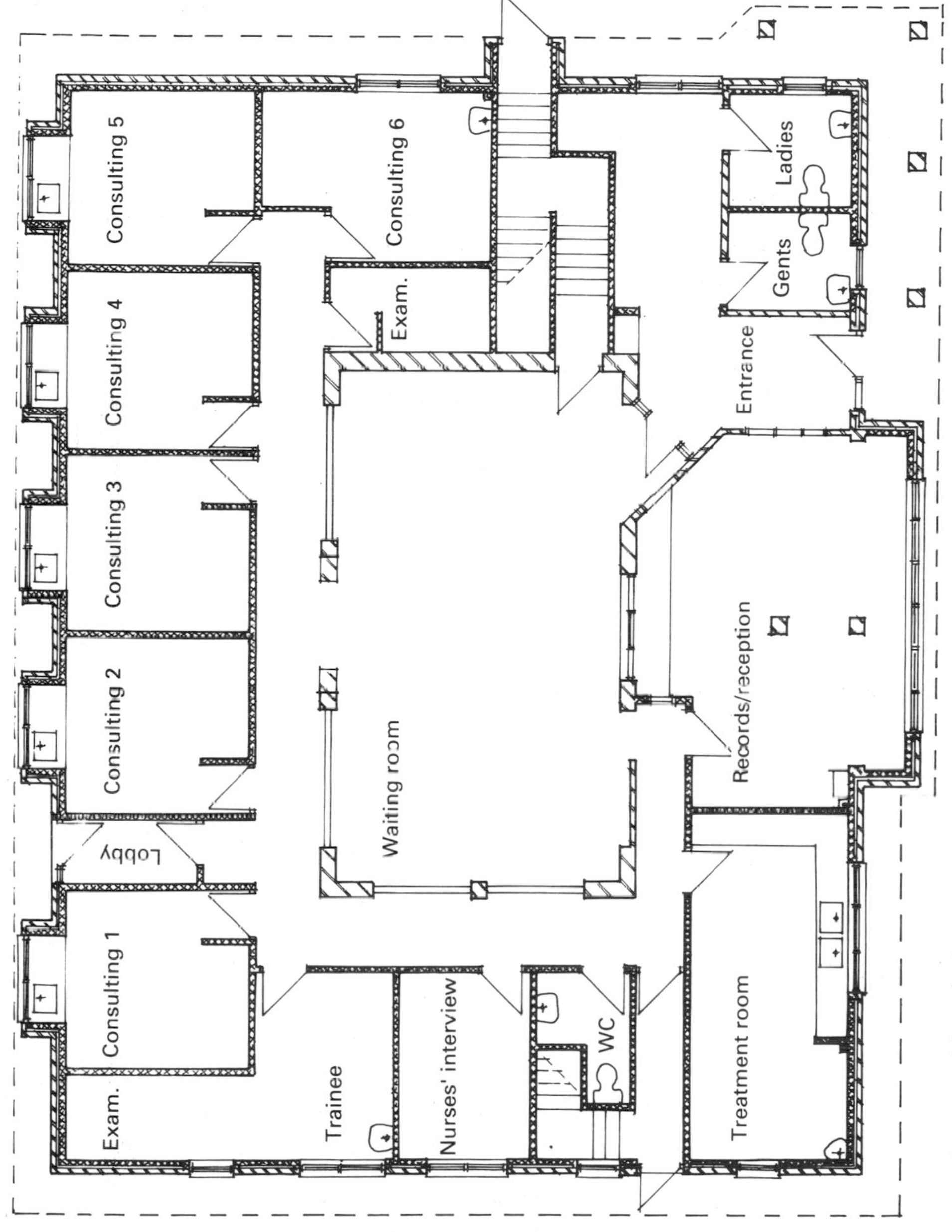

**Fig. 7.23** Case Study 3: ground floor plan.

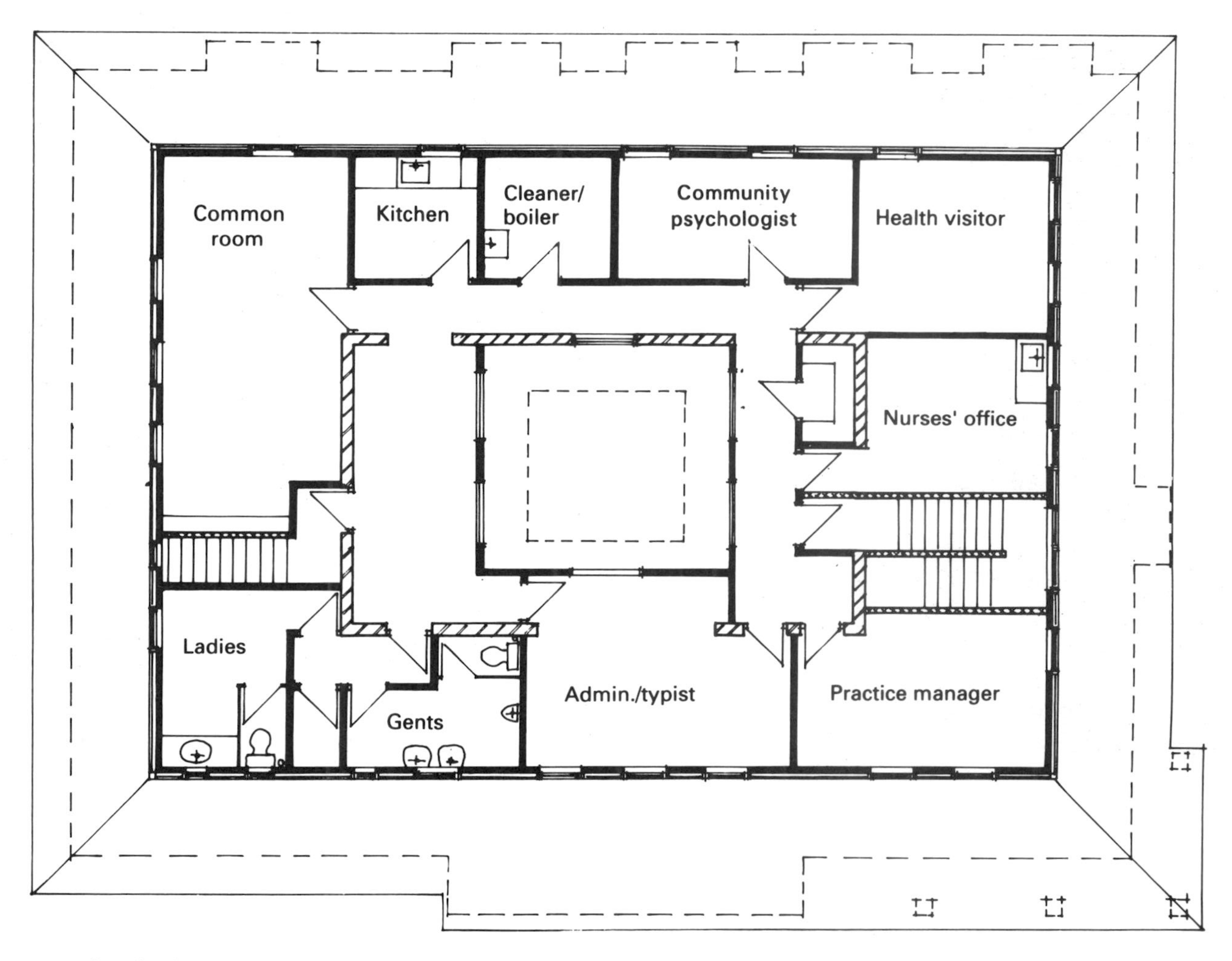


**Fig. 7.24** Case study 3: first floor plan.

**Fig. 7.25** Case study 3: Kimberworth Surgery.

### 7.3.3 *Construction of the building*

*Substructure*

- Foundations
  - Traditional concrete deep strip foundations.
  - Brick and block cavity walls.
- Floor slab
  - Fabric-reinforced concrete bed on DPM on blinded hardcore.

*Superstructure*

- Walls
  - Ground floor: facing brick and block cavity walls.
  - First floor: structural timber stud partitions to external walls supported on laminated structural timber ring beam. Central core internal walls supported on steelwork.
- Windows
  - Powder-coated aluminium set into brickwork to ground floor.
  - Powder-coated aluminium integrated into an aluminium vitrified infill panel cladding system to first floor.
- Partitions
  - Generally brickwork and blockwork at ground-floor level and softwood stud at first-floor level.
  - Walls around central core are brickwork and around staircases are blockwork.

- Roof structure
  - Softwood roof trusses.
- Covering
  - Cement fibre slates.
  - Aluminium lantern light over central waiting area.

*Internal finishes*

- *Brick and block walls*: plastered except waiting and reception areas, which are facing brickwork.
- *Stud partitions*: plasterboard and skim.
- *Ground floors*: screeded, incorporating underfloor heating to waiting area and adjacent corridors, and covered with carpets or vinyl sheeting.
- *Fittings*: mainly worktops, cupboard and wall units and planters.

*Service systems*

- *Heating*: gas central heating incorporating underfloor heating to the waiting area and adjacent corridors.
- *Power, lighting and ventilation*: electric.
- *Other services*: telephone and computer conduits, security fire alarm and patients' call system.

*External works*

- *Paving and access roads*: tarmac, precast concrete kerbs and edgings, brick paving, brick planters.
- *Drainage*: vitrified clay flexible joint pipes and fittings, polypropylene inspection chambers.

### 7.3.4 The contract

- Intermediate Form of Building Contract
- Liquidated damages: £500 per week
- Defects liability period: 12 months
- Insurance cover: minimum £1 000 000.00
- Performance bond required
- The contract period stated in the contract was 30 weeks.

### 7.3.5 The site

The site was located four miles from the contractor's office. It had previously been used for industrial purposes. The original buildings had been demolished, leaving the site reasonably clear. It fronted up to a subsidiary road, which was just off a main road. There were industrial buildings on all other boundaries, including a bus depot on the west boundary.

The site already had some protection as follows:

- On the west boundary there was a concrete retaining wall 3.4 m high with a chain link fence on top.
- On the rear (north) boundary there was a brick wall 2.7 m high.
- To the east boundary there was a chain link fence 2.5 m high.
- To the front there was a chain link fence 2.5 m high which contained a gate with a padlock. The chain link fencing generally was in a poor state of repair and required some attention.

Within the site boundary there were three large advertisement hoardings. Negotiations were in progress regarding their removal, as the leasing period for the use of the hoardings had not expired.

### 7.3.6 Preparing the master programme

*Working week*

It was intended to work a 45-hour week of 5 days.

*Plant*

Items selected a the pre-tender stage were as follows:

- JCB 4: breaker attachment for breaking out existing foundations. Square hole digger for isolated foundations
- Crane for lifting laminated beams and steelwork
- 150/100 mixer for mortar.

*Subcontract work*

Subcontractors were consulted and agreed periods on site and notice required. See schedule of subcontractors (Fig. 7.26).

*Procedure for programming*

(1) A careful study of the drawings was undertaken before starting the master programme.

(2) A list of major operations to be undertaken by the main contractor was prepared, and bill of quantities items were then collected under each programme operation and the labour and plant content established from these figures. See schedule of operations (Fig. 7.27).

(3) A list of all operations to be used in the master programme in approximate order of starting time was prepared incorporating

Kimberworth Surgery - schedule of subcontractors

| Operation | Time required (weeks) | Notice required (weeks) | Remarks |
|---|---|---|---|
| Lantern lights } | 0.50 | 10 | Assumed - awaiting nomination |
| Window units } | 2.00 | 10 | Assumed - awaiting nomination |
| Roof coverings - Main roof } | 2.00 | 3 | |
| Roof coverings - Lower roof } | 0.50 | 3 | |
| Leadwork | 1.00 | 1 | |
| Central heating carcassing | 3.00 | 6 | Assumed - awaiting nomination |
| Plumbing 1st fix | 1.00 | 2 | |
| Mains installation - elect carcass | 3.00 | 8 | Assumed - awaiting nomination |
| Plastering and floor screeds | 4.00 | 2 | |
| Central heating 2nd fix | 3.00 | 2 | Assumed - awaiting nomination |
| Plumbing 2nd fix | 2.00 | 2 | |
| Electrical 2nd fix | 3.00 | 2 | |
| Painting and decorating - External | 1.00 | 4 | |
| Painting and decorating - Internal | 4.00 | 4 | |
| Floor coverings | 1.00 | 4 | |
| Scaffolding | 1.00 | 1 | |
| *External works* | | | |
| Drainage and manholes<br>Excavation for utilities<br>Crossover<br>External foundations to walls } | 5.00 | 3 | |
| Utilities | 2.00 | 5 | |
| Landscaping, topsoil, etc. | 1.00 | 4 | |
| Tarmac | 2.00 | 4 | |
| Final clean and hand over | 0.50 | 2 | |

**Fig. 7.26** Case study 3: schedule of subcontractors.

Kimberworth Surgery - schedule of operations

| Operation | Plant hours | Man hours |
|---|---|---|
| Set up site | | 104 |
| Excavation - Reduce level dig } | 15 | 15 |
| Trenches and pits } | 41 | 41 |
| Foundations | | 210 |
| Brickwork to DPC | | 185 |
| GF const incl intnl drains, etc. | | 268 |
| Walling above DPC | | 960 |
| Struct steel and laminated beams | | 75 |
| First fl const and extnl wall carcass | | 435 |
| Main roof construction inc eaves | | 430 |
| Lower roof construction inc eaves | | 125 |
| RW goods and eaves treatment | | 130 |
| Carpentry 1st fix | | 537 |
| Internal glazing | | 42 |
| Builders work | | 245 |
| Carpentry 2nd fix | | 475 |
| Wall tiling | | 37 |
| Commissioning | | |
| Related work by others | | |
| *External works* | | |
| Walls | | 355 |
| PC curbs and edgings<br>Other paving } | | 360 |

**Fig. 7.27** Case study 3: schedule of operations.

subcontractors' operations. This list contained all the operations shown on the calculation sheet (Fig. 7.28) and the master programme (Fig. 7.29).

(4) Total man-hours and plant-hours were entered for each of the contractor's operations, and the times required by subcontractors were also entered.

Kimberworth Surgery - calculation sheet

| Operation number | Operation | Plant hours | Man hours | Labour and plant | Time required (weeks) | Remarks | |
|---|---|---|---|---|---|---|---|
| 1 | Set up site | | 104 | 2 Carps, 2 labs | 1.00 | Nameboard, temporary buildings, water, electric, telephone, security | |
| 2 | Excavation - Reduce level dig | 15 | 15 | JCB + dr. Lorry + dr, 1 lab | | RL dig, pits, trenches, earthwork support | ***subsequent to prog - sub-c came in cheaper |
| | Trenches and pits | 41 | 41 | Pecker for bk out founds | 1.50 | Square hole digger for pad foundations | |
| 3 | Foundations | | 210 | 3 labs | 1.75 | Trench fill, isolated foundations | |
| 4 | Brickwork to DPC | | 185 | 2 blrs, 1 lab, 150/100 mxr | 1.50 | | |
| 5 | GF const incl intnl drains, etc. | | 263 | 4 labourers | 1.75 | Hardcore filling, sand filling, blinding, compacting, DPM, reinforcement, formwork, fill cavity | |
| 6 | Walling above DPC | | 960 | 4 blrs, 2lab, 150/100 mxr | 4.00 | | |
| 7 | Struct steel and laminated beams | | 75 | 2 blrs, 1lab, 150/100 mxr | 0.75 | Crane required | |
| 8 | First fl const and extnl wall carcass | | 435 | 4 carpenters | 2.75 | Floors, straps, structural timber walls, partition walls, insulation | |
| 9 | Main roof construction inc eaves | | 430 | 4 carpenters | 2.25 | Trusses, wind bracing, fascias, soffits | |
| 10 | Lower roof construction inc eaves | | 125 | 4 carpenters | 0.75 | Structural timbers, fascias, soffits | |
| 11 | RW goods and eaves treatment | | 130 | 4 carpenters | 0.75 | | |
| 12 | Lantern lights | | | Subcontractor | 0.50 | | |
| | Window units | | | | 3.00 | | |
| 13 | Roof coverings - Main roof | | | Subcontractor | 2.00 | | |
| | Roof coverings - Lower roof | | | | 0.50 | | |
| 14 | Leadwork | | | Subcontractor | 1.00 | | |
| 15 | Carpentry 1st fix | | 537 | 3 carpenters | 4.50 | Flooring, windows boards, T & VJ soffits, support timbers and framing, linings, frames, hatches, staircases, fire protection | |
| 16 | Internal glazing | | 42 | 1 carpenter | 1.00 | | |
| 17 | Central heating carcassing | | | Subcontractor | 3.00 | | |
| 18 | Plumbing 1st fix | | | Subcontractor | 1.00 | | |
| 19 | Mains installation - Elect carcass | | | Subcontractor | 3.00 | | |
| 20 | Builders work on ditto | | 245 | 2 labourers | 3.00 | | |
| 21 | Plastering and floor screeds | | | Subcontractor | 4.00 | | |
| 22 | Carpentry 2nd fix | | 475 | 4 carpenters | 3.00 | Toilet cubicles, screens, doors, skirtings, rails, panels, furniture, mirrors, pipe casings, ironmongery, shelving | |
| 23 | Central heating 2nd fix | | | Subcontractor | 3.00 | | |
| 24 | Plumbing 2nd fix | | | Subcontractor | 2.00 | | |
| 25 | Electrical 2nd fix | | | Subcontractor | 3.00 | | |
| 26 | Wall tiling | | 37 | 2 plasterers (ownmen) | 0.50 | | |
| 27 | Painting and decorating - external | | | Subcontractor | 1.00 | | |
| | Painting and decorating - internal | | | Subcontractor | 4.00 | | |
| 28 | Commissioning | | | | 1.00 | | |
| 29 | Related work by others | | | Specialists | 2.00 | Telephones, loose furnishings/fittings, curtains | |
| 30 | Floor coverings | | | Subcontractor | 1.00 | | |
| 31 | Scaffolding | | | Subcontractor | | From first lift until window units complete | |
| | *External works* | | | | | | |
| 32 | Drainage and manholes | | | | | | |
| 33 | Excavation for utilities | | | | | | |
| 34 | Crossover | | | Subcontractor | 5.00 | | |
| 35 | External foundations to walls | | | | | | |
| 36 | Utilities | | | Y. Ele, Y. Water E. Gas, B. Tel | 2.00 | Electric early for site use | |
| 37 | Walls | | 355 | 2 blrs, 1 lab, 150/100 mxr | 3.00 | | |
| 38 | PC curbs and edgings | | | | | | |
| 39 | Other paving | | 360 | 2 blrs, 1 lab, 150/100 mxr | 3.00 | | |
| 40 | Landscaping, topsoil, etc. | | | Subcontractor | 1.00 | | |
| 41 | Tarmac | | | Subcontractor | 2.00 | | |
| 42 | Final clean and hand over | | | 2 labs + contract cleaners | 2.00 | 2 labs, 2 weeks + contract cleaners 0.5 weeks | |

**Fig. 7.28** Case study 3: calculation sheet.

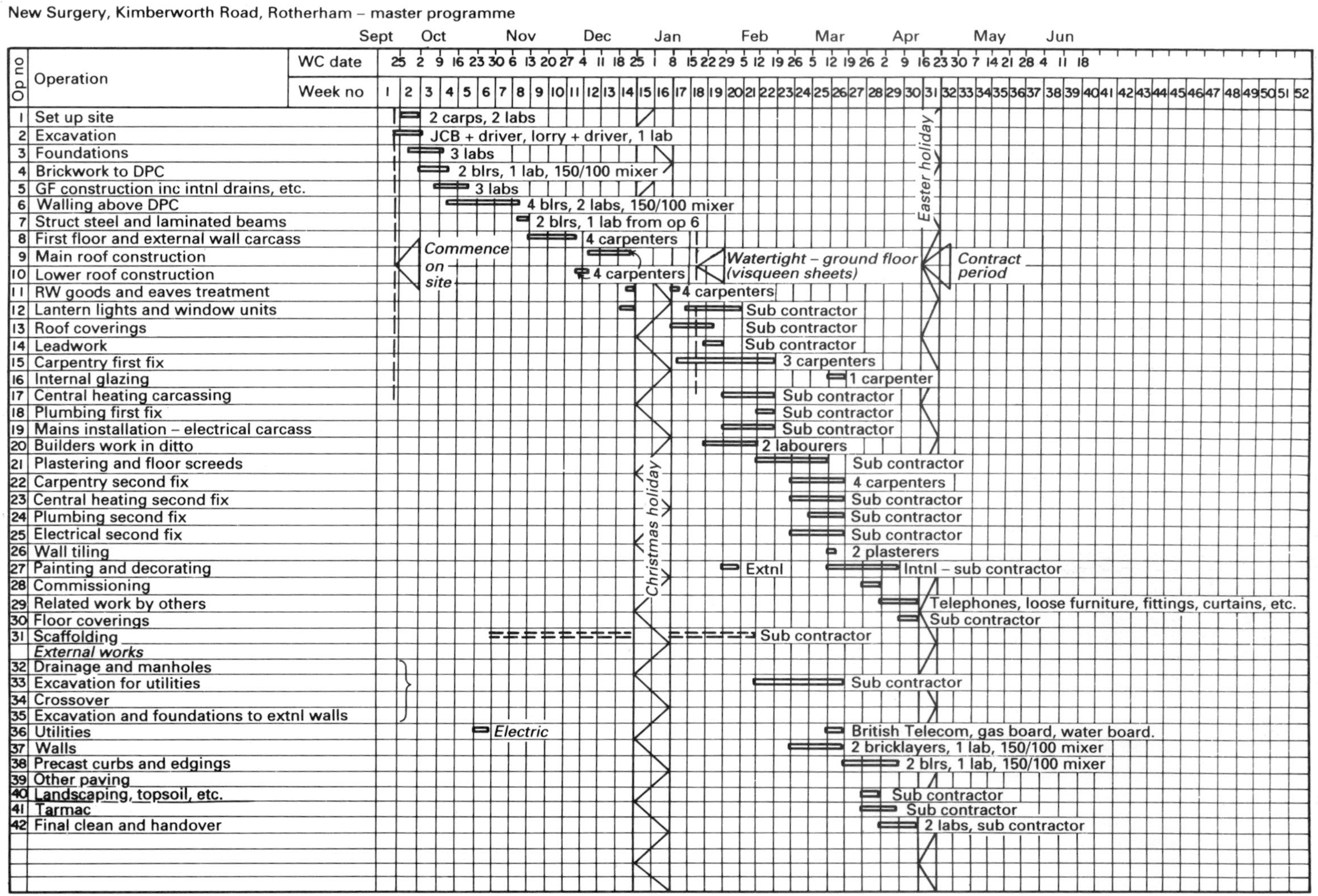

New Surgery, Kimberworth Road, Rotherham – master programme

| Op no | Operation | Resources / notes |
|---|---|---|
| 1 | Set up site | 2 carps, 2 labs |
| 2 | Excavation | JCB + driver, lorry + driver, 1 lab |
| 3 | Foundations | 3 labs |
| 4 | Brickwork to DPC | 2 blrs, 1 lab, 150/100 mixer |
| 5 | GF construction inc intnl drains, etc. | 3 labs |
| 6 | Walling above DPC | 4 blrs, 2 labs, 150/100 mixer |
| 7 | Struct steel and laminated beams | 2 blrs, 1 lab from op 6 |
| 8 | First floor and external wall carcass | 4 carpenters |
| 9 | Main roof construction | |
| 10 | Lower roof construction | 4 carpenters |
| 11 | RW goods and eaves treatment | 4 carpenters |
| 12 | Lantern lights and window units | Sub contractor |
| 13 | Roof coverings | Sub contractor |
| 14 | Leadwork | Sub contractor |
| 15 | Carpentry first fix | 3 carpenters |
| 16 | Internal glazing | 1 carpenter |
| 17 | Central heating carcassing | Sub contractor |
| 18 | Plumbing first fix | Sub contractor |
| 19 | Mains installation – electrical carcass | Sub contractor |
| 20 | Builders work in ditto | 2 labourers |
| 21 | Plastering and floor screeds | Sub contractor |
| 22 | Carpentry second fix | 4 carpenters |
| 23 | Central heating second fix | Sub contractor |
| 24 | Plumbing second fix | Sub contractor |
| 25 | Electrical second fix | Sub contractor |
| 26 | Wall tiling | 2 plasterers |
| 27 | Painting and decorating | Extnl; Intnl – sub contractor |
| 28 | Commissioning | |
| 29 | Related work by others | Telephones, loose furniture, fittings, curtains, etc. |
| 30 | Floor coverings | Sub contractor |
| 31 | Scaffolding | Sub contractor |
| | *External works* | |
| 32 | Drainage and manholes | |
| 33 | Excavation for utilities | Sub contractor |
| 34 | Crossover | |
| 35 | Excavation and foundations to extnl walls | |
| 36 | Utilities | *Electric*; British Telecom, gas board, water board. |
| 37 | Walls | 2 bricklayers, 1 lab, 150/100 mixer |
| 38 | Precast curbs and edgings | 2 blrs, 1 lab, 150/100 mixer |
| 39 | Other paving | |
| 40 | Landscaping, topsoil, etc. | Sub contractor |
| 41 | Tarmac | Sub contractor |
| 42 | Final clean and handover | 2 labs, sub contractor |

| Month | Sept | Oct | | | | Nov | | | | Dec | | | | Jan | | | | | Feb | | | | Mar | | | | Apr | | | | | May | | | | Jun | | | | | | | | | | | | | | | |
|---|---|---|---|---|---|---|---|---|---|---|---|---|---|---|---|---|---|---|---|---|---|---|---|---|---|---|---|---|---|---|---|---|---|---|---|---|---|---|---|---|---|---|---|---|---|---|---|---|---|---|---|
| WC date | 25 | 2 | 9 | 16 | 23 | 30 | 6 | 13 | 20 | 27 | 4 | 11 | 18 | 25 | 1 | 8 | 15 | 22 | 29 | 5 | 12 | 19 | 26 | 5 | 12 | 19 | 26 | 2 | 9 | 16 | 23 | 30 | 7 | 14 | 21 | 28 | 4 | 11 | 18 | | | | | | | | | | | | |
| Week no | 1 | 2 | 3 | 4 | 5 | 6 | 7 | 8 | 9 | 10 | 11 | 12 | 13 | 14 | 15 | 16 | 17 | 18 | 19 | 20 | 21 | 22 | 23 | 24 | 25 | 26 | 27 | 28 | 29 | 30 | 31 | 32 | 33 | 34 | 35 | 36 | 37 | 38 | 39 | 40 | 41 | 42 | 43 | 44 | 45 | 46 | 47 | 48 | 49 | 50 | 51 | 52 |

Commence on site

Watertight – ground floor (visqueen sheets)

Contract period

Christmas holiday

Easter holiday

Fig. 7.29 Case study 3: master programme.

(5) The number of operatives to be used on the first operation was determined, and the time required was calculated. This operation was then entered onto the programme chart (Fig. 7.29) before proceeding to the next operation.
(6) All other operations were then considered. Continuity of work was provided for the operatives wherever possible.

On a project of this size it was sometimes difficult to provide continuity, but the operatives working on this project would be programmed to move to other projects where necessary.

*Flexibility*

To allow for wet time, absenteeism, etc. a 40-hour week was used in the calculations. Flexibility was also provided by reducing the overlap on some operations. Overtime working was used to gain time when this was necessary.

*General notes on the programme*

For operations in the first six weeks see six-week programme (Fig. 7.36).

Initial setting out was carried out in the first week.

At pre-tender stage, excavation was to be carried out by the main contractor, and this was reflected in the overall programme. Just prior to the start of the project a quotation was obtained from a subcontractor to carry out this work, and this was less than the amount included. It was therefore decided to use the subcontractor for this work. This can be seen in the six-week programme (Fig. 7.36).

The laminated beams could not be erected until all the walls supporting them had been raised to the correct levels. The structural timber floor and the structural walls were supported by the laminated beams and the structural steelwork.

In order to progress the internal work, the lower window openings were to be covered in visqueen by the carpenters, and the first floor openings were to be covered by the glazing subcontractor.

External painting was planned to be done early to allow working from the scaffold.

A crane was hired for one day to lift the laminated beams and the steelwork into place.

### 7.3.7 Schedules

Various schedules were produced as shown below. Careful monitoring of these schedules helped to avoid delays due to lack of resources at the time required.

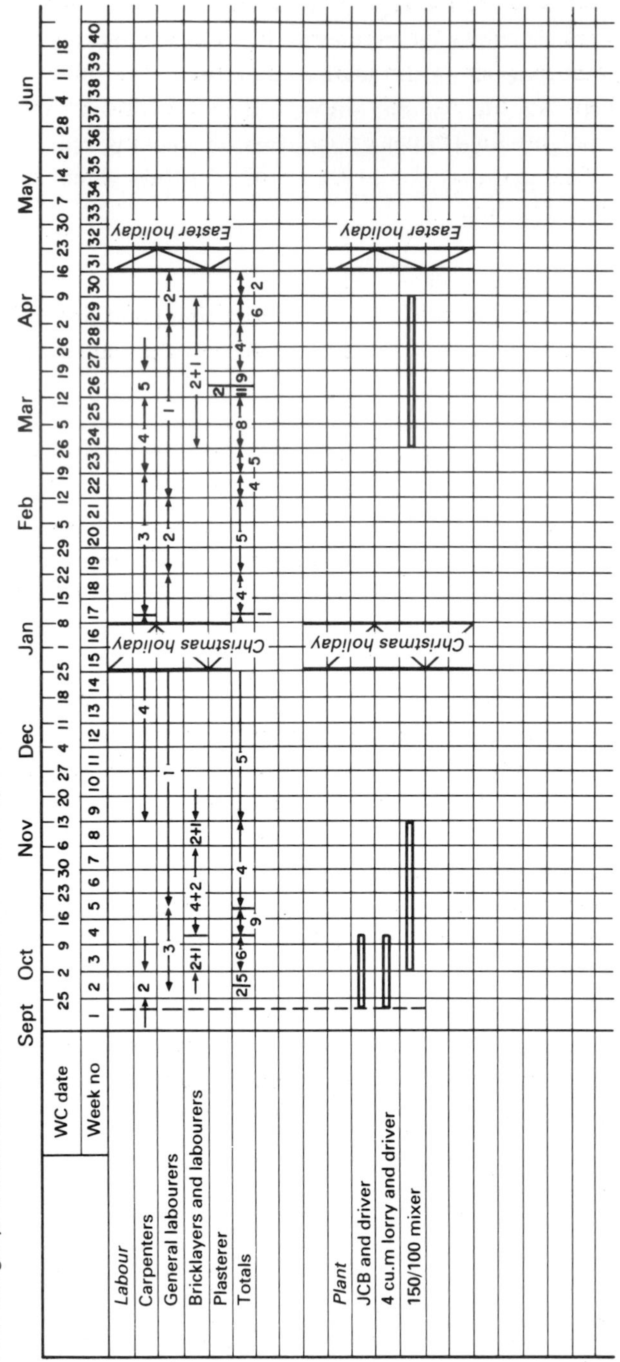


**Fig. 7.30** Case study 3: labour/plant schedule.

*Labour/plant schedule (Fig. 7.30)*

This was produced by adding up the labour and plant required each week. It shows the number of operatives required in each trade, and the total number of the main contractor's men on site. One labourer was kept on site for the whole contract period to ensure that help was available for offloading and general cleaning up.

The plant section of the schedule shows when the major plant items were required.

*Tabular plant schedule (Fig. 7.31)*

This provided information for ordering plant, including supplier's name and starting dates when plant owned by the main contractor was required.

*Sub-contractor schedule (Fig. 7.32)*

This included domestic and nominated subcontractors and the work of statutory authorities. It was shown in tabular form, and gave fairly comprehensive information relating to this work. This assisted in the timing of sending out orders for the work, and provided for comments, which was useful in keeping a record of subcontractors' efficiency.

*Materials schedule (Fig. 7.33)*

This included all major materials, and provided comprehensive information relating to materials. Again, this assisted in the timing of sending out orders and provided a record of the efficiency of suppliers.

Only part of the schedule is shown.

*Components schedule (Fig. 7.34)*

This was similar to the materials schedule, and the same comments apply. Only part of the schedule is shown.

*Information required schedule (Fig. 7.35)*

This included information required from the architect, subcontractors, statutory authorities and all others responsible for supplying information. Only part of the schedule is shown.

Kimberworth Surgery - tabular plant schedule

| Plant | Supplier | Date on site | Date ordered | Order number | Comments |
|---|---|---|---|---|---|
| 150/100 Mixer | Own | 2 Oct | | | |
| Crane | PP Engineering | 6 Nov | | | |
| Dumper | Own | 2 Oct | | | |
| Vibrating roller | Viboplant | 11 Oct | 6 Oct | 256 | |

**Fig. 7.31** Case study 3: tabular plant schedule.

Kimberworth Surgery - subcontractor schedule    Book 5

| BOQ page ref | Work subcontracted | Subcontractor | Date on site | Notice required | Date ordered | Order number | Comments |
|---|---|---|---|---|---|---|---|
| *Domestic subcontractors* | | | | | | | |
| 2/1 | Excavation | Aberford | 21 Sept | 1 week | 14 Sept | 246 | Was to be carried out by main contractor - Subcontractor submitted a lower price |
| 2/2 | Filling, h/c, surface treatment, etc., | Aberford | 9 Oct | 1 week | 14 Sept | 245 | Was to be carried out by main contractor - Subcontractor submitted a lower price |
| 2/17, 2/18 | Roof tiling | Jarvis & Womack | 8 Jan | 1 week | | | |
| 2/18, 2/19 | Leadwork | Poplars | 22 Jan | 1 week | | | |
| 2/54, 2/57 | Plumber 1st fix | Poplars | 12 Feb | 1 week | | | |
| 2/32, 2/25, 5/1 | Plastering, floor screeds, etc. | Lloyd Clough | 12 Feb | 1 week | | | |
| 2/40 | Reception counters, etc. | H. Evans & Sons | 12 March | 6 weeks | | | Was to be carried out by main contractor - Subcontractor submitted a lower price |
| 2/35, 2/38 | Painting and decorating | Gazzards | 29 Jan | 1 week | | | |
| 5/1 | Floor coverings | Shirtcliffes | 9 April | 2 weeks | | | |
| 2/41 | Curtain tracks, etc. | Dearnleys | 2 April | 1 week | | | |
| 2/43, 2/45 | Sanitary appliances | Poplars | 5 March | 3 weeks | | | |
| 2/54, 2/57 | Above-ground drainage | Poplars | 5 March | 3 weeks | | | |
| 4/6 | Landscaping | AWS Landscapes | 26 March | 2 weeks | | | |
| 5/1 | Blinds | Hillarys Blinds | 2 April | 3 weeks | | | |
| 4/3 | Tarmac paving | Worksop Tarmac | 26 March | 2 weeks | | | |
| 1/23, 1/29 | Cleaning | Makebrite | 9 April | 1 week | | | |
| 5/2 | Crossover | Aberford | 12 Feb | 1 week | | | |
| *Nominated subcontractors* | | | | | | | |
| 5/21 | Heating and hot water installation | British Gas | 29 Jan | 3 weeks | | | |
| 5/3 | Electrical installation | R. J. Pickford | 29 Jan | 3 weeks | | | |
| 5/2 | Windows, lantern light | Ashfield Glass | 20 Dec | 6 weeks | | | |
| 5/2 | Artificial plants | Greenspace | 9 Apr | 1 week | | | |
| *Related work by others* | | | | | | | |
| 1/4 | Telephone installation | | 9 Apr | | | | |
| 1/4 | Loose furniture and fittings | | 12 Apr | | | | |
| 1/4 | Curtains | | 12 Apr | | | | |
| *Statutory authorities* | | | | | | | |
| 1/28 | Water supply - for site use | Yorkshire Water | 2 Oct | 1 week | 18 Sept | 240 | |
| 5/2 | Water supply - permanent supply | Yorkshire Water | 19 March | 3 weeks | | | |
| 1/26 | Electric service - for site use | Yorkshire Electric | 2 Oct | 3 weeks | 11 Sept | 244 | |
| 5/2 | Electric service - permanent supply | Yorkshire Electric | 19 March | 3 weeks | | | |
| 5/2 | Sewer connection | Rotherham MBC | 5 March | 3 weeks | | | |
| 5/2 | Gas service | British Gas | 19 March | 2 weeks | | | |
| 1/28 | Telephones - site use | British Telecom | 2 Oct | 3 weeks | 11 Sept | 243 | |
| 5/2 | Telephones - permanent supply | British Telecom | 26 March | 3 weeks | | | |

**Fig. 7.32** Case study 3: subcontractor schedule.

Kimberworth Surgery - materials schedule Book 4

| BOQ page ref | Material | Supplier | Date required | Notice required | Date ordered | Order number | Date received | Comments |
|---|---|---|---|---|---|---|---|---|
| 2/2 | Hardcore and sand | Redland | 5 Oct | 2 days | 25 Sept | 255 | | |
| 2/4 | Reinforcement | Reinforcement Northern | 9 Oct | 7 days | 25 Sept | 256 | | |
| 2/4 | Concrete | RMC | 27 Sept | 2 days | 1 Sept | 235 | | For trenches and pads |
| 2/5 | Mortar - premix | Topmix | 2 Oct | 2 days | 1 Sept | 234 | | |
| 2/5, 4/5 | Common bricks | Ibstock | 2 Oct | 2 wks | 11 Sept | 243 | | |
| 2/5, 2/6, 4/5 | Facings and specials | Aizlewoods | 9 Oct | 2 wks | 18 Sept | 247 | | |
| 4/6 | Red tiles | Maltby Brick Co | 9 Oct | 1 wk | 18 Sept | 248 | | |
| 2/7 | Blue bricks | BBM | 9 Oct | 2 days | 18 Sept | 249 | | |
| 2/7 | Concrete blocks | BBM | 9 Oct | 3 days | 18 Sept | 249 | | |
| 2/7, 2/8 | Lytag blocks | Aizlewoods | 9 Oct | 4 days | 18 Sept | 247 | | |
| 2/8, 2/9 | Cavity ties, jablite, DPCs | Aizlewoods | 2 Oct | 3 days | 18 Sept | 247 | | |
| 2/8, 2/9 | Catnics, etc. | Aizlewoods | 2 Oct | 3 days | 18 Sept | 247 | | |
| 2/9 | Padstones | Procon | 6 Nov | 10 days | 18 Sept | 250 | | |
| 2/12, 2/14 | Structural timbers and plates | Lavers | 8 Nov | 10 days | 18 Sept | 251 | | |
| 2/13, 2/47, 2/51 | Carp 1st fixings | Lavers and GD Woodworking | 15 Jan | 10 days | | | | |
| 2/14 | Straps | Aizlewoods | 29 Nov | 3 days | | | | |
| 2/22, 2/59 | Ceiling timbers, tacboard | Lavers | 19 Jan | 10 days | | | | |
| 2/47 | Insulation | Sheffield Insulation | 19 March | 3 days | | | | |
| 2/47 | Carp 2nd fix timbers | Carrs and GD Woodworking | 26 Feb | 8 days | | | | |
| 2/29, 2/30, 2/62 | Internal glazing | Solaglas | 12 March | 4 days | | | | |
| 2/54 | Rainwater goods | Aizlewoods | 20 Dec | 4 days | | | | |
| | | Stocksigns | | | | | | |

not a complete materials list

**Fig. 7.33** Case study 3: materials schedule.

Kimbleworth Surgery - components schedule

| BOQ page ref | Component | Supplier | Date required | Notice required | Date ordered | Order number | Date received | Comments |
|---|---|---|---|---|---|---|---|---|
| 2/11 | Structural steelwork | Daver Steels | 6 Nov | 2 weeks | 16 Oct | 295 | | |
| 2/12 | Laminated timber beams | Swedspan Glulan | 6 Nov | 4 weeks | 2 Oct | 270 | | |
| 5/1 | Roof trusses | Reeves | 4 Dec | 4 weeks | 30 Oct | 305 | | |
| 2/22 | WC cubicles | Bushboard Parker | 26 Feb | 3 weeks | | | | |
| 2/25 | Borrowed lights | GD Woodworking | 15 Jan | 4 weeks | | | | |
| 2/25 | Screens | GD Woodworking | 26 Feb | 4 weeks | 30 Oct | 304 | | |
| 2/25 | External frames | GD Woodworking | 9 Oct | 2 weeks | 18 Sept | 252 | | |
| 2/25 | Screens to FF void | GD Woodworking | 26 Feb | 4 weeks | 30 Oct | 306 | | |
| 2/26, 2/61 | Doors (2nd fix) | GD Woodworking | 26 Feb | 4 weeks | 30 Oct | 306 | | |
| 2/27, 2/61 | Linings, hatches, etc. | GD Woodworking | 15 Jan | 4 weeks | 30 Oct | 306 | | |
| 2/28 | Stairs | GD Woodworking | 15 Jan | 4 weeks | 30 Oct | 306 | | |
| 2/42 | Fire fighting equipment | Yorkshire Fireguard | 2 April | 1 week | | | | |
| 2/42, 2/43 | Kitchen units, worktops | Magnet Joinery | 12 March | 2 weeks | | | | |
| 2/29, 2/50, 2/63 | Ironmongery | Laidlaws | 26 Feb | 6 weeks | | | | |
| 2/49 | Pendock profiles | Peace & Smith | 12 March | 2 weeks | | | | |

**Fig. 7.34** Case study 3: components schedule.

Kimberworth Surgery - information required schedule

| Information required | From | Start date | Required by | Date of request | Date received | Remarks |
|---|---|---|---|---|---|---|
| Setting out details | Smith Rickard Partnership | 21 Sept | 21 Sept | 14 Sept | 20 Sept | |
| Notice board details | Smith Rickard Partnership | 26 March | 5 March | | | |
| Shelving details | Smith Rickard Partnership | 26 March | 5 March | | | |
| External bin store details | Smith Rickard Partnership | 15 Jan | 1 Jan | | | |
| Decoration schedule - external | Smith Rickard Partnership | 29 Jan | 15 Jan | | | |
| Decoration schedule - internal | Smith Rickard Partnership | 5 March | 19 Feb | | | |
| Floor coverings schedule | Smith Rickard Partnership | 9 April | 12 March | | | |
| Underfloor heating details | Nominated subcontractor | 26 Feb | 12 Feb | | | |
| Setting out details for bldrs' work | Nominated subcontractor | 22 Feb | 15 Feb | | | |
| Structural steel details | Smith Rickard Partnership | 8 Nov | 9 Oct | | | |
| Glulam beam details | Smith Rickard Partnership | 8 Nov | 9 Oct | | | |
| *Nominations* | | | | | | |
| Heating and hot water installation | Smith Rickard Partnership | 29 Jan | ASAP | 14 Sept | 20 Sept | Required to agree programme |
| Electrical installation | Smith Rickard Partnership | 29 Jan | ASAP | 14 Sept | 20 Sept | Required to agree programme |
| Windows, lantern light | Smith Rickard Partnership | 20 Dec | ASAP | 14 Sept | 20 Sept | Required to agree programme |
| Artificial plants | Smith Rickard Partnership | 9 April | 12 March | | | |

**Fig. 7.35** Case study 3: information required schedule.

### 7.3.8 *Short-term planning*

*Six-week programme (Fig. 7.36)*

A detailed programme for the first six weeks was prepared. This was produced to allow closer control, and to ensure that work progressed as shown on the master programme. A review of resource requirements was made based on the detailed programme.

Kimberworth Surgery – six week programme, 18 September – 28 October

| DP no. | Operation | Plant hrs/ man hrs | Labour and plant | |
|---|---|---|---|---|
| | *Set up site* | | | |
| 1 | Repair boundary fence | 66 | 2 labs | 4 |
| 2 | Del steel lock-up | | Supp | |
| 3 | Del portaloo | | Supp | |
| 4 | Site accomm and notice board | 30 | 2 Carps | 2 |
| 5 | Exc for water and elect | 8 | Sub-c 1 lab | 1 |
| 6 | Site telephone | | B/T | |
| 7 | Site water | 1 | YW+sub | |
| 8 | Site electric | 1 | YE+Sub | |
| | *Foundations* | | | |
| 9 | Reduce level dig | 2 | Sub-c | |
| 10 | Exc trenches and bk out | 4 | Sub-c | |
| 11 | Exc isolated piers | 1 | Sub-c | |
| 12 | Support sides | | Sub-c | |
| 13 | Conc trenches and pads | 210 | 4 labs | 6.5 |
| 14 | Internal drains | 50 | 2 labs | 3 |
| 15 | Bwk to DPC | 185 | 2 + 1 | 7¾ |
| 16 | H/C fill and blinding | | Sub-c | |
| 17 | DPM | 8 | 4 labs | 5 |
| 18 | Reinforcement | 16 | | |
| 19 | Concrete floor | 128 | | |
| 20 | Cavity fill | 12 | | |
| | *Superstructure* | | | |
| 21 | Bwk above DPC | 960 | 2 + 1<br>4 + 2 | |

Time scale: September – October; weeks commencing 18, 25, 2, 9, 16, 23 (M T W T F S S). Commence on site.

| Remarks | Labour and subcontractor requirements |
|---|---|
| | General labs: 2, 6, 4, 1 |
| | Bricklayers: 2+1, 4+2 |
| | Carpenters: 2 |
| Originally by main contractor, now sub-let. | Excavation sub |
| | H/C fill sub |

| Remarks | Plant requirements |
|---|---|
| | 150/100 mixer |

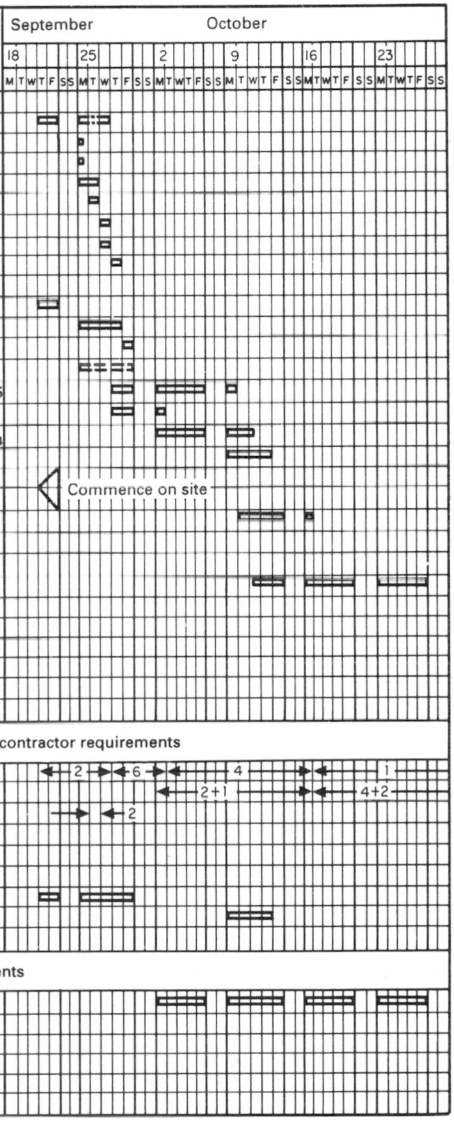

**Fig. 7.36** Case study 3: six-week programme.

*Calculating operation durations*

The items in the bill of quantities were grouped where necessary into each operation in the detailed programme. This particular six-week programme was relatively easy because the operations related directly to bill of quantity items. Some operations were grouped even in this short-term programme, as the work was carried out by one gang and was of short duration.

In some later short-term programmes, quantities for some individual items had to be measured from the drawing as the BOQ did not include sufficient detail. For example, rainwater goods on the upper and lower roof were not separated in the bill.

In some instances it was necessary to use operational records to ascertain durations, as BOQ outputs are average outputs. When breaking down operations, the output will vary depending on the complexity of a particular operation. For example, output of facing brickwork differs depending on the complexity, which can vary in different parts of the project.

*Procedure for programming*

(1) Operations in the master programme were broken down where necessary into more detailed operations, taking care to include all operations.
(2) Man-hours were entered for each operation.
(3) The number of operatives was determined and the time required was calculated. Each operation was then plotted onto the programme as it was considered. The aim was to provide continuity of work and to achieve the progress shown in the master programme. The resources shown on the master programme were generally adhered to, but if necessary this was adjusted.

*Labour requirements*

These were obtained from the six-week programme by adding up the labour required each day. One labourer was retained on site for general duties such as offloading.

*Subcontractor requirements*

These were summarised below the labour requirements.

*Plant requirements*

These were summarised at the bottom of the programme. Apart from small plant and tools, the only plant required was a 150/100 mixer, as the major excavation was carried out by a subcontractor.

*Other requirements*

Requirements for details, nominations, materials, etc. were reviewed using this programme, and a check was made with the schedules produced at the master programming stage to ensure that all requirements were available when required.

### 7.3.9 General notes on six-week programme

*(1) Repair fence*

To start immediately the site was available in order to provide security.

*(2, 3, 4) Lock-up, portaloo and site accommodation*

To start at beginning of first full week.

*(5, 6, 7, 8) Services*

To be laid on immediately site accommodation was ready.

*(9, 10, 11) Excavation*

To start immediately the site was available. After reduce level dig, rear trenches were to be excavated first.

The duration for excavate trenches and break out allowed for the excavator moving to operation 5, excavate for water and electric supplies: i.e. 4 days was the overall time for both operations.

*(12) Support sides*

To be carried out in parallel with trench and pad excavation.

*(13) Concrete trenches and pads*

To overlap the excavation work carried out by a subcontractor.

*(14) Internal drains*

To be carried out in parallel with concrete foundations.

*(15) Brickwork to DPC*

To overlap concrete trenches.

*(16) Hardcore and blinding*

To overlap brickwork to DPC and to finish after brickwork finishes. Rooms to rear were to be prepared first to allow floor construction to proceed as soon as possible.

*(17, 18, 19, 20) Floor construction and cavity fill*
To start as soon as rear rooms were prepared to allow brickwork above DPC to start.

*(21) Brickwork above DPC*
Bricklaying gang to move from brickwork below DPC. The 2 + 1 gang was to be joined by another 2 + 1 gang after the floor was complete.

### 7.3.10 Control of progress

*Progressing the six-week programme (Fig. 7.37)*
The programme was updated at weekly intervals, and a review was made to ascertain whether corrective action was necessary.

As an example, progress is shown at finishing time on Friday 6 October.

*Analysis of progress*

| | |
|---|---|
| (1) | Repair of boundary fences took the programmed time, but was carried out as a continuous operation. This was because laying of the temporary site water service was delayed by 1 day, and consequently the excavation of the trench was also delayed by 1 day. |
| (2, 3, 4) | The steel lock-up, the portaloo and the site accommodation and notice board were all carried out as programmed. |
| (5) | Excavation for site water and electricity was delayed by 1 day. The machine being used on operation 10 was diverted for 3 hours to carry out the excavation work. |
| (6) | The site telephone was installed as programmed. |
| (7, 8) | Temporary site water service was delayed by 1 day (see above) and was carried out on the same day as the temporary electric service. |
| (9) | Reduce level dig took a quarter of a day longer than programmed. |
| (10) | Excavate trench and breaking out took slightly less time than programmed, but it started and finished a quarter of a day late. |
| (11) | Excavating isolated piers took only three quarters of a day and finished as programmed. |
| (12) | The foundations were deep strip foundations, and no trench support was necessary. |
| (13) | Concrete trenches and pads started 1 day late but finished 1 day early: that is, 1 day in front of programme. |

Kimberworth Surgery – six week programme, 18 September – 28 October

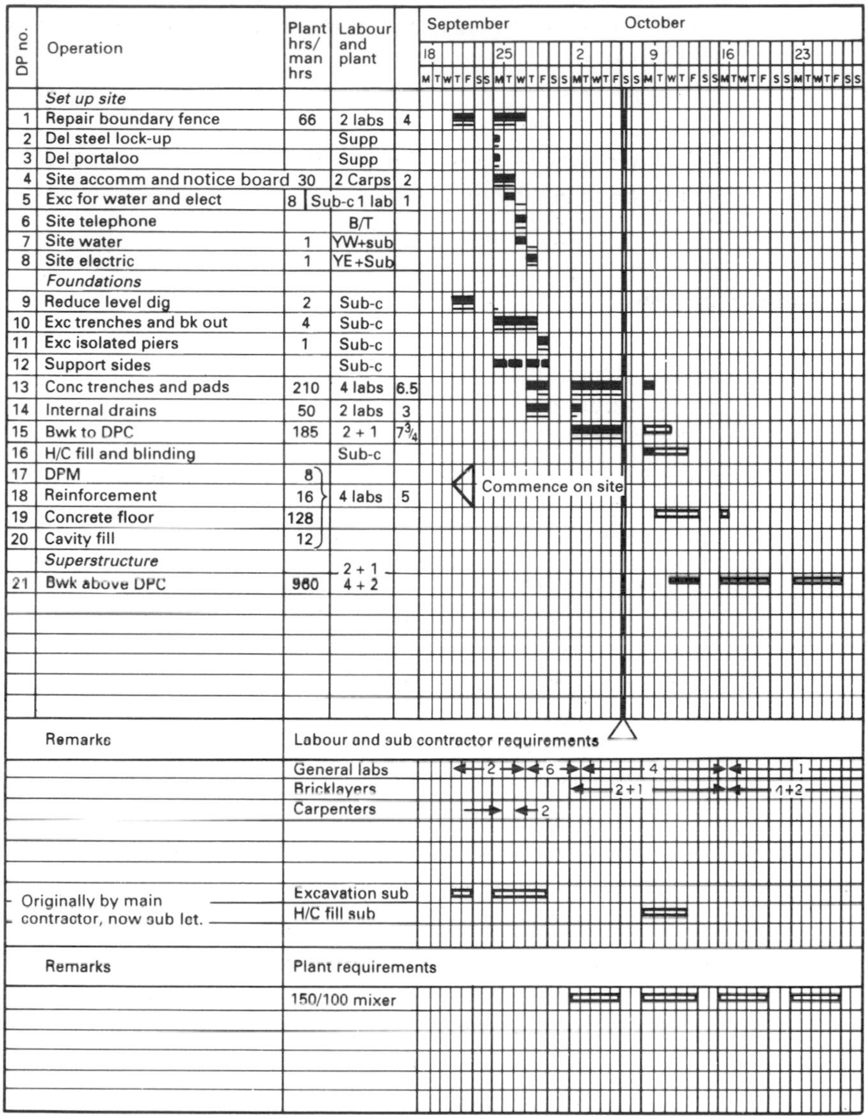

| DP no. | Operation | Plant hrs/ man hrs | Labour and plant | |
|---|---|---|---|---|
| | *Set up site* | | | |
| 1 | Repair boundary fence | 66 | 2 labs | 4 |
| 2 | Del steel lock-up | | Supp | |
| 3 | Del portaloo | | Supp | |
| 4 | Site accomm and notice board | 30 | 2 Carps | 2 |
| 5 | Exc for water and elect | 8 | Sub-c 1 lab | 1 |
| 6 | Site telephone | | B/T | |
| 7 | Site water | 1 | YW+sub | |
| 8 | Site electric | 1 | YE+Sub | |
| | *Foundations* | | | |
| 9 | Reduce level dig | 2 | Sub-c | |
| 10 | Exc trenches and bk out | 4 | Sub-c | |
| 11 | Exc isolated piers | 1 | Sub-c | |
| 12 | Support sides | | Sub-c | |
| 13 | Conc trenches and pads | 210 | 4 labs | 6.5 |
| 14 | Internal drains | 50 | 2 labs | 3 |
| 15 | Bwk to DPC | 185 | 2 + 1 | $7\frac{3}{4}$ |
| 16 | H/C fill and blinding | | Sub-c | |
| 17 | DPM | 8 | 4 labs | 5 |
| 18 | Reinforcement | 16 | | |
| 19 | Concrete floor | 128 | | |
| 20 | Cavity fill | 12 | | |
| | *Superstructure* | | | |
| 21 | Bwk above DPC | 980 | 2 + 1<br>4 + 2 | |

Timeline: September 18, 25; October 2, 9, 16, 23 (M T W T F S S). Commence on site.

| Remarks | Labour and sub contractor requirements |
|---|---|
| | General labs: 2, 6, 4, 1 |
| | Bricklayers: 2+1, 1+2 |
| | Carpenters: 2 |
| Originally by main contractor, now sub let. | Excavation sub |
| | H/C fill sub |

| Remarks | Plant requirements |
|---|---|
| | 150/100 mixer |

**Fig. 7.37** Case study 3: six-week programme, showing progress at finishing time on Friday 6 October.

(14) Internal drains started as programmed and took half a day less than programmed.

(15) Brickwork to DPC was progressing as programmed.

(16) Hardcore fill started 1 day early and was 1 day in front of programme.

*General comments*

At this stage no corrective action was required, as overall progress was approximately as programmed.

The progress from this chart was transferred to the overall programme.

### 7.3.11 Planning and controlling the finishing stages

*Site meetings*

A meeting was called on 8 February to consider the finishing stages of the project. Representatives of the following organisations were present:

- architect
- main contractor
- central heating subcontractor
- electrical subcontractor
- plastering subcontractor.

A review was made of progress to date, and a discussion took place on the likelihood of progress achieving the dates set in the master programme.

The window units subcontractor was not present at the meeting, but had given a written assurance that the ground-floor units would be in place by 20 February.

Owing to a number of variations, and the delay in moving the advertisement hoarding, an extension of 1 week was granted. This extended the contract period to the end of the Easter holiday.

It was agreed by all relevant parties present that completion could be achieved within the revised contract period.

The architect informed the meeting that the hoardings were to be removed on 16 February.

The main contractor stated that a stage programme would be produced to cover the finishing stages of the work.

*Progress at 16 February*

Progress at week ending 16 February was shown on the master programme (Fig. 7.38). This showed that overall progress was generally as programmed. The timelines indicated that the start and finish times of some operations were different from those planned, particularly for the window units. Both upper-floor and lower-floor units were erected later than shown on the master programme.

As originally intended, in order to maintain progress, the ground-floor window openings had been sealed with visqueen by the main

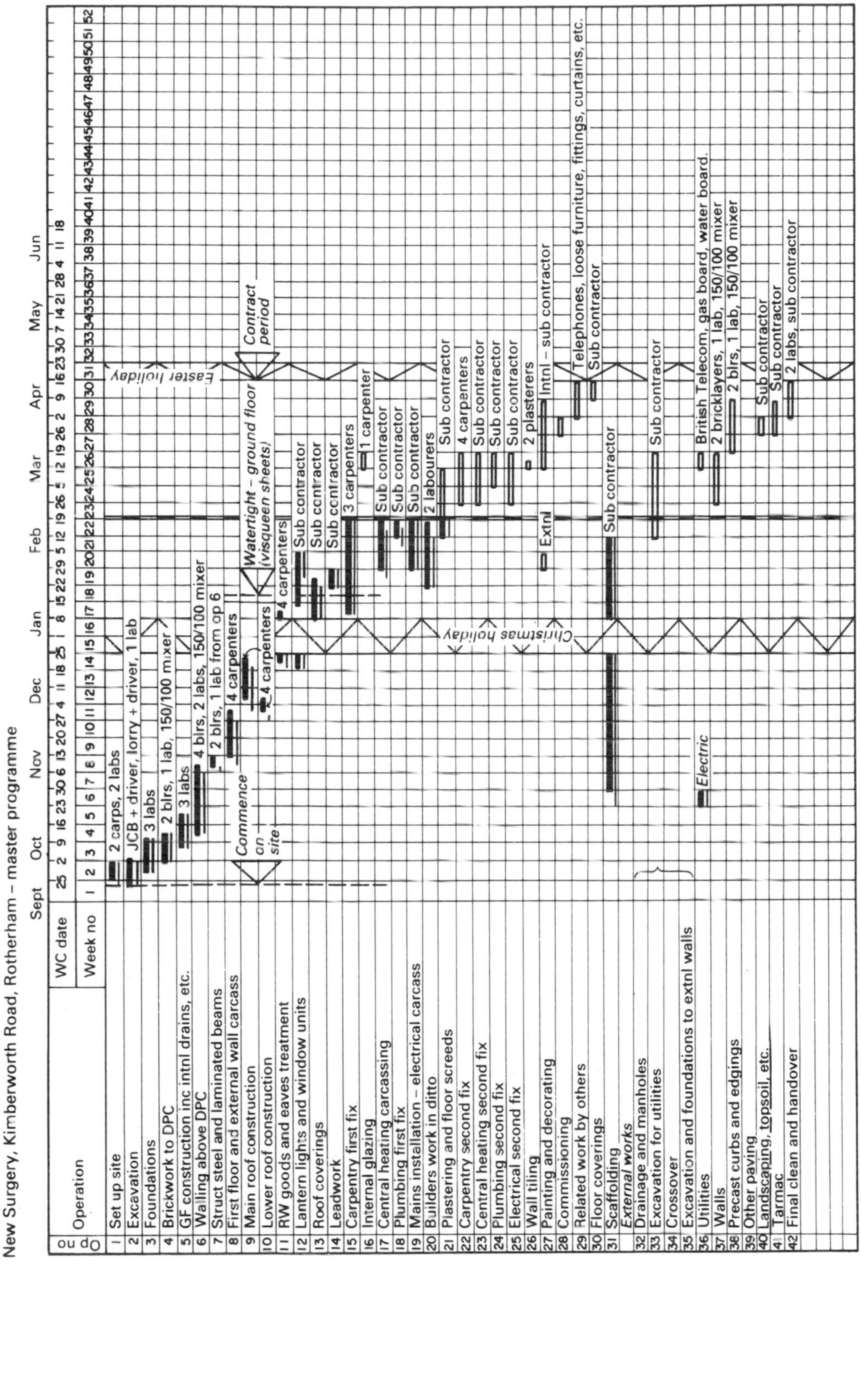

New Surgery, Kimberworth Road, Rotherham – master programme

Sept Oct Nov Dec Jan Feb Mar Apr May Jun

WC date: 25 2 9 16 23 30 6 13 20 27 4 11 18 25 1 8 15 22 29 5 12 19 26 5 12 19 26 2 9 16 23 30 7 14 21 28 4 11 18

Week no: 1 2 3 4 5 6 7 8 9 10 11 12 13 14 15 16 17 18 19 20 21 22 23 24 25 26 27 28 29 30 31 32 33 34 35 36 37 38 39 40 41 42 43 44 45 46 47 48 49 50 51 52

| Op no | Operation | Resources / notes |
|---|---|---|
| 1 | Set up site | 2 carps, 2 labs |
| 2 | Excavation | JCB + driver, lorry + driver, 1 lab |
| 3 | Foundations | 3 labs |
| 4 | Brickwork to DPC | 2 blrs, 1 lab, 150/100 mixer |
| 5 | GF construction inc intnl drains, etc. | 3 labs |
| 6 | Walling above DPC | 4 blrs, 2 labs, 150/100 mixer |
| 7 | Struct steel and laminated beams | 2 blrs, 1 lab from op 6 |
| 8 | First floor and external wall carcass | 4 carpenters |
| 9 | Main roof construction | Commence on site; Watertight – ground floor (visqueen sheets); Contract period |
| 10 | Lower roof construction | 4 carpenters |
| 11 | RW goods and eaves treatment | 4 carpenters |
| 12 | Lantern lights and window units | Sub contractor |
| 13 | Roof coverings | Sub contractor |
| 14 | Leadwork | Sub contractor |
| 15 | Carpentry first fix | 3 carpenters |
| 16 | Internal glazing | 1 carpenter |
| 17 | Central heating carcassing | Sub contractor |
| 18 | Plumbing first fix | Sub contractor |
| 19 | Mains installation – electrical carcass | Sub contractor |
| 20 | Builders work in ditto | 2 labourers |
| 21 | Plastering and floor screeds | Sub contractor |
| 22 | Carpentry second fix | 4 carpenters |
| 23 | Central heating second fix | Sub contractor |
| 24 | Plumbing second fix | Sub contractor |
| 25 | Electrical second fix | Sub contractor |
| 26 | Wall tiling | 2 plasterers |
| 27 | Painting and decorating | Extnl; Intnl – sub contractor |
| 28 | Commissioning | |
| 29 | Related work by others | Telephones, loose furniture, fittings, curtains, etc. |
| 30 | Floor coverings | Sub contractor |
| 31 | Scaffolding | Sub contractor |
| | *External works* | |
| 32 | Drainage and manholes | |
| 33 | Excavation for utilities | Sub contractor |
| 34 | Crossover | |
| 35 | Excavation and foundations to extnl walls | |
| 36 | Utilities | *Electric*; British Telecom, gas board, water board. |
| 37 | Walls | 2 bricklayers, 1 lab, 150/100 mixer |
| 38 | Precast curbs and edgings | 2 blrs, 1 lab, 150/100 mixer |
| 39 | Other paving | |
| 40 | Landscaping, topsoil, etc. | Sub contractor |
| 41 | Tarmac | Sub contractor |
| 42 | Final clean and handover | 2 labs, sub contractor |

*Christmas holiday* (weeks 15–16); *Easter holiday* (week 31)

**Fig. 7.38** Case study 3: master programme, showing progress at week ending 16 February.

contractor's joiners prior to starting work on carpentry first fix. The subcontractor had temporarily sealed the upper windows with visqueen. The subcontractor had, however, fitted all these units by 16 February.

The only other significant variations were as follows:

- External painting had not been carried out. This was because the appointed subcontractor could not carry out the work, and a new subcontractor had to be appointed.
- Drainage and excavation had not started. This was because the advertisement hoardings had not been removed, and this interfered with the external works. Arrangement had been made for these to be removed on 16 February.

*Stage programme*

A stage programme was produced covering the finishing stages and the external works.

*Preparing the stage programme*

The procedure followed was similar to that for the master programme, but the breakdown was in much more detail. The stage programme calculation sheet is shown in Fig. 7.39 and the programme is shown in Fig. 7.40.

*General notes*

The programme generally followed that laid down in the master programme, and again continuity of work was maintained wherever possible.

Throughout the project, targets were set for the operatives, and these targets related directly to operations in the current six-week programme. This practice was therefore followed during the progress of the items on the stage programme. For example, carpentry second fix (first floor) was broken down into three targets based on two carpenters per gang. Each of these targets represented one operation in the current six-week programme and covered the period shown on the stage programme. In the case of items such as joinery fittings one target was given. The six-week programme prepared at this stage is not considered here.

A check was made on requirements for materials and subcontractors, and action was taken to ensure that these would be on site when required.

### 7.3.12 General notes on the project

The project was completed and handed over a 20 April, the revised completion date.

Kimberworth Surgery - stage programme calculation sheet Book 11

| Operation | Plant hours | Man hours | Labour and plant | Time required (days) | Remarks |
|---|---|---|---|---|---|
| Plastering - first floor | | | Subcontractor | 5.00 | First floor 50% complete |
| ground floor | | | Subcontractor | 10.00 | |
| Carpentry 2nd fix - first floor | | 235 | 3/4 Carpenters | 9.00 | Including variations |
| ground floor | | 295 | 3/4 Carpenters | 10.25 | Including variations |
| Floor screeds - ground floor only | | | Subcontractor | 2.00 | |
| Plumbing 2nd fix - first floor | | | Subcontractor | 5.00 | |
| ground floor | | | Subcontractor | 5.00 | |
| Heating 2nd fix - first floor | | | Subcontractor | 4.00 | |
| ground floor | | | Subcontractor | 11.00 | |
| Electrical 2nd fix - first floor | | | Subcontractor | 7.00 | |
| ground floor | | | Subcontractor | 9.00 | |
| Painting and decorating - external | | | Subcontractor | 5.00 | Was to be carried out earlier but original contractor gave back-work and could not carry out the work |
| first floor | | | Subcontractor | 10.00 | |
| ground floor | | | Subcontractor | 10.00 | |
| Internal glazing | | 41 | 1 Carpenter | 5.00 | |
| Wall tiling - first floor | | 19 | 1 Plasterer (Own employee) | 2.50 | |
| ground floor | | 21 | 1 Plasterer (Own employee) | 2.50 | |
| Joinery fittings | | 15 | 1 Carpenter | 2.00 | |
| Pendock floor ducts - trays | | 32 | 1 Carpenter | 4.00 | |
| Pendock floor ducts - covers | | 30 | 1 Carpenter | 4.00 | |
| Pendock casings | | 34 | 1 Carpenter | 4.00 | |
| Curtain tracks | | | Subcontractor | 2.00 | |
| Blinds | | | Subcontractor | 2.00 | |
| Floor coverings | | | Subcontractor | 3.00 | |
| Artificial flowers | | | Subcontractor | 2.00 | |
| Telephones | | | Subcontractor | 2.00 | |
| Curtains | | | Subcontractor | 2.00 | |
| Commissioning | | | Subconts. + Main contractor | 1.00 | |
| Clear up | | 34 | 2 Labourers | 2.00 | |
| Contact clean | | | Subcontractor | 2.00 | |
| *External works* | | | | | |
| Excavate to reduce levels, fill and compact | | | Subcontractor | 3.00 | |
| Excavate trenches for wall foundations | | | Subcontractor | 2.00 | |
| Excavate for drains and manholes | | | Subcontractor | 9.00 | |
| Excavate trenches for services | | | Subcontractor | 3.00 | |
| External walls | | 280 | 2 Bricklayers-1 Labourer | 11.50 | |
| Drainage and manholes | | | Subcontractor | 9.00 | |
| Kerbs, edgings and pavings | | 345 | 2 Bricklayers-1 Labourer | 14.50 | |
| Tarmac paving and parking lines | | | Subcontractor | 4.00+1.00 | |
| Sewage connection and crossover | | | Subcontractor | 5.00 | |
| Water, gas and telecom mains | | | Utilities | 3.00 | |

Note - All man-hours and subcontract times shown include variation up to 23rd February

**Fig 7.39** Case study 3: finishings and external works stage programme calculation sheet.

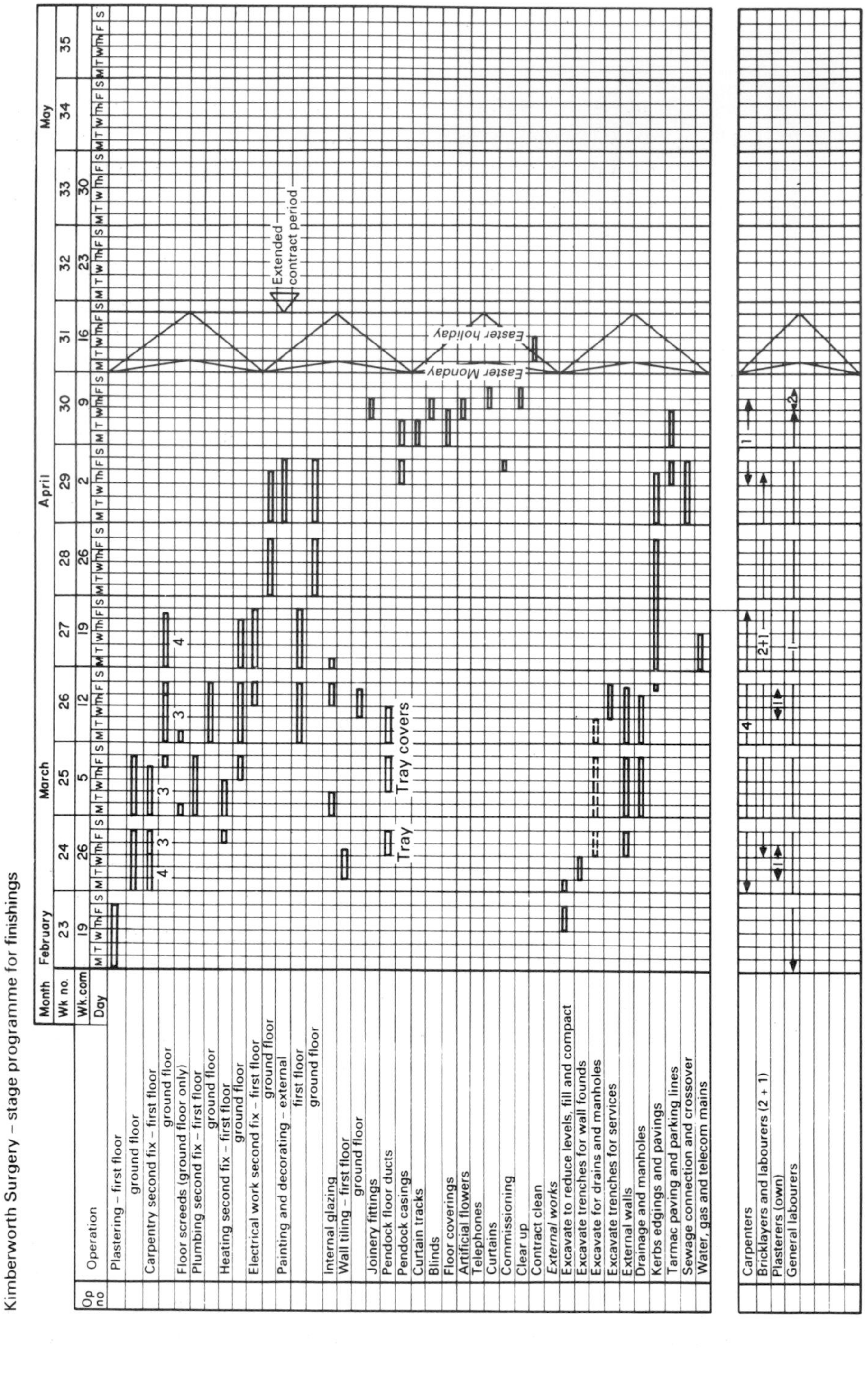

Kimberworth Surgery – stage programme for finishings

Month: February, March, April, May

Wk no.: 23, 24, 25, 26, 27, 28, 29, 30, 31, 32, 33, 34, 35

Wk.com: 19, 26, 5, 12, 19, 26, 2, 9, 16, 23, 30

Day: M T W Th F S (each week)

Op no | Operation

Plastering – first floor
ground floor
Carpentry second fix – first floor
ground floor
Floor screeds (ground floor only)
Plumbing second fix – first floor
ground floor
Heating second fix – first floor
ground floor
Electrical work second fix – first floor
ground floor
Painting and decorating – external
first floor
ground floor
Internal glazing
Wall tiling – first floor
ground floor
Joinery fittings
Pendock floor ducts
Pendock casings
Curtain tracks
Blinds
Floor coverings
Artificial flowers
Telephones
Curtains
Commissioning
Clear up
Contract clean
*External works*
Excavate to reduce levels, fill and compact
Excavate trenches for wall founds
Excavate for drains and manholes
Excavate trenches for services
External walls
Drainage and manholes
Kerbs edgings and pavings
Tarmac paving and parking lines
Sewage connection and crossover
Water, gas and telecom mains

4 3 3 3 4

Tray

Tray covers

Easter Monday

Easter holiday

Extended contract period

Carpenters
Bricklayers and labourers (2 + 1)
Plasterers (own)
General labourers

4 1 2+1 1 1 1 2

Fig 7.40 Case study 3: stage programme for finishings.

## 7.4. Case study 4: Multi-project planning

### 7.4.1 *Purpose*

The purpose of this case study is to illustrate a method of carrying out the planning and control process when a number of small projects are to be carried out concurrently and are sharing the same resources. The bar charts shown in the previous examples and case studies are for one project only.

Firms carrying out smaller projects, where operatives will be on site for short periods of time, need to assess the demand that each project will place on the total resources of the firm. In small firms, operatives tend to undertake a wider variety of work: for example, carpenters will carry out carcassing, first fix, second fix and finishings, whereas in larger firms they tend to specialise in one or more aspects of work. In practice, many projects will be proceeding simultaneously, and it is obvious that the total available resources and the demand for such resources by all current and future known projects should be considered when drawing up the programmes for new projects.

The following outlines a method that allows planning of a number of projects, taking account of the total available resources of the firm or section of the firm, and facilitates the coordination of a number of projects on the one chart.

### 7.4.2 *Method of presentation*

Using this planning technique, all operations that follow each other within one particular project are on the same line of the chart. This saves considerable space on the chart, thus allowing more projects to be presented. For clarity, each block on the chart was coloured using a particular colour for each trade, and arrows of the same colour were used to indicate the transfer of men from one project to another. As the following case study has not been shown in colour, the arrows have also been omitted. Where no resources are shown, the operations were to be carried out by subcontractors.

### 7.4.3 *Procedure*

As each project was awarded to the firm it was integrated with the previously planned projects, which were all displayed on a master chart. As stated earlier, care was taken when deciding on the timing of operations to see that available resources were not exceeded, where this was possible. When a new project was added to the master chart, the programme for each projects manager and trade foreman was taken off and presented separately.

### 7.4.4 *Description of the six projects*

As an example of the use of this technique, six projects are considered. The six projects were under the supervision of the projects manager.

(1) flat refurbishment over retail shops, London Road, Sheffield;
(2) extension and refurbishment of doctors' surgery, Manchester Road, Sheffield;
(3) alterations to a health centre, Maltby;
(4) refurbishment of terrace houses, Rotherham (two properties);
(5) alterations and extensions to Herringthorpe United Reformed Church, Rotherham;
(6) new flats, Bawtry.

Photographs of some of these projects are shown in Figs 7.41–7.44.

### 7.4.5 *The programme (Fig. 7.45)*

Projects 1–5 were all in progress at the start of this multi-project programme, and were scheduled to make maximum use of the resources employed by the firm, supplemented where necessary by labour-only subcontractors.

**Fig. 7.41** Case study 4: flat coversion, London Road, Sheffield.

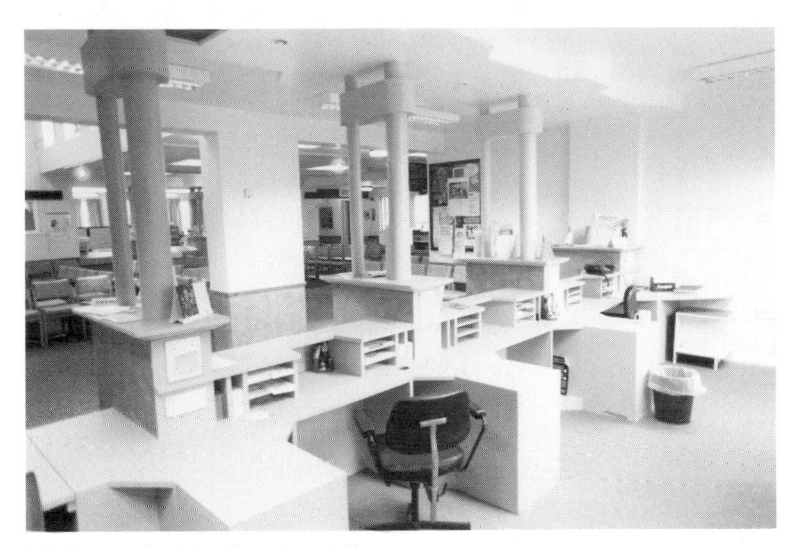

**Fig. 7.42** Case study 4: health centre, Maltby.

**Fig. 7.43** Case study 4: refurbishments, Rotherham.

**Fig. 7.44** Case study 4: new flats, Bawtry.

To illustrate how continuity of work was achieved, the trade of carpenter will be used.

There were three gangs of carpenters, and their deployment was scheduled as follows:

- Gang A
  - Project 1, first fix then move to
  - Project 4A, first fix then to
  - Project 4B, first fix then to
  - Project 4A, second fix then to
  - Project 4B, second fix then to
  - Project 5, main roof construction.
- Gang B
  - Project 2, first fix followed by second fix and finishings.
- Gang C
  - Project 3, second fix then move to
  - Project 1, second fix and finishings then to
  - Project 5, ceiling battens then to
  - Project 5, porch then to
  - Project 5, main roof construction (with gang A).

*Introduction of project 6*

At the start of week 6, project 6 was to commence, and this had to be coordinated with the other projects. At the start of project 6, two labourers from project 1 were scheduled to strip out internally and then

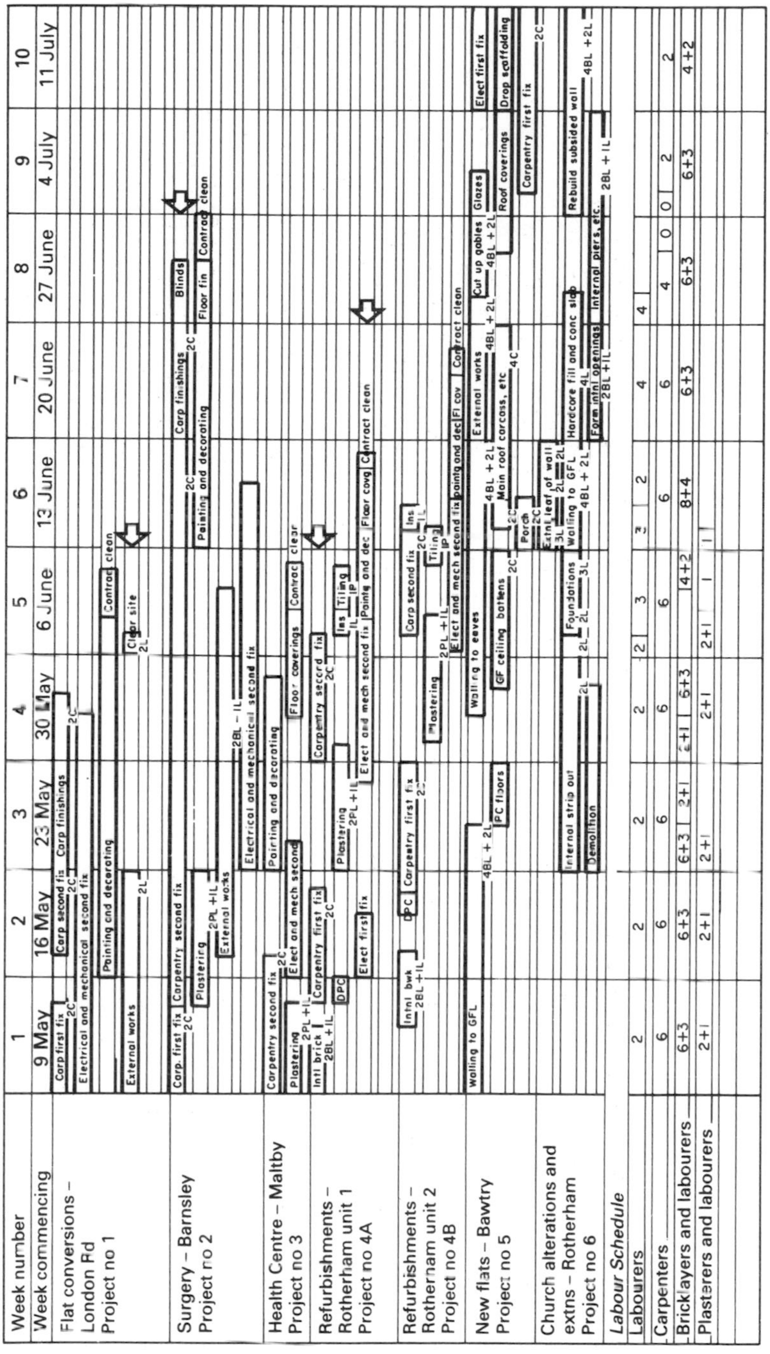

| Week number | 1 | 2 | 3 | 4 | 5 | 6 | 7 | 8 | 9 | 10 |
|---|---|---|---|---|---|---|---|---|---|---|
| Week commencing | 9 May | 16 May | 23 May | 30 May | 6 June | 13 June | 20 June | 27 June | 4 July | 11 July |
| Flat conversions – London Rd Project no 1 | Carp first fix 2C; Electrical and mechanical second fix; External works | Carp second fix 2C; Painting and decorating; 2L | Carp finishings | 2C | Contract clean; Clear site 2L | | | | | |
| Surgery – Barnsley Project no 2 | Carp. first fix 2C; Carpentry second fix | Plastering 2PL +1L; External works | Electrical and mechanical second fix | 2BL – 1L | | 2C; Painting and decorating | Carp finishings 2C | Blinds; Floor fin; Contract clean | | |
| Health Centre – Maltby Project no 3 | Carpentry second fix 2C; Plastering 2PL +1L | Elect and mech second | Painting and decorating | Floor coverings | Contract clear | | | | | |
| Refurbishments – Rotherham unit 1 Project no 4A | Intl brick 2BL + 1L | Carpentry first fix 2C; DPC; Elect first fix | Plastering 2PL +1L | Carpentry second fix 2C; Elect and mech second fix | Ins 1L; Tiling 1P; Paintg and dec | Floor covg; Contract clean | | | | |
| Refurbishments – Rotherham unit 2 Project no 4B | Intnl bwk 2BL +1L | DPC; Carpentry first fix 2C | | Plastering | 2PL +1L; Carp second fix 2C; Elect and mech second fix | Ins 1L; Tiling 1P; paintg and dec | Fl cov; Contract clean | | | |
| New flats – Bawtry Project no 5 | Walling to GFL | 4BL + 2L | PC floors | Walling to eaves; GF ceiling battens 2C | | 4BL + 2L; Main roof carcass, etc; 2C; Porch 2C | External works 4BL + 2L; 4C | Cut up gables 4BL + 2L | Glazes; Roof coverings; Carpentry first fix | Elect first fix; Drop scaffolding; 2C |
| Church alterations and extns – Rotherham Project no 6 | | | Internal strip out; Demolition | 2L | 2L; Foundations 2L; 3L | Extnl leaf of wall 3L 2L 2L; Walling to GFL 4BL + 2L | Hardcore fill and conc slab 4L; Form intnl openings 2BL +1L | Internal piers, etc. | 2BL + 1L; Rebuild subsided wall | 4BL +2L |
| *Labour Schedule* | | | | | | | | | | |
| Labourers | 2 | 2 | 2 | 2 | 2 / 3 | 3 / 2 | 4 | 4 | | |
| Carpenters | 6 | 6 | 6 | 6 | 6 | 6 | 6 | 4 / 0 | 0 / 2 | 2 |
| Bricklayers and labourers | 6+3 | 6+3 | 6+3 / 2+1 | 2+1 / 6+3 | 4+2 | 8+4 | 6+3 | 6+3 | 6+3 | 4+2 |
| Plasterers and labourers | 2+1 | 2+1 | 2+1 | 2+1 | 2+1 / 1 | 1 | | | | |

Fig. 7.45 Case study 4: multi-project programme.

to continue with foundation work. They were then to be joined by a labour-only labourer, and the three were then to move on to taking down the external leaf of an existing cavity wall. The labour-only labourer was to move to project 4B to carry out insulation and then leave site as he would no longer be required. The other two were to carry on taking down the wall and then join two more labour-only labourers laying the hardcore fill and the concrete slab.

On week 6, two 2 + 1 labour-only bricklaying gangs were to be taken on for one week, and this was to be reduced to one gang on week 7. This remaining gang was to be retained until the end of week 9 forming internal piers. The 4 + 2 gang from project 5 was to rebuild the wall affected by mining subsidence on project 6.

The programmed movement of operatives would achieve continuity of work for these gangs and at the same time keep the demand on the resources to a minimum.

### 7.4.6 *Exceeding available resources*

Where the total available resources within the firm were exceeded, adequate warning that further recruitment of labour or hire of labour-only subcontractors or plant was given. Occasionally the demand for tradesmen such as bricklayers and carpenters exceeds supply, and in such circumstances this planning technique is invaluable in making the best use of available resources and providing them with continuity of work.

### 7.4.7 *Labour schedules*

As each new project was introduced, labour schedules were drawn up below the programme to show the level of each resource.

The aim was to keep the labour force as stable as possible while allowing operatives to complete an operation once they had started it. This allows bonusing to work more effectively.

## 7.5 Case study 5: Rossington bungalows

### 7.5.1 *Purpose*

The purpose of this case study is to illustrate the initial planning stages of a design-and-build contract.

### 7.5.2 *Description of project*

The project consisted of the erection of four pairs of two-bedroom detached bungalows for the elderly, including landscaping, paths, road and paved areas, drainage and fencing (see Figs 7.46–7.48).

This project was featured in the RIBA Yorkshire Region Directory, 1996.

- *Employer's agent*: T. Sleath & Partners
- *Design and build contractors*: O & P Construction Services Ltd
- *Architects*: Milnes and Associates
- *Structural engineers*: Hutter Jennings & Titchmarsh.

The employer's agent, the architect and the structural engineer had been utilised in the preparation of the scheme design for the project, and in obtaining planning permission. Building Regulations approval had not

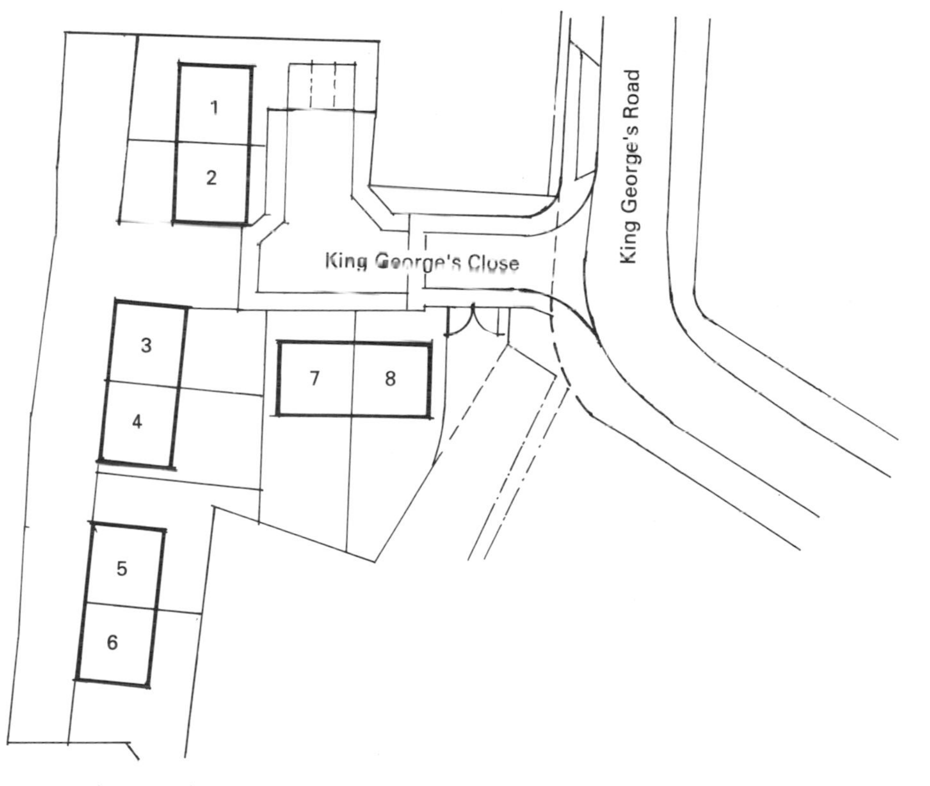


**Fig. 7.46** Case study 5: site plan.

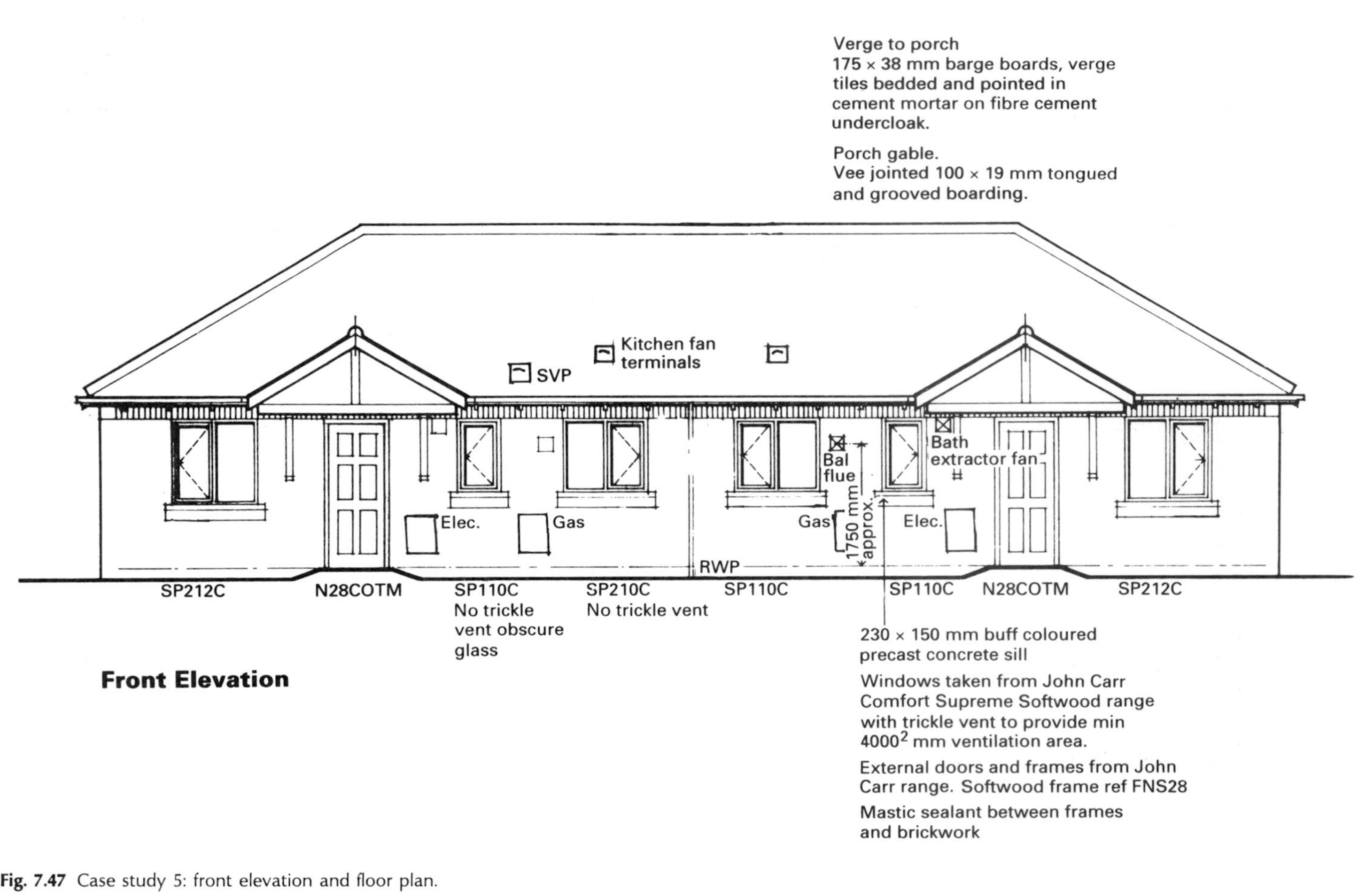

**Fig. 7.47** Case study 5: front elevation and floor plan.

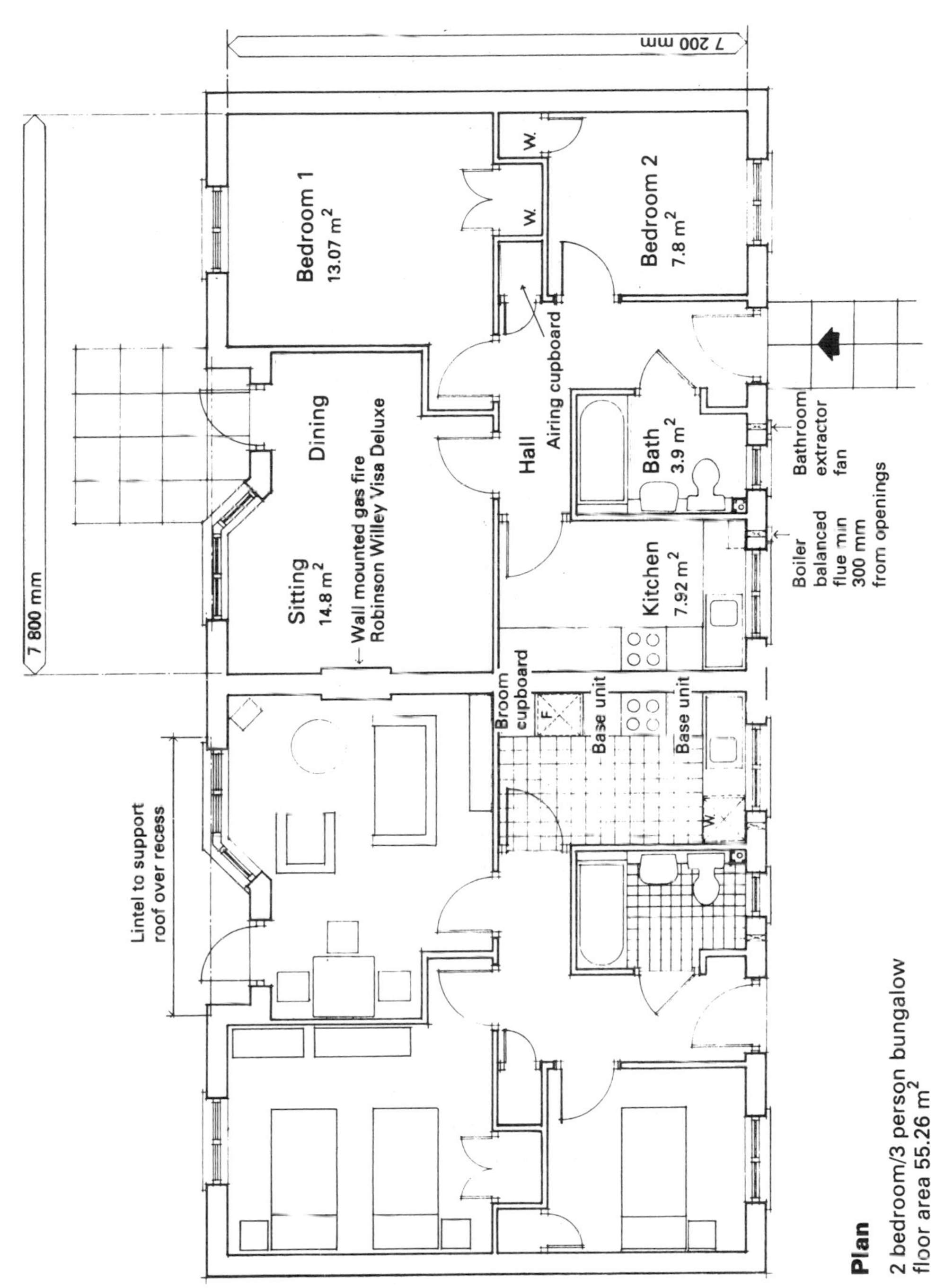


**Plan**

2 bedroom/3 person bungalow
floor area 55.26 m$^2$

**Fig. 7.47** (continued).

**Fig. 7.48** Case study 5: Rosssington bungalows.

of course been obtained at this stage. The contract was of the novated design and build type. It was the wish of the employer that the architect be retained by the successful tenderer.

The contractor was to be responsible for the whole of the design of the works.

In this contract, contractors had to submit contractors proposals and a contract sum analysis, as these were to be taken into account during tender evaluation together with other factors.

### *7.5.3 Contractor's proposals*

These had to include, in addition to other documents, the following:

(1) a completed form of tender;
(2) a contract sum analysis in the format as described in the documents and accompanied by the supporting information listed therein;
(3) a detailed general construction and design programme;
(4) a cash flow detailing the anticipated rate of expenditure throughout the project, split between design works and contract works;
(5) details of the quality assurance scheme and procedures that the contractor proposed for the project;
(6) form of contract;
(7) approved contract drawings;
(8) approved specification.

Items 2, 3 and 4 will be covered by this case study.

### 7.5.4 Contract sum analysis

This had to include a breakdown of the contract sum on a unit-by-unit lump-sum basis, identifying separately any specific costs not related to units. The tenderers also had to submit a detailed analysis of their tender price, together with a separate breakdown of preliminaries elements. The design costs and any statutory fees and charges included in the tender had also to be clearly identified separately.

### 7.5.5 Construction of the buildings

*Substructure*

- *Foundations*: flat slab reinforced raft, on sand and hardcore; brick block cavity wall up to DPC.
- *Floor slab*: DPM, Jablite and concrete fill on raft.

*Superstructure*

- *Walls*: facing brick and block cavity walls with insulation.
- *Windows*: standard softwood, double glazed.
- *External doors*: standard hardwood.
- *Partitions*: 100 mm blockwork.
- *Roof structure*: softwood roof trusses.
- *Roof covering*: interlocking concrete tiles.

*Internal finishes*

- *Walls*: two coats plaster.
- *Ceilings*: plasterboard and skim.
- *Floor*: latex and vinyl tiles and vinyl sheeting.

*Service systems*

- *Heating*: gas central heating.

*External works*

- *Drainage*: clay supersleeve. Precast concrete manhole.
- *Road*: tarmac and block paving.
- *Parking*: block paving.
- *Paths*: paving flags.
- *Landscaping*: grassed areas, shrubs and hedges.
- *Street lights*: lamp standards.
- *Fencing*: timber screen fencing and wrought iron railings.

### 7.5.6 The contract: JCT Standard Form of Contract with contractor's design

- ❑ *Liquidated damages*: £400.00 per week.
- ❑ *Defects liability period*:
  - ■ 6 months for the building work;
  - ■ 12 months for electrical and mechanical works, landscaping, external works and adopted works.
- ❑ *Insurance cover*: £2 000 000.00.
- ❑ *Contract period*: 26 weeks.
- ❑ *Tender submission date*: 13 December 1993.
- ❑ *Date of possession of site*: 28 February 1994.

### 7.5.7 The site

The site was located 12 miles (19 km) from the contractor's office. It was previously wasteland, which contained a commercial filling station. Only the garage slab remained. The site fronted up to an estate road.

The north boundary consisted of bollards, beyond which was a Working Mens' Club and a car park.

The west boundary consisted of a post-and-wire fence, which separated the site from the gardens of existing properties.

The south boundary was a post-and-wire fence, which divided the site from a school playing field and a community hall. The entrance to the community hall had to be left open at all times, and was to be relocated as part of the contract.

### 7.5.8 Requirements prior to signing contract

After the contractor had been selected, a detailed construction and design programme had to be prepared and all information brought up to a sufficient level to enable the employer to enter into a contract with the tenderer. These included:

- ❑ contract drawings
- ❑ detailed specification
- ❑ contract sum analysis
- ❑ detailed programme (construction and design)
- ❑ working drawings sufficient to enable preliminary client approval to detail and standards of finishes and fittings.

### 7.5.9 Planning at the tender stage

It was decided at the tender stage to use Pertmaster Advance, a computer project planning package, as a basis for the planning of the project.

| Rossington Bungalows | | 1 | |
|---|---|---|---|
| Tender analysis | | | |
| House type | no | Cost per unit | Total |
| Semi-detached bungalows | 8 | 16581.53 | 132652.20 |
| (1) *Specific costs not relating to units* | | | |
| Landscaping maintenance contract | | | 750.00 |
| Siteworks | | | 40106.32 |
| Drainage | | | 8294.42 |
| External services and lighting | | | 1899.73 |
| Adopted siteworks | | | 15269.78 |
| Adopted drainage | | | 5238.89 |
| Adopted external lighting | | | 2265.00 |
| Design costs, statutory fees, preliminaries and OH & P | | | 67430.00 |
| | | Total contract sum | 273906.34 |

**Fig. 7.49** Case study 5: tender analysis.

The contract sum analysis had to be presented in the format requested in the tender documents. This required the breakdown to be shown as:

(1) tender analysis
(2) elemental analysis.

### 7.5.10 *Tender analysis (Fig. 7.49)*

This required an overall cost per dwelling and a breakdown of specific costs not related to units.

### 7.5.11 *Elemental analysis (Fig. 7.50)*

This required a detailed breakdown of the cost of:

(1) the dwellings
(2) the external works
(3) post-completion maintenance works
(4) preliminaries analysis
(5) contract sum analysis final cost summary.

Other items, the cost of which were requested by the employer's agent, were shown after the preliminaries analysis.

### 7.5.12 *Pre-tender programme*

The logic was first set out using a precedence diagram. The procedure followed in preparing the programme is set out below.

6.3 Elemental analysis

| | | |
|---|---|---|
| 6.3.1.1 | ***Dwellings*** | |
| 1.1A | Substructure | 18847.04 |
| 1.2 | ***Superstructure*** | |
| 1.2A | Frame | |
| 1.2B | Upper floors | 0.00 |
| 1.2C | Roof | 20613.36 |
| 1.2D | Stairs/ramps | 0.00 |
| 1.2E | External walls | 16255.36 |
| 1.2F | Windows/external doors | 12374.60 |
| 1.2G | Internal walls | 6508.36 |
| 1.2H | Internal doors | 4975.68 |
| 1.3 | ***Finishes*** | |
| 1.3A | Wall finishes | 9118.28 |
| 1.3B | Floor finishes | 3960.36 |
| 1.3C | Ceiling finishes | 3006.56 |
| 1.4 | Fittings/special items | 6718.00 |
| 1.5 | ***Services*** | |
| 1.5A | Sanitary fittings | 2915.92 |
| 1.5B | Services equipment | 0.00 |
| 1.5C | Disposal installations | 2686.44 |
| 1.5D | Water installations | 1493.48 |
| 1.5E | Mechanical installation | 12276.00 |
| 1.5F | Electrical installations | 7008.00 |
| 1.5G | Gas installations | 343.00 |
| 1.5H | Lift installations | 0.00 |
| 1.5I | Protective installations | 0.00 |
| 1.5J | Communication installations | 51.76 |
| 1.5K | Special installations | 3500.00 |
| 1.5L | BWIC | 0.00 |
| | 6.3.1.1 Dwellings total to final cost summary | £132652.20 |

6.3.1.2 External works

| | | |
|---|---|---|
| 2.1A | Site clearance and preparatory earthworks | 8706.90 |
| 2.2A | Roads and hardstandings | 8609.45 |
| 2.2B | Paths and paving | 15806.27 |
| 2.2C | Landscaping and planting | 3443.60 |
| 2.2D | Walls and fences | 18809.88 |
| 2.3A | Foul and surface drainage | 13533.31 |
| 2.4A | External lighting | 3065.00 |
| 2.4B | Site water disposal | 678.05 |
| 2.4C | Site electricity distribution | 421.18 |
| 2.4D | Site BT distribution | 0.00 |
| 2.4E | Site gas distribution | 0.00 |
| 2.4F | Fire hydrant installation | 0.00 |
| 2.4G | Fire main installation | 0.00 |
| 2.4H | Water capital installation | 0.00 |
| 2.4I | Electrical capital installation | 0.00 |
| 2.4J | BT capital installation | 0.00 |
| 2.4K | Gas capital installation | 0.00 |
| 2.4L | Cable TV installation | 0.00 |
| | 6.3.1.2 External works total to final cost summary | £73073.64 |

Preliminaries analysis

| | |
|---|---|
| Scaffolding | 1898.00 |
| General clearing up | 1000.00 |
| Snag | 1000.00 |
| Transporting materials (most delivered direct to site) | 500.00 |
| Plant and mixer | 800.00 |
| Contract clean | 400.00 |
| Water | 350.00 |
| Electric | 500.00 |
| Telephone | 220.00 |
| Site accommodation, etc. | 1010.00 |
| Security fencing | 1200.00 |
| Sign board | 100.00 |
| Supervision charges for roads and sewer | 460.00 |
| Performance bond | 405.00 |
| Divert power cables | 5000.00 |
| | 14843.00 |
| Allowance in contract sum for smoke detectors | 360.00 |
| Extra cost for grass cutting front gardens | 360.00 |
| Extra costs for UPVC windows | 650.00 |

**Fig. 7.50** Case study 5: elemental analysis.

| | | |
|---|---|---|
| 6.3.1.3 | Post-completion maintenance works | |
| 3.1A | Landscaping maintenance contract | 750.00 |
| | 6.3.1.2 Post completion maintenance works to final cost summary | 750.00 |
| 6.3.1.3 | Contract sum analysis final cost summary | |
| | Dwellings (from 6.3.1.1) | 132652.20 |
| | External works (from 6.3.1.2) | 73073.64 |
| | Post completion maintenance works (from 6.3.1.3) | 750.00 |
| | Additional costs: | |
| | (a) Design costs | 7153.00 |
| | (b) All statutory fees and other charges | |
| | Gas | 887.00 |
| | Water | 6984.00 |
| | Sewer | 5424.00 |
| | Electric | 2800.00 |
| | Telephones | 0.00 |
| | (c) Preliminaries and management costs during the progress of the works | 14843.00 |
| | (d) Overhead and profit | 29339.50 |
| | (e) Inflation and fixed price | 0.00 |
| | (f) Interest charges | 0.00 |
| | (g) Land acquisition cost | 0.00 |
| | Total to form of tender - section 7 | 273906.34 |

**Fig. 7.50** (continued).

### 7.5.13 *Calendar*

A contract calendar was created, which related the day numbers to calendar dates.

### 7.5.14 *Resources*

Resources were defined and a code name given to each resource. The type of loading was stated; for example, in the case of monetary resources, the charge could be made at the beginning or end of an activity, or spread throughout the activity. In the case of manpower and plant the number available was also stated.

### 7.5.15 *Creating the plan*

General information about the plan was first entered. This included:

- a plan name
- the title of the project, type of programme and name of company
- other general information about the project, such as start and finish dates
- the calendar code name.

### 7.5.16 *Programme information – calculation sheet (Fig. 7.51)*

Before the programme could be generated, the activity durations had to be ascertained.

Programme information — Eight bungalows at Rossington, Doncaster

| Ref | Description | Hours in estimate (per pair) | Target hours (85%) | Programme hours (80%) | Gang size | Gang hours | Duration 1 pair (days) | Duration 4 pairs (days) | Duration 1 bungalow (days) | |
|---|---|---|---|---|---|---|---|---|---|---|
| | | For a typical pair of bungalows | | | | | | | | |
| | Prepare all design proposals | | | | | | | 39.0 | | |
| | Obtain preliminary NHER rating | | | | | | | 15.0 | | |
| | Obtain planning permission | | | | | | | 10.0 | | |
| | Prepare working drawings | | | | | | | 29.0 | | |
| | Prepare precontract programme | | | | | | | 14.0 | | |
| | Prepare cashflow | | | | | | | 3.0 | | |
| | Prepare schedules | | | | | | | 3.0 | | |
| | Engineering drawings and calculations | | | | | | | 5.0 | | |
| | Construction detailing | | | | | | | 20.0 | | |
| | Obtain regulations approval | | | | | | | 10.0 | | |
| Setup | Set up site | | | | | | | 2.0 | | |
| Exc rl | Excavate topsoil, retain, RL dig and CA | | | | | | | 13.0 | | |
| Adj | Foul, SW and serv trenches adj building | | | | 2.0 | | | 12.0 | | |
| Hardcore | Consolidated hardcore under raft | 33.8 | 27.0 | 21.6 | 3 | 7 | 0.9 | 3.6 | | |
| Sleeve | Service sleeves and drain entry | 8.0 | 6.4 | 5.1 | 3 | 2 | 0.2 | 0.9 | | |
| Sand | Sand bed | 20.0 | 16.0 | 12.8 | 3 | 4 | 0.5 | 2.1 | | |
| Formwork | Formwork to edge of raft | 12.3 | 9.8 | 7.8 | 1 | 8 | 1.0 | 3.9 | | |
| Raft | Concrete raft, mesh and slipsheet | 91.4 | 73.1 | 58.5 | 3 | 19 | 2.4 | 9.7 | | |
| Walldpc | Walling up to DPC | 35.7 | 28.6 | 22.9 | 3 | 8 | 1.0 | 3.8 | | |
| Conc | Concrete fill and insulation | 68.0 | 54.4 | 43.5 | 3 | 15 | 1.8 | 7.3 | | |
| Wallsup | Walling in superstructure | 482.8 | 386.2 | 309.0 | 3 | 103 | 12.9 | 51.5 | | 2 gangs |
| Infra | Infrastructure and connection charges | | | | | | | 1.0 | | |
| Scaff | Scaffolding - erect and strip | | | | | | | 10.0 | | |
| Roofcon | Roof construction and eaves | 235.6 | 188.5 | 150.8 | 2 | 75 | 9.4 | 37.7 | | 2 gangs |
| Roofcov | Roof coverings | | | | | | 2.0 | 8.0 | | |
| Car1 | Carpentry first fix | 120.4 | 96.3 | 77.0 | 2 | 39 | 4.8 | 19.3 | 2.4 | |
| Plumb1 | Plumbing and heating first fix | | | | | | 2.0 | 8.0 | 1.0 | |
| Elec1 | Electrical first fix | | | | | | 4.0 | 16.0 | 2.0 | |
| Plas | Plastering | 266.6 | 213.3 | 170.6 | 3 | 57 | 7.1 | 28.4 | 3.6 | |
| Elec2 | Electrical second fix | | | | | | 4.0 | 16.0 | 2.0 | |
| Car2 | Carpentry second fix | 172.3 | 137.9 | 110.3 | 2 | 55 | 6.9 | 27.6 | 3.4 | |
| Plumb2 | Plumbing and heating second fix | | | | | | 6.0 | 24.0 | 3.0 | |
| Conn | Service connect | | | | | | | 5.0 | | |

**Fig. 7.51** Case study 5: programme information.

For a typical pair of bungalows

| Ref | Description | Hours in estimate (per pair) | Target hours (85%) | Programme hours (80%) | Gang size | Gang hours | Duration 1 pair (days) | Duration 4 pairs (days) | Duration 1 bungalow (days) |
|---|---|---|---|---|---|---|---|---|---|
| Tile | Tiling | 14.8 | 11.9 | 9.5 | 1 | 9 | 1.2 | 4.7 | 0.6 |
| Comm | Commissioning | | | | | | | | |
| Ins | Roof insulation | 26.2 | 21.0 | 16.3 | 1 | 17 | 2.1 | 8.4 | 1.0 |
| Decor | Decoration to bungalows | | | | | | 8.0 | 32.0 | 4.0 |
| Floor | Floor coverings | | | | | | 3.0 | 12.0 | 1.5 |
| Snag | Snagging and finishing off | | | | | | 2.0 | 8.0 | 1.0 |
| Clean | Final clean up and scrub out | | | | | | 2.0 | 8.0 | 1.0 |
| Drnl | Remaining SW drainage, serv trenches | | | | 2.0 | | | 10.0 | |
| Light | Street lights | | | | | | | 2.0 | |
| Topsoil | Topsoil and hardcore to roads and pavg | | | | JC3 + 1 man | | | 16.0 | |
| Bfdns | Boundary wall foundations | | | | 2.0 | | | 3.0 | |
| Kerb | Kerb edgings | | | | 2.0 | | | 5.0 | |
| Pavg | Paving flags | | | | 2.0 | | | 18.0 | |
| Bwalls | Boundary walls | | | | 3.0 | | | 14.0 | |
| Rail | Railings and fencing | | | | Sub | | | 8.0 | |
| Blockpav | Block paving | | | | 2.0 | | | 23.0 | |
| Tarmac | Tarmac roads and footpaths | | | | Sub | | | 5.0 | |
| Inspect | Final inspection and handover | | | | | | | 1.0 | |
| Landsc | Landscaping | | | | | | | 2.0 | |
| Warden | Warden call | | | | | | | 1.0 | |

**Fig. 7.51** (continued).

The estimate for this project was prepared using MASTER (see section 6.3 and case study 1), and consequently times included for the activities were readily available.

For programming purposes, the expected activity durations were calculated from the hours in the estimate.

The same information was used ultimately as a basis for targets, which were set at 85% of the hours in the estimate. Operatives were generally expected to take a maximum of 75% of the target, but the programme was based on 80% to allow some flexibility. Lotus 1-2-3 was used to calculate the activity durations. At the pre-tender stage some of the activities on the calculation sheet were grouped.

### 7.5.17 *Entering activities*

Activities were entered by typing the abbreviated name, description, duration, a code to assist in sorting the activities into order, and links to succeeding activities including duration between links.

### 7.5.18 *Scheduled day numbers*

These were used for the start of the project, the end of the project and the commencement of operations on site.

### 7.5.19 *Entering resources*

Resources were entered using the resource code, followed by a number to represent the quantity of the particular resource. These entries generally represent the resource requirements per unit of time spread evenly over the activity: for example, carpenters. Some monetary resources were entered as point resources: for example, service connections and infrastructure charges.

### 7.5.20 *Generating reports*

Numerous reports can be generated using Pertmaster Advance. Before producing reports, pertmaster analyses the data and loads the calendar and abbreviations file. It then sorts the activities into the order required by the first report to be prepared.

The power of the package is the wealth of information that can be obtained after initial entry. Only the reports directly relevant to this case study are considered here.

O & P CONSTRUCTION SERVICES LTD
PERTMASTER - BARCHART REPORT

ROSSINGTON
BUNGALOWS

PRE TENDER PROGRAMME
VERSION 1

Task Description

Jan '94 | Feb '94 | Mar '94 | Apr '94 | May '94 | Jun '94 | Jul '94 | Aug '94 | Sep '94 | Oct
27 | 3 | 10 | 17 | 24 | 31 | 7 | 14 | 21 | 28 | 7 | 14 | 21 | 28 | 4 | 11 | 18 | 25 | 2 | 9 | 16 | 23 | 30 | 6 | 13 | 20 | 27 | 4 | 11 | 18 | 25 | 1 | 8 | 15 | 22 | 29 | 5 | 12 | 19 | 26 | 3

***PLANNING AND DESIGN WORK***
***SET UP SITE***
***FOUNDATIONS***
EXC TOPSOIL,RETAIN,RL DIG REM F
TEMPORARY ROAD
RAFT FOUNDATIONS
WALLING UP TO DPC
CONCRETE FILL
***SUPERSTRUCTURE***
WALLING IN SUPERSTRUCTURE
SCAFFOLDING
INFRASTRUCTURE AND CONNECTIO
ROOF CONSTRUCTION
ROOF COVERING
***INTERNAL WORK***
CARPENTRY FIRST FIX
PLUMBING AND HEATING FIRST FIX
ELECTRICAL FIRST FIX
PLASTERING
CARPENTRY SECOND FIX
PLUMBING AND HEATING SECOND FI
SERVICE CONNECTIONS
ELECTRICAL SECOND FIX
WALL TILING
ROOF INSULATION
DECORATION
FLOOR COVERINGS
WARDEN CALL SYSTEM
SNAGGING AND FINISHING OFF
FINAL CLEAN UP AND SCRUB OUT
***EXTERNAL WORK***
DRAINAGE AND SERVICE TRENCHES
BOUNDARY WALL FOUNDATIONS
STREET LIGHTS
BOUNDARY WALLS
TOPSOIL AND HARDCORE
RAILINGS AND FENCING
KERB EDGINGS
BLOCK PAVINGS
PAVING FLAGS
TARMAC ROADS AND FOOTPATHS
LANDSCAPING
***FINAL INSPECTION AND HANDOVER***

Float, free | Float, total | Float, start | Normal Task | Normal, Critical | Normal, Stretched

**Fig. 7.52** Case study 5: abbreviations file.

### 7.5.21 Screen reports

Bar charts, histograms and activity listings were displayed on the screen. It was then simple to move around the project examining the information in detail as required.

### 7.5.22 Bar chart (Fig. 7.52)

A pre-tender bar chart was produced. This contained the heading, with dates, and was divided into weeks. Critical and non-critical activities and float could easily be seen. At this stage the project was programmed to take the full contract period and some of the activities shown on the calculation sheet were grouped. All activities were shown at their earliest times in this programme.

### 7.5.23 Cash flow (Fig. 7.53)

The net cost of each operation was ascertained using MASTER. Lotus 1-2-3 was then used to calculate:

- the 'marked up' costs
- the 'marked up' costs less retention.

When all the figures had been calculated, a check was carried out to ensure that no errors had been made. This information was then entered into Pertmaster, and a spending report was generated (see Fig. 7.54).

### 7.5.24 Contract budget (Fig. 7.55)

A budget was then generated in the form of a graph. The budget was in fact based on the shorter contract period shown on the pre-contract programme (Fig. 7.56) because this reflects the expected cash flow more accurately.

### 7.5.25 Contract programme (Fig. 7.56)

After acceptance of the tender, a pre-contract programme was developed from the pre-tender programme. All the operations on the calculation sheet were included at this stage. The overall duration was less than the contract period to allow for flexibility. Again, all activities were shown at their earliest times.

8 bungalows at Rossington, Doncaster

| Description | Net costs<br>Mat | Lab | Sub | Plt | Total marked up costs for 4 pairs | Total marked up costs for 4 pair less 5% retention | |
|---|---|---|---|---|---|---|---|
| Prepare all design proposals | | | | | | | |
| Obtain preliminary NHER rating | | | | | | | |
| Obtain planning permission | | | Client | | | | |
| Prepare working drawings | | | 4000.00 | | 4000.00 | 3800.00 | |
| Prepare precontract programme | | | | | | | |
| Prepare cash flow | | | | | | | |
| Prepare schedules | | | | | | | |
| Engineering drawings and calculations | | | 1600.00 | | 1600.00 | 1520.00 | |
| Construction detailing | | | 1155.00 | | 1155.00 | 1097.25 | |
| Obtain regulations approval | | | 809.00 | | 809.00 | 768.55 | |
| Set up site | 190.00 | 1350.00 | 670.00 | 700.00 | 2910.00 | 2764.50 | |
| Excavate topsoil, retain, RL dig & CA | | | 10811.39 | | 12452.56 | 11829.93 | |
| Foul, SW and serv trenches adj building | 4990.34 | 1326.19 | 2118.61 | | 9715.59 | 9229.81 | |
| Hardcore to raft | 243.00 | 159.64 | | 37.80 | 2029.20 | 1927.74 | |
| Service sleeves and drain entry | 26.00 | 39.92 | | | 303.71 | 288.52 | |
| Sand bed | 80.00 | 94.60 | | 22.40 | 907.62 | 862.24 | |
| Formwork to edge of raft | 19.60 | 67.01 | | | 399.03 | 379.08 | |
| Concrete raft, mesh and slipsheet | 1158.75 | 432.32 | | | 7330.38 | 6963.86 | |
| Infrastructure and connection charges | | | 4023.75 | | 18538.22 | 17611.31 | |
| Walling up to DPC | 260.98 | 195.42 | | | 2102.73 | 1997.59 | |
| Concrete fill on insulation | 629.68 | 241.23 | | | 4012.46 | 3811.83 | |
| Walling in superstructure | 2973.50 | 2636.33 | | 300.00 | 27230.07 | 25868.57 | |
| Scaffolding erect and strip | | 100.00 | | 1708.00 | 1808.00 | 1717.60 | |
| Roof construction and eaves | 1619.93 | 1284.33 | | | 13382.81 | 12713.67 | |
| Roof coverings | 90.50 | 34.73 | 1570.00 | | 7810.26 | 7419.75 | |
| Carpentry first fix | 2436.82 | 686.76 | | | 14390.96 | 13671.41 | |
| Plumbing and heating first fix | | | 1120.88 | | 5164.12 | 4905.91 | |
| Electrical first fix | | | 926.00 | | 4266.27 | 4052.95 | |
| Plastering | 969.17 | 1279.49 | | | 10360.03 | 9842.03 | |
| Electrical second fix | | | 926.00 | | 4266.27 | 4052.95 | |
| Carpentry second fix | 2246.98 | 831.91 | | | 14415.42 | 13694.65 | |
| Plumbing and heating second fix | | | 3362.64 | | 15492.36 | 14717.74 | |
| Service connections | | | | | | | |
| Tiling | 121.04 | 31.17 | | | 931.62 | 885.04 | |
| Commissioning | | | | | | 0.00 | |
| Roof insulation | 330.12 | 123.93 | | | 2091.90 | 1987.30 | |
| Decoration to bungalows | | | 1714.50 | | 7899.04 | 7504.09 | |
| Floor coverings | | | 990.09 | | 4561.54 | 4333.47 | |
| Snagging and finishing off | | 1780.00 | | | 1780.00 | 1691.00 | |
| Final clean up and scrub out | | | 400.00 | | 400.00 | 380.00 | |
| Remaining drainage and service trenches | 4990.34 | 1326.19 | 2118.61 | | 9715.59 | 9229.81 | |
| Street lights and private lights | | | 2665.00 | | 3069.55 | 2916.07 | |
| Topsoil and hardcore | 3693.60 | 615.00 | 2625.10 | | 7987.39 | 7588.02 | |
| Boundary wall foundations | | | 448.20 | | 516.24 | 490.42 | |
| Kerb edgings | 1723.44 | 625.71 | 425.25 | | 3202.46 | 3042.34 | |
| Paving flags and clothes posts | 4340.73 | 2899.77 | | | 8339.61 | 7922.63 | |
| Boundary walls | 3273.80 | 3095.64 | 7336.55 | | 7336.55 | 6969.72 | |
| Railings and fencings | | 116.48 | 10991.68 | | 11683.88 | 11099.68 | |
| Block paving | 2787.94 | 3447.14 | | | 7181.57 | 6822.49 | |
| Tarmac roads and footpaths | | | 3069.8 | | 3535.80 | 3359.01 | |
| Final inspection and handover | | | | | | 6847.66 | (Half retention) |
| Landscaping | | | 2422.5 | | 2790.24 | 2650.72 | |
| Warden call system | | | 875.00 | | 4031.30 | 3829.74 | |
| | | | | | 273906.32 | Check 267058.66 | |
| | | | | | | 6847.66 | (Half retention) |
| | | | | | | 273906.32 | OK |

Fig. 7.53 Case study 5: calculation sheet for cash flow.

| O & P Construction Services Ltd<br>Spending Report | ROSSINGTON BUNGALOWS | PRE CONTRACT PROGRAMME<br>VERSION 1 |
|---|---|---|

| Task Description | Budget Costs |
|---|---|
| ***PRELIMINARY WORK*** | ***£7,185.90*** |
| PREPARE WORKING DRAWINGS | £3,800.00 |
| ENGINEERING DRAWINGS AND CALCULATIONS | £1,520.00 |
| OBTAIN PRELIMINARY NHER RATING | £0.00 |
| OBTAIN PLANNING PERMISSION | £0.00 |
| PREPARE PRE-CONTRACT PROGRAMME | £0.00 |
| PREPARE ALL DESIGN PROPOSALS | £0.00 |
| PREPARE CASH FLOW | £0.00 |
| PREPARE SCHEDULES | £0.00 |
| CONSTRUCTION DETAILING | £1,097.25 |
| OBTAIN REGULATIONS APPROVAL | £768.65 |
| ***SET UP SITE*** | ***£2,764.50*** |
| ***FOUNDATION*** | ***£37,290.60*** |
| EXC TOPSOIL,RETAIN,RL DIG, REM FROM SITE | £11,829.93 |
| FOUL, SW & SERVICE TRENCHES ADJ BUILDING | £9,229.81 |
| HARDCORE TO RAFT | £1,927.74 |
| SERVICE SLEEVES AND DRAIN ENTRIES | £288.52 |
| SAND BED | £862.24 |
| FORMWORK TO EDGE OF RAFT | £379.08 |
| CONCRETE RAFT, MESH AND SLIPSHEET | £6,963.86 |
| WALLING UP TO DPC | £1,997.59 |
| CONCRETE FILL ON INSULATION | £3,811.83 |
| ***SUPERSTRUCTURE*** | ***£65,330.90*** |
| WALLING IN SUPERSTRUCTURE | £25,868.57 |
| INFRASTRUCTURE AND CONNECTION CHARGES | £17,611.31 |
| SCAFFOLDING - ERECT AND STRIP | £1,717.60 |
| ROOF CONSTRUCTION AND EAVES | £12,713.67 |
| ROOF COVERING | £7,419.75 |
| ***INTERNAL WORK*** | ***£85,548.28*** |
| CARPENTRY FIRST FIX | £13,671.41 |
| PLUMBING AND HEATING FIRST FIX | £4,905.91 |
| ELECTRICAL FIRST FIX | £4,052.95 |
| PLASTERING | £9,842.03 |
| CARPENTRY SECOND FIX | £13,694.65 |
| PLUMBING AND HEATING SECOND FIX | £14,717.74 |
| SERVICE CONNECTIONS | £0.00 |
| ELECTRICAL SECOND FIX | £4,052.95 |
| WALL TILING | £885.04 |
| ROOF INSULATION | £1,987.30 |
| DECORATION | £7,504.09 |
| FLOOR COVERINGS | £4,333.47 |
| WARDEN CALL SYSTEM | £3,829.74 |
| COMMISSIONING | £0.00 |
| SNAGGING AND FINISHING OFF | £1,691.00 |
| FINAL CLEAN UP AND SCRUB OUT | £380.00 |
| ***EXTERNAL WORK*** | ***£62,090.91*** |
| REMAINING DRAINAGE AND SERVICE TRENCHES | £9,229.81 |
| BOUNDARY WALL FOUNDATIONS | £490.42 |
| STREET LIGHTS | £2,916.07 |
| BOUNDARY WALLS | £6,969.72 |
| TOPSOIL AND HARDCORE TO ROADS AND PAVING | £7,588.02 |
| RAILINGS AND FENCING | £11,099.68 |
| KERB EDGINGS | £3,042.34 |
| BLOCK PAVINGS | £6,822.49 |
| PAVING FLAGS | £7,922.63 |
| TARMAC ROADS AND FOOTPATHS | £3,359.01 |
| LANDSCAPING | £2,650.72 |
| ***FINAL INSPECTION AND HANDOVER*** | ***£6,847.66*** |
| **TOTALS** | **£267,058.75** |

Report Produced by Pertmaster Project Planning Sofware

**Fig. 7.54** Case study 5: spending report.

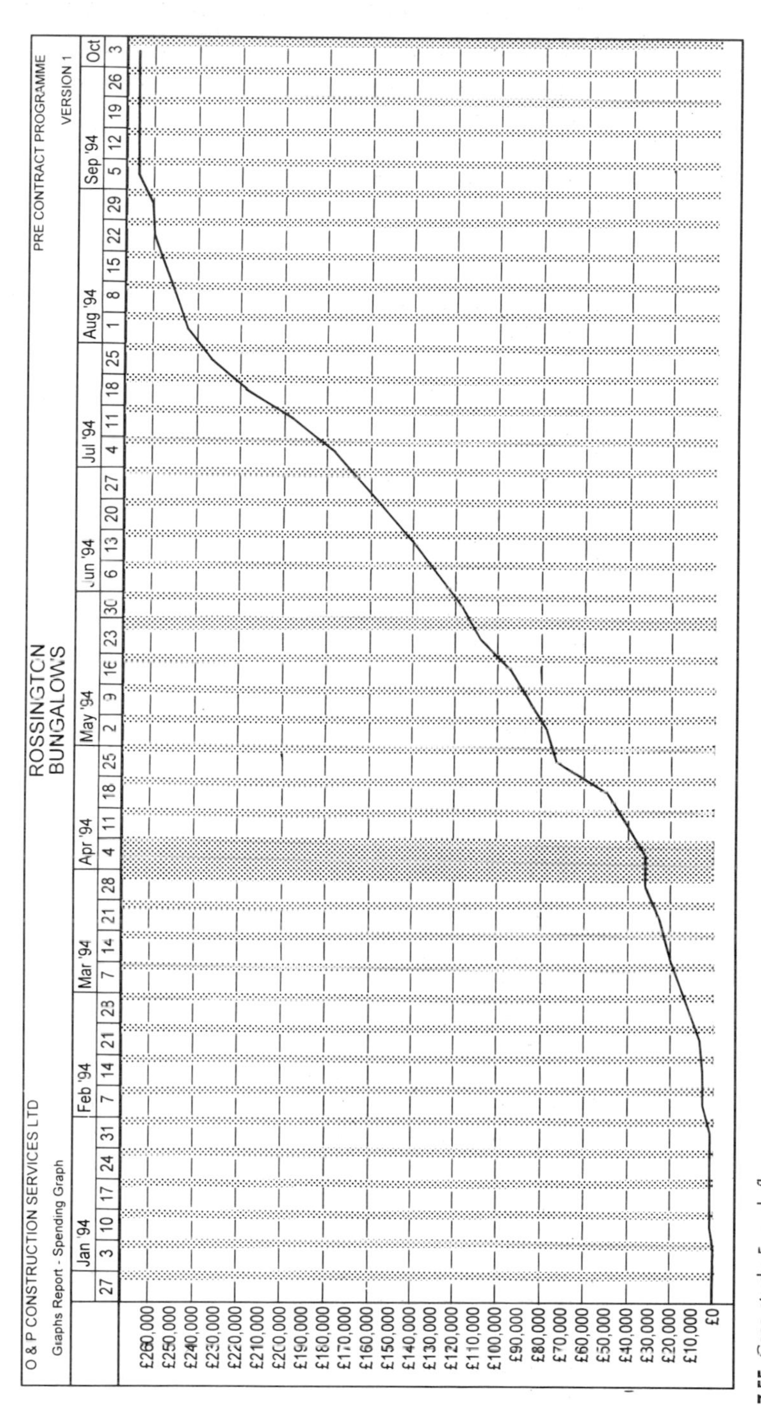


**Fig. 7.55** Case study 5: cash flow.

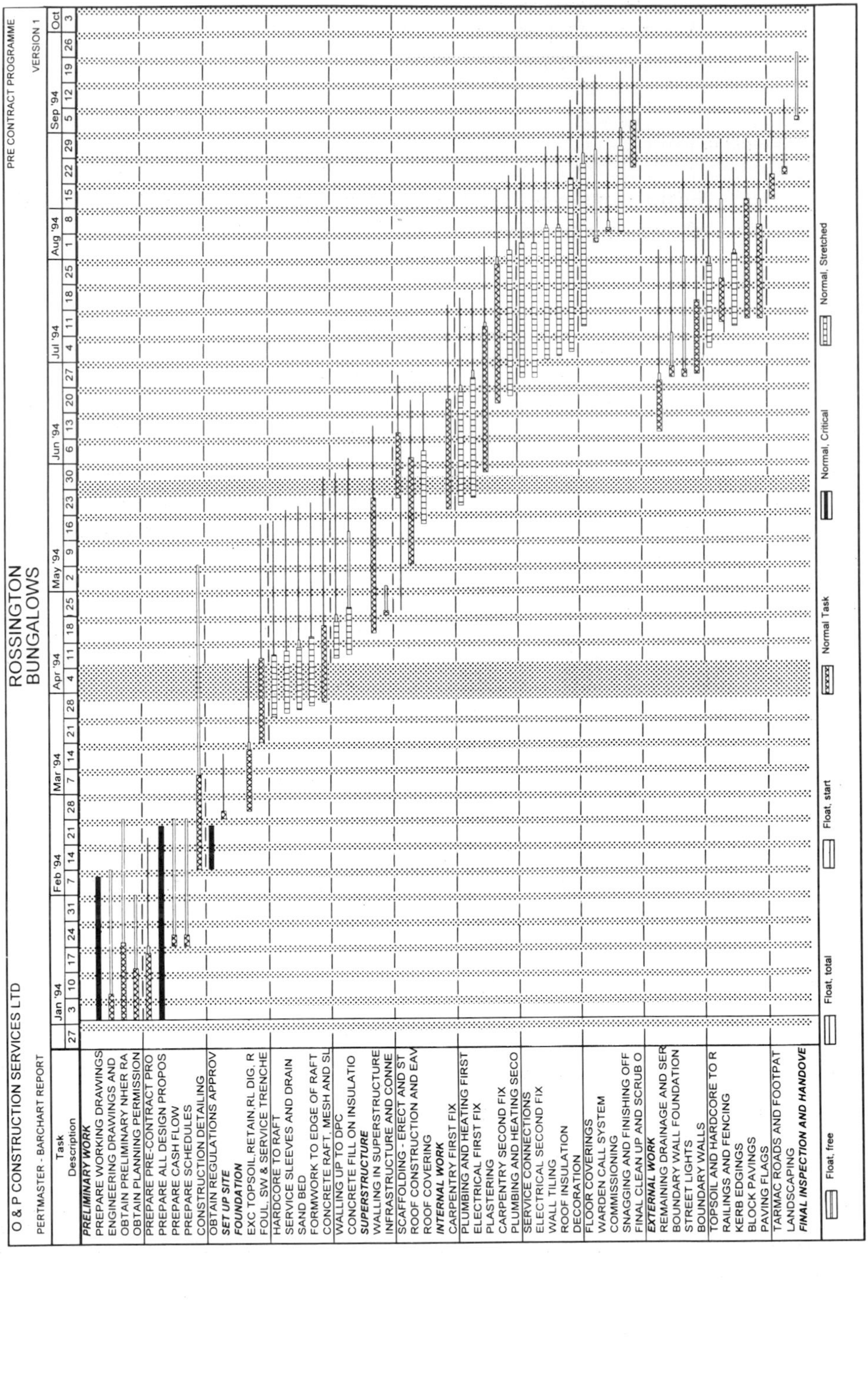

O & P CONSTRUCTION SERVICES LTD
PERTMASTER - BARCHART REPORT
ROSSINGTON BUNGALOWS
PRE CONTRACT PROGRAMME
VERSION 1

Task Description
Jan '94 | Feb '94 | Mar '94 | Apr '94 | May '94 | Jun '94 | Jul '94 | Aug '94 | Sep '94 | Oct
27 3 10 17 24 31 7 14 21 28 7 14 21 28 4 11 18 25 2 9 16 23 30 6 13 20 27 4 11 18 25 1 8 15 22 29 5 12 19 26 3

***PRELIMINARY WORK***
PREPARE WORKING DRAWINGS
ENGINEERING DRAWINGS AND
OBTAIN PRELIMINARY NHER RA
OBTAIN PLANNING PERMISSION
PREPARE PRE-CONTRACT PRO
PREPARE ALL DESIGN PROPOS
PREPARE CASH FLOW
PREPARE SCHEDULES
CONSTRUCTION DETAILING
OBTAIN REGULATIONS APPROV
***SET UP SITE***
***FOUNDATION***
EXC TOPSOIL,RETAIN,RL DIG, R
FOUL, SW & SERVICE TRENCHE
HARDCORE TO RAFT
SERVICE SLEEVES AND DRAIN
SAND BED
FORMWORK TO EDGE OF RAFT
CONCRETE RAFT, MESH AND SL
WALLING UP TO DPC
CONCRETE FILL ON INSULATIO
***SUPERSTRUCTURE***
WALLING IN SUPERSTRUCTURE
INFRASTRUCTURE AND CONNE
SCAFFOLDING - ERECT AND ST
ROOF CONSTRUCTION AND EAV
ROOF COVERING
***INTERNAL WORK***
CARPENTRY FIRST FIX
PLUMBING AND HEATING FIRST
ELECTRICAL FIRST FIX
PLASTERING
CARPENTRY SECOND FIX
PLUMBING AND HEATING SECO
SERVICE CONNECTIONS
ELECTRICAL SECOND FIX
WALL TILING
ROOF INSULATION
DECORATION
FLOOR COVERINGS
WARDEN CALL SYSTEM
COMMISSIONING
SNAGGING AND FINISHING OFF
FINAL CLEAN UP AND SCRUB O
***EXTERNAL WORK***
REMAINING DRAINAGE AND SER
BOUNDARY WALL FOUNDATION
STREET LIGHTS
BOUNDARY WALLS
TOPSOIL AND HARDCORE TO R
RAILINGS AND FENCING
KERB EDGINGS
BLOCK PAVINGS
PAVING FLAGS
TARMAC ROADS AND FOOTPAT
LANDSCAPING
***FINAL INSPECTION AND HANDOVE***

Float, free | Float, total | Float, start | Normal Task | Normal, Critical | Normal, Stretched

**Fig. 7.56** Case study 5: spending report.

### 7.5.26 *Summary report (Fig. 7.57)*

A summary report was generated for management information showing the major phases of the work.

### 7.5.27 *Histograms*

Histograms were generated for all trades. The one for carpenters is shown in Fig. 7.58.

### 7.5.28 *Conclusion*

A computer-planning package was used on this project because of the amount of information requested at the pre-tender and pre-contract stages. The package was used throughout the project for control purposes. It was generally accepted that the company benefitted from the use of the package. While all programmes illustrated show the activities at their earliest times, when the project started adjustments were made to smooth out requirements.

However, the greatest benefits of computers for planning are gained on large and/or complex projects when a number of reports are required, because they can all be produced from the same basic data. Alternative strategies can be explored and the results seen immediately, thus saving considerable sums of money on site.

O & P CONSTRUCTION SERVICES LTD — ROSSINGTON BUNGALOWS — PRE CONTRACT PROGRAMME

PERTMASTER SUMMARY REPORT — VERSION 1

| Task Description | Start Date | Finish Date | 1994 Jan Feb Mar Apr May Jun Jul Aug Sep Oct (weeks 1 5 9 13 17 21 25 29 33 37 41) | Budget Costs |
|---|---|---|---|---|
| PRELIMINARY WORK | 04/Jan/94 | 11/Mar/94 | | £7,185.90 |
| SET UP SITE | 28/Feb/94 | 01/Mar/94 | | £2,764.50 |
| FOUNDATION | 02/Mar/94 | 26/Apr/94 | | £37,290.60 |
| SUPERSTRUCTURE | 20/Apr/94 | 13/Jun/94 | | £65,330.90 |
| INTERNAL WORK | 24/May/94 | 07/Sep/94 | | £85,548.28 |
| EXTERNAL WORK | 14/Jun/94 | 25/Aug/94 | | £62,090.91 |
| FINAL INSPECTION AND HANDOVER | 08/Sep/94 | 08/Sep/94 | | £6,847.66 |
| TOTALS | | | | £267,058.75 |

Plan Timenow 04/Jan/94 Plan Finish Date 26/Sep/94 — Report produced by Pertmaster Project Planning Software

**Fig. 7.57** Case study 5: contract budget.

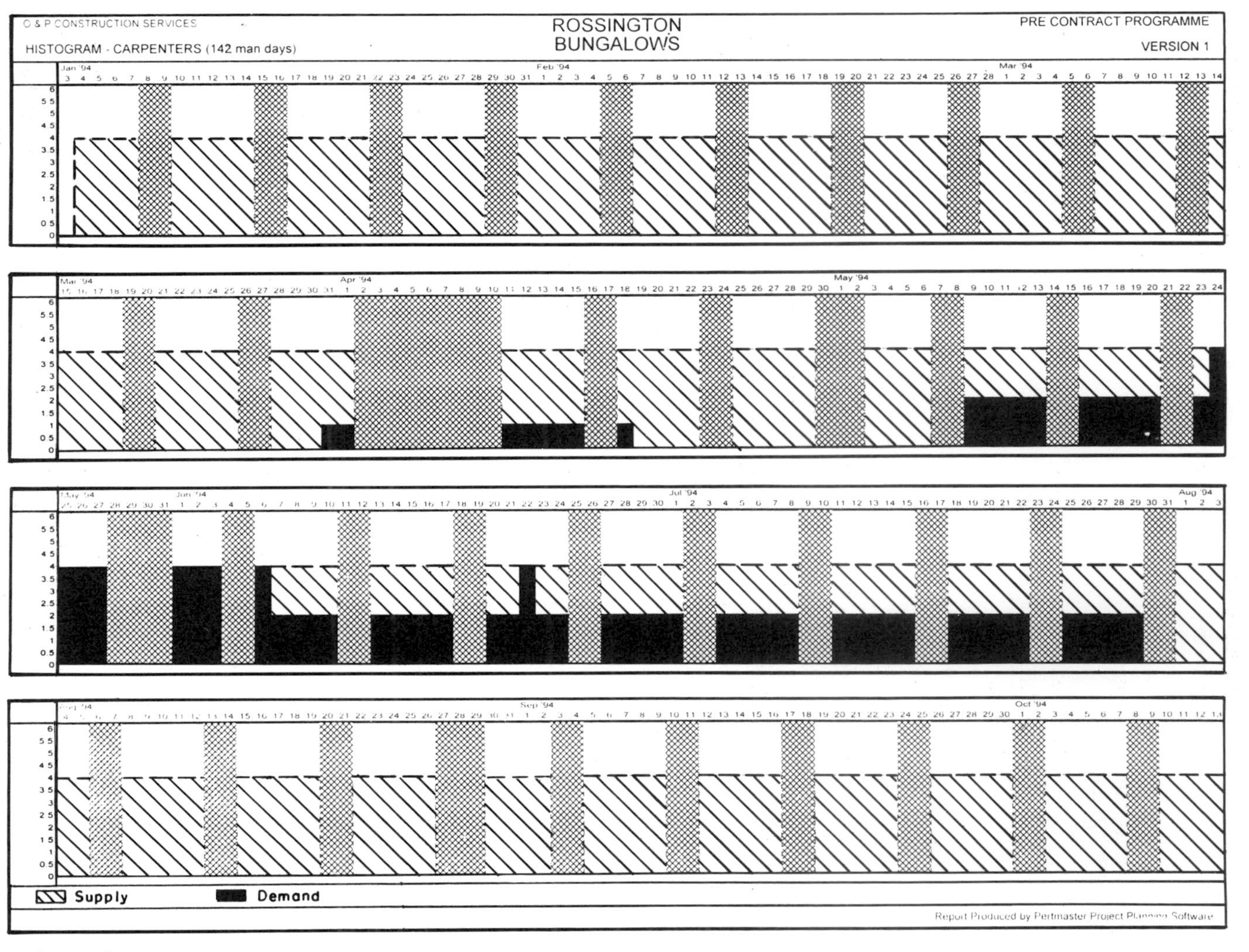


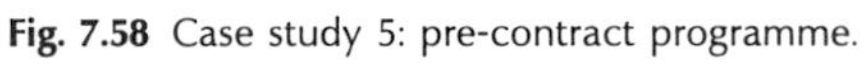

**Fig. 7.58** Case study 5: pre-contract programme.

The use of office and site-based microcomputers, and the direct computer links that can now be achieved, make computer-based planning a very attractive alternative to manual methods.

### 7.5.27 *Other features of Pertmaster*

Many features are available using Pertmaster in addition to those illustrated in the case study.
These include:

- scheduled day numbers within the programme: e.g. delivery of materials and components in short supply;
- milestones: key events within the programme;
- updating plans in order to control the project;
- setting up a library of skeleton plans for similar projects;
- testing the plan against different calendars (e.g. working a 6-day week) to see what will happen;
- providing a histogram for the entire programme;
- printing the actual figures on the cash flow curves;
- listing the activities and building up plans on the screen;
- producing the precedence or arrow diagram on the printer;
- carrying out a plan fit to smooth resources;
- the use of sub-resources under parent resources;
- spending reports.

## 7.6 Case Study 6: Herringthorpe United Reformed Church

### 7.6.1 *Purpose*

The purpose of this case study is to illustrate the use of pre-tender method statements and detailed method statements, and the production of site layout plans.

### 7.6.2 *Description of project*

The project consisted of alterations and extensions to and refurbishment of an existing church building. The works involved the reinstatement of mining subsidence damage to the worship room and the toilet block, and other remedial work and alterations to the existing building. In addition, the existing flat-roofed kitchen and main toilet area, which

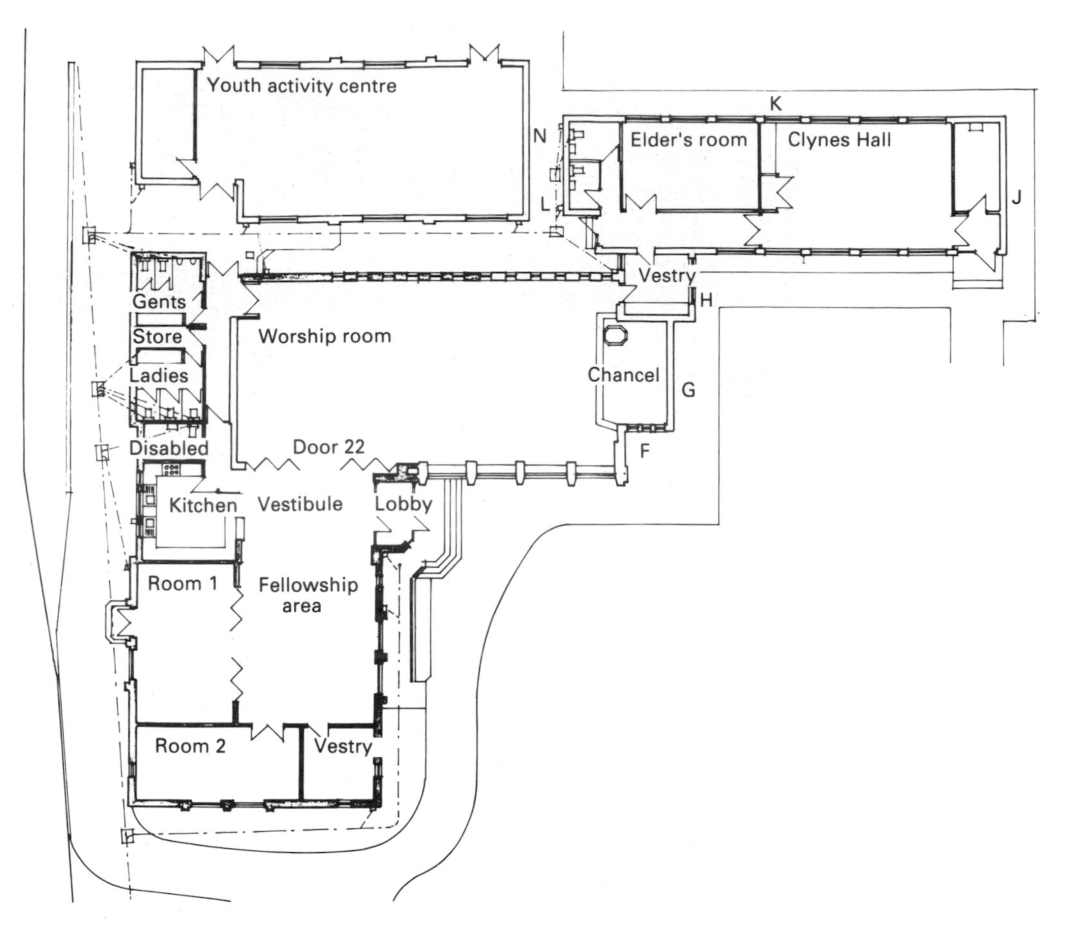


**Fig. 7.59** Case study 6: floor plan.

was also badly damaged by mining subsidence, had to be demolished and replaced by a single-storey pitched roof extension that contained new toilets, a kitchen, two meeting rooms, fellowship area, vestry and vestibule (see Figs 7.59 and 7.60).

- *Architect*: Neville Brooksbank
- *Main contractor*: O & P Construction Services Limited.

This project was featured in *Church Building* July/August 1995.

### 7.6.3 Construction of the building

*Substructure*

- *Foundations*: traditional reinforced concrete strip foundations, solid brick and brick cavity walls.

**Fig. 7.60** Case study 6: Herringthorpe United Reformed Church.

- *Floor slab*: fabric-reinforced concrete bed on styrofoam insulation on DPM on blinded limestone hardcore.

*Superstructure*

- *Walls*: facing brick and block cavity walls, but a number of areas had facing bricks both sides. The rear wall of the worship room was rebuilt with bricks reclaimed from the demolition work. All had insulated cavities.
- *Windows and doors*: purpose-made, double-glazed, high-performance made from Douglas fir.
- *Partitions*: most were blockwork, but some were fair-faced brick-work.
- *Roof structure*: softwood roof trusses with some traditional infill areas, using sawn timbers. The pyramid area was strengthened with horizontal wind girders connected to stainless steel wind posts built into piers in the brickwork.
- *Covering*: Redland Regent concrete single lap tiles.
- *Rainwater goods*: Powder-coated extruded aluminium.

*Internal finishes*

- *Walls*: all exposed surfaces of blockwork were plastered. A number of the partition walls and internal skins of cavity walls were fair-faced brickwork. The upper part of the walls in the fellowship area were clad in American red oak.
- *Ceilings*: plasterboard and skim in the toilet and kitchen area but all other new ceilings were plasterboard covered with a proprietary finish (RMC Alltek).

- *Floors*: screeded with either carpet or ceramic tile finish. The worship room floor area was a woodstrip floor, which had to be taken up and replaced with an American red oak strip floor.
- *Doors and frames*: frame was Douglas fir, doors American red oak.
- *Fittings*: fitting out the kitchen was carried out by an outside contractor.

*Service systems*

- *Heating*: balanced flue, gas-fired, wall-mounted blown-air units.
- *Power, lighting and ventilation*: electric.
- *Other services*: telephone, security and fire alarm systems.

*External works*

- *Paving and access drive*: tarmac, precast concrete kerbs and edgings. Concrete block paving and staircase feature.
- *Drainage*: Vitrified clay flexible joint pipes and fittings, precast concrete inspection chambers. (The main 300 mm diameter public sewer down the west side of the new building had to be repaired, owing to subsidence damage, and diverted to miss the corner of the new building.)

### 7.6.4 The contract

The form of contract was the JCT Agreement for Minor Building Works 1992.

- *Defects liability period*: 6 months.
- *Insurance cover*: £1 000 000
- *Contract period*: 27 weeks.

### 7.6.5 The site

The site was 2 miles (3 km) from the contractor's office. The site fronted onto a main road on the south and mainly onto domestic properties on the north, east and west; its boundaries were, in the main, secure.

### 7.6.6 Preparing the site layout (Figs 7.61 and 7.62)

*Access to the site*

Ideally there should be an entrance and an exit to the site, as this has the effect of stimulating the flow of traffic. In this case the site was situated on a busy road, very close to a large traffic island. An additional access

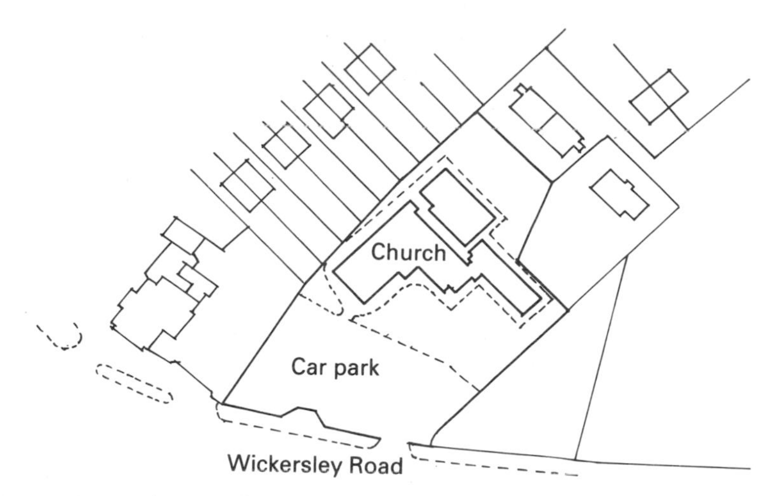


**Fig. 7.61** Case study 6: site plan.

point even closer to the island than the existing one would not therefore have been appropriate. The existing access therefore had to serve as both entrance and exit. Access to the contractor's working area was via the car park, which was still being used by people using the youth activity building and Clynes Hall. The existing footpath crossover was found to be adequate to take the loads imposed by the site traffic.

*Hoarding*

A proprietary metal mesh hoarding was provided to the south and east of the contractor's working area for security. The gate in the hoarding was formed from two standard mesh panels.

*Temporary roads*

There was no necessity to construct a temporary road, as the existing car park, which was finished with blast furnace slag, was found to be capable of taking all site traffic, including the heavy mobile crane. It was necessary, however, to mark out the road areas so that material storage areas would not encroach upon this area.

*Plant*

No heavy plant was used. Lifting of some steel beams and roof trusses was carried out by mobile crane with telescopic jib. This was brought onto site as required and parked on the temporary road area. The crane used to offload the toilet container unit was taken along the temporary road with the lorry up the west side of the existing building to offload the unit at the back of the youth activity building (location requested by the client).

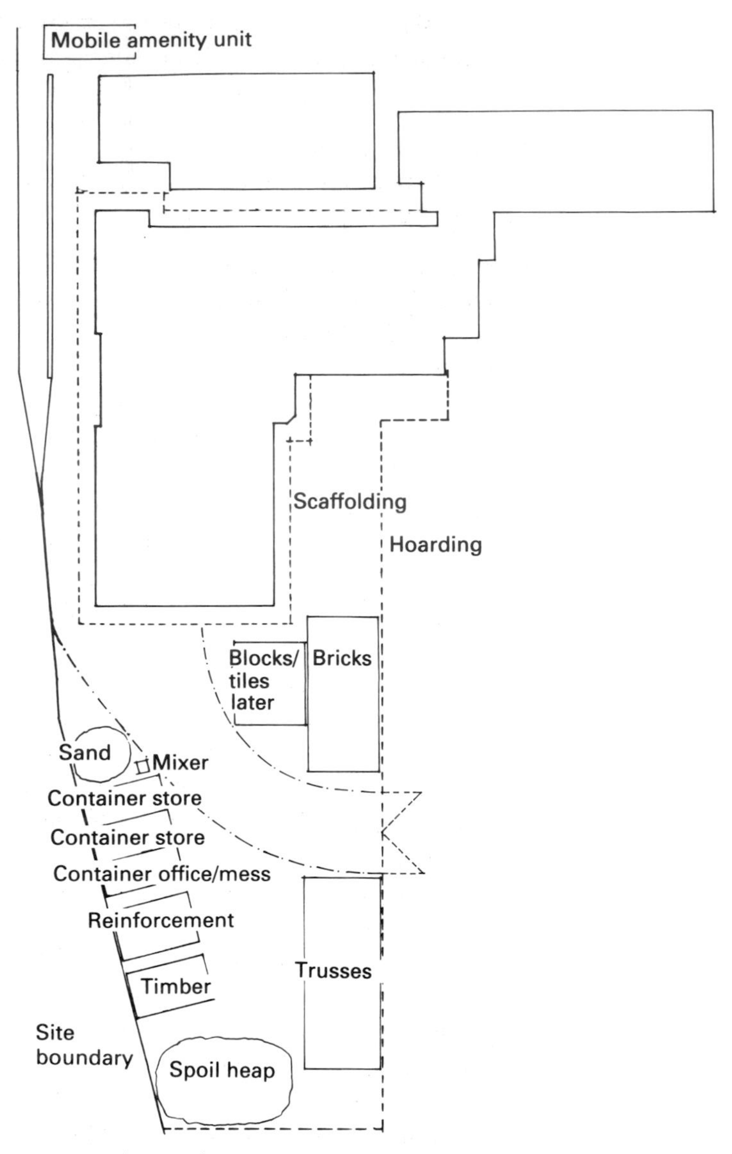

**Fig. 7.62** Case study 6: site layout.

The concrete pump and ready-mixed concrete lorries were also parked on the temporary road area when the concrete floor was laid.

The mortar mixer was stored in the container overnight and brought out every morning and located as shown on the site layout.

The scaffolding was of the independent type to give access to the bricklaying operation and the construction of the roof. Independent scaffolding was also used for the demolition and rebuilding of the rear wall of the worship room; scaffold towers were also erected to support the rear of the worship room roof while work was carried out on this wall.

*Materials*

It is good practice to mark out all storage areas on site and then ensure that they are strictly adhered to.

- *Mesh reinforcement*: stored vertically alongside steel containers, cut into strips as required and placed in foundations. Mesh reinforcement for the floor slabs was stored in the same location but used as full sheets.
- *Bricks and blocks*: all bricks and blocks were delivered packaged. The area set aside had to be within the reach of the lorry-mounted crane. All reclaimed bricks were taken down, cleared and then stored alongside the rear wall of the worship room, ready for reuse.
- *Hardcore*: limestone hardcore used for filling under floors was tipped directly into the area where it was required, spread and levelled by JCB site master and compacted by vibrating roller.
- *Insulation for floor*: styrofoam slab insulation was stored in a container prior to being used under the concrete slab.
- *Artificial stone*: the artificial stone sills, heads, etc., were stored on the floor of the worship room and covered over to protect delicate arrises. They were moved out onto the scaffold just prior to fixing.
- *Trussed rafters*: these were all stored vertically in the rack provided, and covered.
- *Other roof timbers*: there were other sawn timbers required for the infill parts of the roof as well as PSE timbers for soffits and fascias. These were stored in the area indicated and covered with tarpaulins.
- *Roof tiles*: these were stored along with hiptiles as close as possible to the scaffold, for ease of transportation onto the roof slopes.
- *Rainwater goods*: the rainwater goods needed very special care, as they were powder-coated cast and extruded aluminium. They were stored in a section of one of the containers and covered at all times for added protection.
- *Plaster and plasterboard*: delivered to site as required and stored inside building just prior to being used.
- *Floor screed*: premixed screed was used throughout. This was delivered to site the day it was required and deposited outside the entrance to the building.
- *Window and door frames*: during the construction of the walls, templates were made to form all the openings. This was to prevent any of the high-quality door and window frames from getting

damaged. When delivered to site all door and window frames were stored inside the building and chained together for security. The main reason for this was that they were too large to be stored in the containers.

- *Double-glazed units*: stored vertically in racks on foam cushions in the containers. Stored in the order in which they would later be required.
- *Second-fix timbers*: stored in racks inside the containers.
- *Doors*: these were delivered only when the building had been thoroughly dried out. They were taken into the building and fixed straight after delivery.

*Accommodation*

- *Location*: the area to be used for the accommodation was selected primarily because: (a) it was possible to have all the accommodation in one area, which helped with communication between personnel; (b) it made the control of materials much simpler.
- *Construction and fittings*: all units were based on steel containers, one of which was partitioned to provide office accommodation at one end and a messroom at the other. This unit was insulated to reduce heat loss, and was provided with windows and a door, all of which had steel shutters for security at night. The office was fitted out with desk, chairs and filing cabinets. The messroom was fitted out with a table, benches, a sink unit and a gas-fired oven/grill. All were heated by Calor gas.
- *Water closet*: this was situated away from the mess room and office areas. The facility was in the form of a sectional timber cabin with insulated walls fitted out with WC and wash basin. It was located as close as possible to the mobile amenities unit provided for the client, so that it could be connected to the same services.

*Services*

- *Water*: supplied to the mixer plant (via a hose from a standpipe), the messroom and the toilets.
- *Electricity*: originally electricity was supplied by generator, but when the electricity supply was reintroduced, the 240 V supply was transferred down to 110 V before being used on site to power handtools. The distribution board was housed inside the building in a secure, watertight enclosure.
- *Telephone*: this was connected as soon as the site accommodation was located on site.

Herringthorpe United Reformed Church, Wickersley Road, Rotherham, South Yorkshire
Date: Compiled by:

| Item | Quantity | Remarks | Labour and Plant | Time Required |
|---|---|---|---|---|
| Set up temporary toilets (for client) | – | Excavate into bank to form level standing for toilet unit (steel container). Spoil deposited for reuse | JCB Sitemaster operator and two labourers | 1 day |
| | | Toilet delivered on flat back lorry - lifted into place by crane | 15 t crane (telescopic jib) operator and two labourers | 1 day |
| | | Connect toilet unit to services (water, drains and electricity) | One plumber One labourer One Electrician | 1 day |
| Hoarding | 80 mts.lin | Standard UK fence hire units into concrete base blocks on south and east of working area | Two labourers | 1 day |
| Access | – | Vehicle and pedestrian access via existing entrance and then across car park to compound | | |
| Accommodation | 4 no | All accommodation to be in steel container type units. Toilets for workmen to be built from our standard sectional units | Two labourers | 1 day |
| Temporary services | – | Provide electrical, water and WC services as shown on site layout | Two labourers One plumber One electrician | 2 days |
| Demolition | – | Existing single storey toilet/kitchen area - JCB Sitemaster using front bucket and backacter Break up concrete roof slab, remove from site | JCB Sitemaster operator and two labourers. Two tool compressors, 16t lorry and driver | 3 days |
| Excavation | $278 m^3$ | Excavate to reduced levels - strip foundations and internal drainage | JCB Sitemaster operator and two labourers, 16 t lorry and driver | 4 days |
| General hoisting and transportation | – | All bricks and blocks - lorry crane. Off load in compound area. Structural steel and trusses lifted by crane | Mobile crane/telescopic jib and operator | |
| Concreting | $76 m^3$ | Ready-mixed concrete | | |
| | | *Transportation* - placed directly into trenches for foundations. Pumped into slab - hit and miss method of construction | Ready-mix lorry Lorry-mounted concrete pump Two labourers | 2 days |
| | | *Placing* - poker vibrator for foundations. Vibrating screed for slabs | Vibrating poker Four labourers Vibrating screed | 2 days 2 days |
| Mortar mixing | – | Mortar for bricklaying will be mixed in a 150/100 mixer located on a hard standing. Bagged cement will be stored in the steel container, directly adjacent | One labourer | |
| Scaffolding | – | Scaffolding will be of independent type (traditional tubular steel) 1.2 wide, two lifts on extension, three lifts on rear wall of worship room | Two scaffolders | |
| | | Lattice beams supporting roof of worship room to be supported on scaffold tube tower during demolition and reconstruction of rear wall | Two scaffolders | |

**Fig. 7.63** Case study 6: pre-tender method statement.

Detailed method statement
Herringthorpe United Reformed Church, Wickersley Road, Rotherham, South Yorkshire
Date: Compiled by:

| Item | Quantity | Method | Output (man hrs) | Plant | Labour | Time required |
|---|---|---|---|---|---|---|
| Installation of beams over D22 | 4 no | 1. Form holes through 225/50/225 cavity wall to receive steel beams as needles at 1200 mm centres | 4 hrs each | Kango | Two labourers | 1 day |
| | 4 no | 2. Insert steel beam needles, supported on Acrow props - bolt Acrow props to beams and tighten up props to engage beams against brickwork - brace props with scaffold tubes | 2 hrs each | 8 Acrow props | Two labourers | ½ day |
| | | 3. Remove all brickwork under needles, including existing beams spanning existing opening | 8 hrs | Kango | Two labourers | ½ day |
| | 2 no | 4. Cut out a 375 high × 180 deep slot in the back face of two brick columns | 4 hrs each | Kango | Two labourers | ½ day |
| | | 5. Insert inner beam | | subcontract | | |
| | | 6. Insert slate wedges between top flange of beam and underside of brickwork and point up | 8 hrs | | Two labourers | ½ day |
| | | 7. Cut out a 375 high × 350 deep slot in the front face of two pilaster faced brick columns | 8 hrs | Kango | Two labourers | ½ day |
| | | 8. Insert outer beam, spacer tubes and bolts | | subcontract | | |
| | 8 mts.lin | 9. Insert slate wedges between top flange of beam and underside of brickwork and point up | 8 hrs | | Two labourers | ½ day |
| | | 10. Demolish and cart away all brickwork in columns below beam level | 16 hrs | Kango | Two labourers | 1 day |
| | 4 no | 11. Remove needles and props | 1 hr each | | Two labourers | ¼ day |
| | 4 no | 12. Make good holes in 225/50/225 cavity walls (this area of wall in new roof one side - plastered on other side) | 2 hrs each | | Two labourers | ½ day |

**Fig. 7.64** Case study 6: detailed method statement for forming of opening for door 22.

### 7.6.7 *Pre-tender method statement*

The pre-tender method statement shown in Figure 7.63 was prepared for Herringthorpe United Reformed Church to assist in tendering.

### 7.6.8 *Detailed method statement*

The example that has been used is the forming of the opening for door 22. This entailed the removal of two existing steel beams over a small existing opening, and the removal of the bottom section of two brick columns. Two 8 m long 356 × 171 × 67 UBs were then inserted to support the 225/50/225 cavity wall and the tops of the brick columns.

The structural engineering consultant proved that the beams did not require padstones, as the existing solid dense bricks were of adequate strength to carry the beams.

# Chapter 8

# Exercises

## 8.1 Refurbishment project: Case study 1

The following variations occurred early in the above contract, and the estimated man-hours were determined as follows:

| | Man-hours |
|---|---|
| Alterations to boundary wall at the rear of the house | 36 |
| Straps to tie the wall at rear of the house | 38 |
| Additional drainage | 14 |
| Rebuild brick wall between bedroom and bathroom | 15 |
| Replace floor in the front room | 40 |

(1) (a) Adjust the targets to take account of the variations.
    (b) Adjust the programme.
    (c) State the action you would take regarding any extension of time as a result of variations.
(2) Carpentry first fix was completed in 90 man-hours. 6 hours of unclaimable daywork occurred during the operation.

Place entries on the cost sheets for this operation and calculate the saving/loss.

## 8.2 Line of balance applied to a refurbishment project: Case study 2

(1) Draw up the overlap sheets for the operations not covered in the line of balance schedule.
(2) Assuming 75 units have to be refurbished at the same handover rate:
    (a) Draw up a calculation sheet
    (b) Draw up a line of balance schedule.

## 8.3 Kimberworth Surgery: Case study 3

The project described in case study 3 and shown in Figs 7.23 and 7.24 is to be used for the following site layout exercise.

On Fig. 8.1 draw your recommended layout for the following:

(1) All static plant required on the site
(2) Temporary buildings and services
(3) Materials stock piles and compounds
(4) Site security and means of access onto and about the site.

In your solution, state your reasons for your layout in each case.

*Note:* In formulating your solution you are to assume the worst case: that is, when the main contractor and all his subcontractors are on the site at the same time.

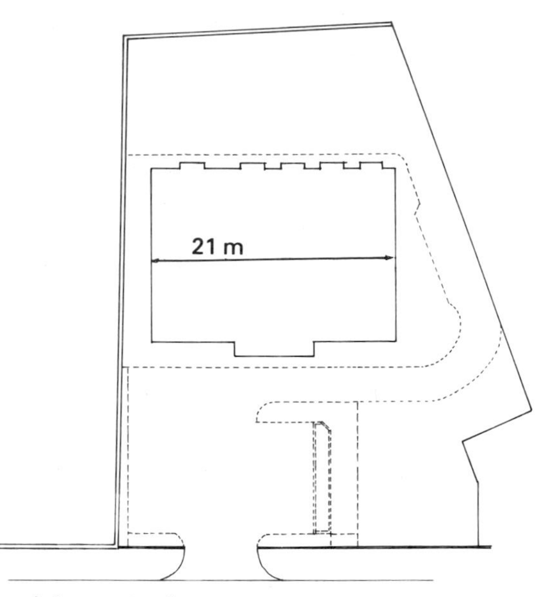


**Fig. 8.1** Kimberworth Surgery site plan.

## 8.4 Multi-project scheduling: Case study 4

At the end of week 3, progress was as follows:

- Flat conversions, London Road
  - carpentry finishings 25% complete
  - electrical and mechanical second fix 75% complete

  - painting and decorating 30% complete
  - external works complete.
- Surgery, Barnsley
  - carpentry second fix 35% complete
  - plastering complete
  - external works 50% complete.
- Health Centre, Maltby
  - painting and decorating 80% complete
  - electrical and mechanical complete.
- Refurbishment, Rotherham
  - Unit 1: plastering complete; electrical and mechanical 20% complete
  - Unit 2: carpentry first fix complete.
- New flats, Bawtry
  - PC floors complete.
- Church alterations and extensions, Herringthorpe
  - internal strip-out 75% complete
  - demolition 70% complete.

(1) (a) Update the programme
(b) Comment on the effect of progress to date on the remainder of the programme.

(2) Assuming that Kimberworth Surgery (Case study 3) was to commence on 13 June, include the early operations of Kimberworth Surgery on the multi-project schedule.

## 8.5 Herringthorpe United Reformed Church: Case study 6

The project described in case study 6 is to be used for the following programming exercise.

The project was planned based on a 45-hour, 5-day week.

Quantities were taken off, and from these the total man-hours were calculated for all operations to be carried out by the contractor's own labour. See Table 8.1.

Subcontractors were consulted, and indicated their required time on site and the periods of notice required following the placement of orders (Table 8.2). Initial orders were placed by 17 May.

Draw up a pre-contract programme for the project, commencing on 23 May.

**Table 8.1** Main contractor's operations

| Operation | Man-hours |
|---|---|
| *Demolition and construction of extension* | |
| Set up site | 80 |
| Remove fitted units/cupboards | 110 |
| Form temporary path to rear of Clynes building | 55 |
| Remove outer leaf of brickwork and clean | 56 |
| Concrete foundations | 34 |
| Walling to underside of slab | 110 |
| Hardcore, insulation and concrete floor slab | 199 |
| Drainage | 311 |
| Form opening to D22 including new foundation and pier | 266 |
| Remove stage vestibule wall and mezzanine floor | 174 |
| Form openings in existing walls | 65 |
| Walling to eaves | 650 |
| Scaffolding | |
| Roof carcassing | 406 |
| Aluminium rainwater goods | 96 |
| Carpentry first fix | 167 |
| Plastering | 255 |
| Carpentry second fix | 366 |
| Quilt insulation | 26 |
| WC cubicles | 32 |
| Church cross | 8 |
| *Work to existing building* | |
| Screed and DPM | 92 |
| Replace windows and glaze to worship room | 64 |
| Replace panelling with brickwork below feature window | 35 |
| Scaffold support | |
| Take down and rebuild rear wall | 686 |
| Re-lay rear external paving | 81 |
| Elevations F, G, H, I including paving | 40 |
| Elevations L, M including paving | 82 |
| Elevation N | 48 |
| Make good cracks in brickwork generally | 8 |
| Extend worship room ceiling over old stage area | 75 |
| Re-plaster | 122 |
| Carpentry second fix | 9 |
| *External* | |
| Form ramp and steps | 97 |
| Paving | 276 |
| Boundary fencing | 18 |
| Clear up on completion | 32 |

**Table 8.2** Subcontractors' work

| Operation | Period required (working days) | Period of notice required (weeks) |
|---|---|---|
| *Demolition work and construction of extension* | | |
| Demolish single-storey building | 3 | 1 |
| Excavate to reduce levels and trench exc. | 4 | 2 |
| Roof covering and leadwork | 13 | 3 |
| Plumbing and heating first fix | 5 | 2 |
| Electrical first fix | 20 | 2 |
| Plumbing and heating second fix | 15 | 2 |
| Electrical second fix | 15 | 2 |
| Serving hatch shutters | 2 | 4 |
| Ceramic wall and floor tiles | 4 | 3 |
| Painting and decorating | 10 | 3 |
| Textured spray finish to ceilings | 5 | 4 |
| French polishing | 4 | 2 |
| *Work to existing building* | | |
| Ceramic tile floor | 3 | 3 |
| Painting and decorating | 5 | 3 |
| *External* | | |
| Tarmac paving | 3 | 2 |

# Bibliography

## Construction planning and control

Anthony, R. N. (1965). *Planning and Control Systems – a Framework for Analysis.* Harvard University Press.

Calvert, R. E. (1995) *Introduction to Building Management*, 6th edn. Butterworth-Heinemann, Oxford.

The Chartered Institute of Building, *Programmes in Construction – A guide to good practice.* CIOB.

The Chartered Institute of Building, *The Practice of Site Management*, Vols 1, 2 and 3. CIOB.

Cooke, B. (1988) *Contract Planning Case Studies.* Macmillan.

Cooke, B. (1992) *Contract Planning and Contractual Procedures*, 3rd edn. Macmillan.

Illingworth, J. R. (1993) *Construction Methods and Planning.* E & FN Spon.

Pilcher, R. (1993) *Principles of Construction Management*, 3rd end. McGraw-Hill, London.

## Project network techniques

Antill, J. M. (1990) *Critical Path Methods in Construction Practice*, 4th edn. Wiley, New York.

BS 4335 (1972) *Glossary of Terms used in Project Network Techniques.* British Standards Institution, London.

Burman, P. J. *Precedence Networks for Project Planning and Control.* McGraw-Hill, London.

Cooke, B. (1989) *Contract Planning Case Studies.* Macmillan.

Harris, F. & McCaffer, R. (1995) *Modern Construction Management*, 4th edn. Blackwell Science, Oxford.

Lockyer, K. G. (1991) *Critical Path Analysis and other Project Network Techniques.* Pitman.

O'Brien, J. J. (1993) *CPM in Construction Management*, 4th edn. McGraw-Hill, London.

Pilcher, R. (1993) *Principles of Construction Management*, 3rd edn. McGraw-Hill, London.

## Work study

BS 3138 *Glossary of Terms in Work Study.* British Standards Institution, London.

Calvert, R. E. (1995) *Introduction to Building Management*, 6th edn. Butterworth-Heinemann.

Currie, R. M. & Faraday, J. E. (1993) *Work Study*. Pitman.
Fellows, R., Langford, D., Newcombe, R. & Urry, S. (1988) *Construction Management in Practice*. Longman.
Fryer, B. (1990) *The Practice of Construction Management*. Blackwell Science, Oxford.
Geary, R. *Work Study Applied to Building*. George Godwin.
Harris, F. & McCaffer, R. (1995) *Modern Construction Management*, 4th edn. Blackwell Science, Oxford.
Herzberg, F. (1968) *Work and the Nature of Man*. Staples.
Herzberg, F. (1993) *The Motivation to Work*. Transaction.
International Labour Office (1986) *Introduction to Work Study*. ILO.
Neale, R. H. & Neale, D. E. (1989) *Construction Planning*. Thomas Telford.
*Outline of Work Study*, Parts 1, 2, 3. The British Institute of Management.
Pilcher, R. (1992) *Principles of Construction Management*. McGraw-Hill.
Rougvie, A. (1987) *Project Evaluation and Development*. CIOB.
Rutter, P. A. & Martin, A. S. (1990) *Management of Design Offices*. Thomas Telford.
*Spon's Architects' and Builders' Price Book*. E & F Spon. Annual.

## Budgetary and cost control

Barnes, M. (1990) *Financial Control*. Thomas Telford.
Cooke, B. & Jepson, W. B. *Cost and Financial Control for Construction Firms*. Macmillan.
Drury, A. (1990) *Costing*. Chapman-Hall.
Gobourne, J. *Cost Control in the Construction Industry*. Newnes Butterworth.
Harris, F. & McCaffer, R. (1995) *Modern Construction Management*, 4th edn. Blackwell Science, Oxford.
Institute of Cost and Works Accountants, *Introduction to Budgetary Control, Standard Costing, Material Control and Production Control*. Gee & Co.
Pilcher, R. (1992) *Principles of Construction Management*. McGraw-Hill, London.
Pilcher, R. (1994) *Project Cost Control*, 2nd edn. Blackwell Science, Oxford.
Scott, J. A. *Budgetary Control and Standard Costs*. Pitman.

## Computer applications

Barton, P. (1985) *Information Systems in Construction Management*. Batsford.
The Chartered Institute of Building, *Construction Computing* (monthly journal). CIOB.
Cooke, B. (1989) *Contract Planning Case Studies*. Macmillan.
Flowers, R. (1996) *Computing for Site Managers*. Blackwell Science, Oxford.

## General

Ashby, C. J. (1992) *Quality Management System Implementation*. CCTA, The Government Centre for Information Systems.
Ashford, J. L. (1989) *The Management of Quality in Construction*. E and FN Spon.
European Construction Institute (1992) *Total Quality in Construction–Stage 2*. ECI.
European Construction Institute (1993) *Total Quality in Construction–Measurement Matrix and Guidelines for Improvement*. ECI.
Garvin, D. A. (1988) *Management Quality*. Collier Macmillan.
Hellard, R. B. (1993) *Total Quality in Construction Projects*. Thomas Telford.
Waller, J. & Allen, D. (1993) *The Quality Management Manual*. Kogan Page.

# Index